D1797502

WEATHER AND CLIMATE EXTREMES

Changes, Variations and a Perspective from the Insurance Industry

WEATHER AND CLIMATE EXTREMES

*Changes, Variations and a Perspective
from the Insurance Industry*

Edited by

Thomas R. Karl

Neville Nicholls

and

Anver Ghazi

Reprinted from Climatic Change
Volume 42, No. 1, 1999

Sponsored by DG XII
Science Research and Development

Kluwer Academic Publishers
Dordrecht / Boston / London

Library of Congress Cataloging-in-Publication Data

Weather and climate extremes : changes, variations, and a
  perspective from the insurance industry / edited by Thomas R.
  Karl, Neville Nicholls, and Anver Ghazi.
       p.   cm.
  Includes Index.
  ISBN 0-7923-5711-6 (hc. : alk. paper)
  1. Climatic extremes. 2. Climatic changes. 3. Insurance.
  I. Karl, Thomas. II. Nicholls, N. (Neville) III. Ghazi, A.,
  1940-   .
  QC981.8.C53W4   1999
  551.6--dc21   99-25780

Published by Kluwer Academic Publishers,
P.O. Box 17, 3300 AA Dordrecht, The Netherlands

Sold and distributed in North, Central and South America
by Kluwer Academic Publishers,
101 Philip Drive, Norwell, MA 02061, U.S.A.

In all other countries, sold and distributed
by Kluwer Academic Publishers,
P.O. Box 322, 3300 AH Dordrecht, The Netherlands

Printed on acid-free paper

Printed in The Netherlands

Contents

Some Critical Aspects of Extreme Events

OVERVIEW

In recent years, more climate data has become available to enable climatologists to analyze these data for trends and variations related to extreme weather and climate events. At the same time, the insurance industry has experienced a significant increase in weather and climate related insurance claims. The set of papers in this volume represent a major step toward developing a comprehensive set of information about worldwide changes in climate extremes. Nonetheless, there are still many difficulties in analysis, data availability, data quality and consistency that make it difficult to derive a clear picture of changes in worldwide extreme weather and climate events. Clearly, in some areas we know much more than in other areas. For example, analysis of temperature extremes are much more comprehensive than other variables, although there is now emerging a number of new analyses related to precipitation extremes. What is more striking, however, is the complexity of linking current information about climate extremes to the needs of the insurance industry. As the reader of this volume will quickly learn, however, there are many opportunities for clever scientists and insurance industry analysts to leverage various types of information from both fields to develop better statistics on climate variability that have a direct effect on insurance claims and rates.

On average, as the climate has warmed, there has been a reduced frequency of extremely cold days leading to a reduction in frost and freeze frequency, although this general trend can be temporally reversed, as described in two separate analyses of severe freezes that have affected Florida and the southeast United States as global temperatures warmed during the 1980s. The reduction in the number of freezes and frosts is exaggerated by the fact that the minimum temperature is rising faster than the daily maximum temperature. There has also been an tendency toward an increase in the frequency of extremely warm days, but a strong trend has yet to emerge.

Analyses of changes in precipitation extremes suggest that for some regions and times of the year there is clear evidence of an increase in extreme and heavy precipitation events, e.g., USA, PRC, and during summer in Australia, Canada, Norway, Mexico, Poland, and the former Soviet Union. There is now a solid foundation to expect increases in these types of events as the climate warms and precipitation increases. There are some areas, however, where differences in the way in which the data have been analyzed make it difficult to interpret the results, e.g., Canada, and can lead to apparent contradictions. Another difficulty relates to the inability of scientists to get access to data that is residing within national archives.

The analysis of changes in tropical storms reveal intriguing relationships between large-scale climate variables and the number of storms that march across the oceans each year. Although there are important decadal variations evident in the data, no clear trends emerge, except for some regional changes noted in the Austral-Australian region. Changes in extratropical storms have important consequences related to wind damage and other factors, but little evidence is available to suggest a real increase in damaging winds, although the record is woefully underanalyzed.

Finally, there are some regions where local severe weather events such as hail frequency have been analyzed. For some parts of the world, existing insurance claims provide a means to cross-validate weather and climate data for consistency of trends and variations. This is a potentially fruitful area of research in regions of the world that have a history of insurance claims.

NOAA, U.S. Department of Commerce/NOAA   THOMAS KARL
National Climatic Data Center
151 Patton Avenue, Room 120
Asheville, NC 28801-5001, U.S.A.

CLIVAR/GCOS/WMO WORKSHOP ON INDICES AND INDICATORS FOR CLIMATE EXTREMES

WORKSHOP SUMMARY

THOMAS R. KARL

National Climatic Data Center, 151 Patton Ave., Asheville, NC 28801-5001, U.S.A.

NEVILLE NICHOLLS

Bureau of Meteorology Research Center, GPO Box 1289K, Melbourne Vic 3001, Australia

ANVER GHAZI

European Commission, 200, rue de la Loi, Brussels B-1049, Belgium

There is general agreement that changes in the frequency or intensity of extreme weather and climate events are likely to have profound impacts on society and the environment (Karl et al., 1997). A Workshop on Indices and Indicators for Climate Extremes was held in Asheville, North Carolina, 3-6 June 1997, to encourage the development of data sets, and analysis techniques, to determine whether such extreme events are becoming more extreme or variable. Over 100 participants, from 23 countries, including representatives from 15 insurance and re-insurance countries (which have a clear interest in extreme weather and climate), examined the following questions:

- What needs to be done to improve data sets and analyses for extreme weather monitoring?
- Can we establish priorities for specific data set development and improvement?
- Can we establish indices and indicators of extreme weather and climate?
- What are the impediments to improving the monitoring of climate extremes?

The incentive for holding this workshop came largely from the difficulties the Intergovernmental Panel on Climate Change (IPCC) has had in answering the question: *Has the climate become more variable or extreme?* The Second Assessment Report (SAR) of the IPCC was completed in 1995 and determined that (Nicholls et al., 1996):

> Overall, there is no evidence that extreme weather events, or climate variability, has increased, in a global sense, through the 20th century, *although data and analyses are poor and not comprehensive*. On regional scales there is clear evidence of changes in some extremes and climate variability indicators. Some of these changes have been toward greater variability; some have been toward lower variability.

Climatic Change **42**: 3–7, 1999.
© 1999 *Kluwer Academic Publishers. Printed in the Netherlands.*

The SAR further noted:

- *The data on climate extremes and variability are inadequate to say anything about global changes,* but in some regions, where data are available, there have been decreases or increases in extreme weather events and variability.
- Other than the few areas with longer term trends to lower rainfall (eg., the Sahel), little evidence is available of changes in drought frequency or intensity.
- There have been few studies of variations in extreme rainfall events and flood frequency. In some areas with available data there is evidence of increase in the intensity of extreme rainfall events, but no clear large-scale pattern has emerged.
- There is some evidence of recent (since 1988) increases in extreme extra-tropical cyclones over the North Atlantic. Intense tropical cyclone activity in the Atlantic has decreased over the past few decades although the 1995 season was more active than recent years. Elsewhere, changes in observing systems and analysis methods confound the detection of trends in the intensity or frequency of extreme synoptic systems.
- There has been a clear trend to fewer extremely low minimum temperatures in several widely separated areas in recent decades. Widespread significant changes in extreme high temperature events have not been observed.
- There have been decreases in daily temperature variability in recent decades, in the Northern Hemisphere mid-latitudes.

Despite the difficulties in assessing whether there had been recent trends in climate extremes and variability, the SAR examined model projections, to assess the likelihood of future changes in extremes and variability (Kattenberg et al., 1996) and concluded that:

- A general warming would tend to lead to an increase in extremely high temperature events and a decrease in winter days with extremely low temperatures (eg., frost days in some areas).
- With increasing greenhouse gas concentration, many models suggest an increase in the probability of intense precipitation. A number of simulations also show an increase in the probability of dry days and the length of dry spells (consecutive days without precipitation).
- New results reinforce the view that variability associated with the enhanced hydrological cycle translates into prospects for more severe droughts and/or floods in some places and less severe droughts and/or floods in other places.
- In the few analyses available, there is little agreement between models on changes in storminess that might occur in a warmer world. Conclusions regarding extreme storm events are obviously even more uncertain. The formation of tropical cyclones depend not only on sea surface temperature

(SST) but also on a number of atmospheric factors. Although some models now represent tropical storms with some realism for present day climate, the state of the science does not allow assessment of future changes.

The lack of certainty with regard to past and future extremes evident in the SAR indicates that further work is necessary to develop data bases capable of monitoring changes in climate extremes and variability, and to develop systems for such monitoring. This will require daily weather data to be made available over long periods, and for problems arising from changes in instrumentation, exposure, siting, and even analysis and meteorological interpretation, to be overcome. The Third Assessment Report (TAR) of the IPCC will be completed at the end of this century. Considerable efforts are needed before then, to ensure that the data and analyses are available to allow the TAR to provide a more complete answer to the question: *Has the climate become more variable or extreme?* CLIVAR provides a mechanism to coordinate studies aimed at answering questions raised by IPCC assessments. The WCRP CLIVAR program (CLIVAR, 1995) includes amongst its objectives:

- To extend the record of climate variability over the time-scales of interest through the assembly of quality-controlled paleoclimatic and instrumental data sets.
- To understand and predict the response of the climate system to increases of radiatively active gases and aerosols and to compare these predictions to the observed climate record in order to detect the anthropogenic modification of the natural climate signal.

These objectives will be pursued through one of the 12 Principal Research Areas of CLIVAR: *Climate Change Detection and Attribution.* Detection of climate change is the process of demonstrating that an observed variation in climate is highly unusual in a statistical sense. Detection of climate change requires demonstrating that the observed change is larger than would be expected to occur by natural internal fluctuations. Attribution of change to human activity requires showing that the observed change cannot be explained by natural causes, forced or unforced. It is the process of establishing cause and effect relations, including the testing of competing hypotheses. CLIVAR will aim to improve the current state of studies of climate change detection and attribution through, *inter alia*, encouraging the development of historical time series of relevant climate data, with the Global Climate Observing System (GCOS).

The first step in the detection/attribution of climate change is the assembly of high-quality time-series of key variables. We need to ensure that credible historical time-series are available, through the correction for time-varying biases caused by changes in observing practices, instrumentation, and location.

As well, continuation of high-quality climate observations into the future are a prime requirement. The Global Climate Observing System (GCOS) and the Global Ocean Observing System (GOOS) have a leading role in ensuring the continuation of such observations, as does WMO with its Reference Climate Station program.

The preparation of high-quality, historical climate data sets for climate change detection will be pursued through the CLIVAR/CCL Working Group on Climate Change Detection (WGCCD). This group has the expertise and access to data and station documentation necessary for the detection and removal of inhomogeneities in the historical data. The GCOS Data and Information Management Panel (GCOS-DIMP) too has a role in this area. Close coordination between the WGCCD and the GCOS-DIMP will ensure that duplication of effort is avoided, and that the best quality data sets are produced for climate change detection studies.

The Workshop on Indices and Indicators for Climate Extremes, it was hoped, could address many of the questions IPCC and CLIVAR need to answer. The production of time series of global and regional indicators and indices of climate extremes by exploiting daily weather data, and the development of a mechanism to update these indicators and indices on a regular basis, can go far to ensuring that the IPCC TAR will be able to determine whether the climate is becoming more extreme or variable. The Workshop was intended as a first step towards the production of such time-series, with their necessary mechanisms for updating.

The Workshop commenced with views from IPCC, CLIVAR, GCOS, agriculture, and the insurance and re-insurance industry with regard to extreme climate and weather. Scientific papers then examined the available data sets, and methods for analysing climate extremes. Regional papers examined the quality of data and analyses for different parts of the world.

After the scientific papers, three breakout groups (Storms; Precipitation; Temperature) met to consider the requirements for the three areas of extremes. On the final day of the Workshop, a plenary session discussed the results of the deliberations of the breakout groups and reached the following conclusions:

- It was agreed that a small group of climate extremes indices should be selected, to simplify the analysis and presentation of climate extreme trends. These indices would be selected by the Chairs of the breakout groups.
- Regional rapporteurs were selected to encourage improved access to data and coordination of analyses, and the development of time-series of the selected climate extremes indices.
- A small task group should be established to facilitate inter-regional consistency of analyses and to improve data access and liaison with the insurance industry.

- The regional rapporteurs and the task group would decide on improved methods for archiving global climate extremes data.
- Global and regional bodies would be approached to encourage regional projects to develop data sets and analyses of climate extremes.
- A further Workshop would be held in approximately two years to determine progress and any impediments to determining whether the global climate was becoming more extreme or variable.

It was also agreed that the presentations at the Workshop, and summaries of the breakout group discussions, should be published as a special edition of *Climatic Change*. Some of the presentations were combined, to reduce the number of papers, to provide a more comprehensive regional coverage, and to promote "cross-fertilisation" between authors approaching the subject from different perspectives. The papers and discussions are included in this volume, and provide the most comprehensive description of what we can ascertain regarding changes in climate and weather extremes, the data we need to improve the monitoring of these extremes, and the analyses that still need to be undertaken.

References

CLIVAR Scientific Steering Group: 1995, 'CLIVAR. A study of climate variability and predictability', MO TD No. 690, August, 157 pp.

Karl, T. R., Nicholls, N. and Gregory, J.: 1997, 'The coming climate', *Scientific American*, May, 54-59.

Kattenberg, A., Giorgi, F., Grassl, H., Meehl, G.A., Mitchell, J.F.B., Stouffer, R.J., Tokioka, T., Weaver, A.J., and Wigley, T.M.L.: 1996. Climate models – projections of future climate. In *Climate Change 1995. The Science of Climate Change: Contribution of Working Group I to the Second Assessment Report of the Intergovernmental Panel on Climate Change*. Houghton, J.T., Meira Filho, L.G., Callander, B.A., Harris, N., Kattenberg, A., and Maskell, K., eds., 285-357. Cambridge: Cambridge University Press, 572 pp.

Nicholls, N., Gruza, G.V., Jouzel, J., Karl, T.R., Ogallo, L.A., and Parker, D.E., 1996. Observed climate variability and change. In *Climate Change 1995. The Science of Climate Change: Contribution of Working Group I to the Second Assessment Report of the Intergovernmental Panel on Climate Change*. Houghton, J.T., Meira Filho, L.G., Callander, B.A., Harris, N., Kattenberg, A., and Maskell, K., eds., 133-192. Cambridge: Cambridge University Press, 572 pp.

(Received 5 May 1998; in revised form 20 November 1998)

WORKSHOP ON INDICES AND INDICATORS FOR CLIMATE EXTREMES, ASHEVILLE, NC, USA, 3-6 JUNE 1997

BREAKOUT GROUP A: STORMS

KEVIN E. TRENBERTH

National Center for Atmospheric Research, Box 3000, Boulder, CO 80307 U.S.A.

TIMOTHY W. OWEN

NOAA/National Climatic Data Center, 151 Patton Avenue, Asheville, NC 28801 U.S.A.

Abstract. The Working Group on Storms considered tropical cyclones, extratropical cyclones, thunderstorms and their associated winds and effects other than on temperatures and precipitation (which are dealt with by the other working groups) to be in their purview. Changes in observing systems and distribution of observers and people impacted by these phenomena confound trend analysis. In light of the difficulty of assembling homogeneous time series of small-scale phenomena such as thunderstorms, tornadoes and hail, and also the problems in wind measurements, the working group recommends that indices of wind be developed by taking advantage of long surface (or sea-level) pressure measurements and analyses. Because wind is a vector, two pairs of readings that are orthogonal are desirable. Instantaneous values over about 1000 km scales are desirable to generate statistics relevant to wind extremes. Recommendations are given on how the data might profitably be processed. Several other recommendations are made concerning data acquisition and processing, some of which apply to reanalysis of past data and some apply to future processing of data. Various "extremes indices" are also suggested.

1. Introduction

When society is impacted by an extreme meteorological event other than drought or heat wave, the event is almost certainly some type of storm. Storms, loosely defined here to include tropical cyclones, extratropical cyclones and thunderstorms (with associated small-scale phenomena), can result in substantial property losses and, in some cases, loss of life. Unfortunately, climate researchers have long been hampered in their ability to quantify long-term changes in the frequency or intensity of extreme events directly related to storms for a number of reasons. These include the spatial and temporal variability of the phenomena and acute limitations in the number of observations and thus data availability and quality (i.e, period of record, numbers and types of parameters, consistency), e.g., see Doswell and Burgess (1988). For instance, the annual number of tornadoes in the United States is steadily increasing, but this is simply because there are more people who are more widely distributed to observe them (a more comprehensive discussion of tornadoes and problems with the database are given by Church et al. (1993-- see especially pp. 459-466). The Working Group on Storms at the June, 1997 Workshop on Indices and Indicators for Climate Extremes was charged with making recommendations to the climate data and research community in order to mitigate these limitations and propose potential extremes indices. The following report summarizes key findings of the working group in each of the three aforementioned categories of storms, in addition to its major recommendation.

The main thrust of the recommendations here is on building databases suitable for analysis of how phenomena that cause disruptions and extremes have changed. We note, in passing, the desirability to also include information wherever possible about the processes involved and the linkages among different extremes. In particular, heavy rainfall originates from the same storms considered here, but is dealt with by a separate working group. Similarly, temperature extremes often occur with extratropical storms. Therefore, extremes of wind, hail and other phenomena are often related to extremes in these other categories.

2. Major Recommendation

In light of the difficulty of assembling homogeneous time series of small-scale phenomena such as thunderstorms, tornadoes and hail, and also the problems in wind measurements (as discussed below), the working group recommends that *indices of wind be developed by taking advantage of long surface (or sea-level) pressure measurements and analyses.* Because wind is a vector, two pairs of readings that are orthogonal are desirable. Sea-level pressure readings are robust to small station shifts in location and elevation, and also instrument changes. For reliability it is desirable to choose station pairs when available, although pairs of grid points from sea-level pressure analyses can and should also be used. While there is considerable interest in pairs of station pressure readings over large distances that represent major circulation features, such as the North Atlantic Oscillation (which uses normalized pressure differences between Lisbon, Portugal and Stykkisholmur, Iceland for instance (Hurrell, 1995)), or large-scale zonal and meridional indices of wind, the need here addresses scales of about 1000 km and instantaneous values. Thus, it more directly addresses local wind speed, which is linked to wind damage and ocean waves. The latter are relevant for mariners and coastal erosion. We note that data provided at scales of 1000 km once or twice daily will not capture extremes from smaller-scale phenomena such as tropical cyclones. Therefore, where possible, higher spatial resolution and temporal sampling should be used.

The following steps can be taken to implement the major recommendation:

1. Determine a geostrophic wind vector at least once daily, and preferably four times daily, with a 1000 km (about 10 degrees latitude) resolution globally or at least throughout the extratropics.
2. Relate this for limited periods to observed winds for calibration and interpretation purposes. Also perform, where possible, correlative analysis with

ocean wave data (roughness, swell, wave height, etc).

3. Generate wind statistics of all kinds, especially for wind speeds exceeding certain thresholds, such as the main Beaufort categories.

Table I

Beaufort categories

Beaufort Category	Description	Threshold Wind Speed (knots/meters per second)
0 to 4	Calm to Moderate Breeze	up to 16 / 7.9
5 and 6	Fresh and Strong Winds	up to 27 / 13.8
7 and 8	Moderate and Fresh Gales	up to 40 / 20.7
9 and 10	Whole Gale Force	up to 55 / 28.4
11	Storm Force Winds up to 63 / 33.5	
12 and greater	Hurricane Force Winds	> 63 / > 33.5

Include monthly statistics such as wind run (cumulative wind), scalar speed, prevailing direction, vector component monthly means and standard deviations, and indices such as the number of "gale days", and estimated percentiles and return periods.

4. Analyze wind variability, how the statistics have changed with time, and whether the changes are significant (physically and statistically). In particular, compare estimates of return periods based on statistics for different decades and see how these are changing.

These analyses can most readily be performed from gridded sea-level pressure analyses which are available for the extratropical northern hemisphere from 1899 and regionally further back in time. However, the reliability is questionable in some places in earlier years (Trenberth and Paolino, 1980), and so cross validation is desirable. In making use of the results of these analyses, it should be recognized that mean wind speed is only one factor in wind damage, in particular, turbulence and gustiness of wind contributes substantially and allowance for gust factors in assessing damage can and should be done (e.g., Krayer and Marshall, 1992).

3. Tropical Cyclones

Tropical cyclones are warm core, convective, oceanic systems with a mesoscale (10-100 km) radius of maximum winds. This category of storms regularly impacts tropical coastal locations along the Atlantic, Pacific and Indian Oceans. In light of their ability to cause many deaths (e.g., ~300,000 from a single cyclone in Bangladesh in 1970) and massive property losses (e.g, $26.5 billion from Hurricane Andrew in Florida in 1992) (Holland, 1993; Hebert et al., 1996), better quantification of tropical cyclone variability is particularly desirable to the meteorological, re-insurance and emergency management communities.

In addition, there is considerable interest in how tropical cyclones may change as the climate changes, yet considerable uncertainty remains. While global warming from increasing greenhouse gases in the atmosphere is apt to produce increases in sea surface temperatures, which would favor more and stronger tropical cyclones, changes in atmospheric stability and circulation may produce offsetting tendencies. General circulation models of the atmosphere do not resolve the scales required to properly address this issue (for instance, the stability of the atmosphere in the tropics is dominated by moist convection and thus sub-grid scale processes which are parameterized in models, and hurricanes can not be resolved adequately either). Therefore, reliable estimates of observed trends in tropical storms are important, although it should be recognized that decreases in one area may be offset by increases in another area because of the global connectivity of the tropical atmospheric circulation, such as is manifested during El Niño events. During an El Niño, for example, the incidence of hurricanes typically decreases in the Atlantic and far western Pacific and Australian regions, but increases in the central Pacific (Gray, 1984; Nicholls, 1979; Revell and Goulter, 1986; Lander, 1994; Jose and Manalo, 1997).

Globally, there are a number of tropical cyclone data sets that are stratified by regional basin (including the Atlantic, N.E. Pacific, N.W. Pacific, N. Indian, S. Indian and S. Pacific basins). The Global Tropical/Extratropical Cyclone Data Set (NCDC, 1997) includes two parameters for all basins: six-hourly cyclone positions and intensities (measured by maximum sustained winds (for one or ten minutes) at ten meters and minimum sea-level pressure). The period of record for reliable data is only three decades in most basins --satellite observations began only in the mid-1960s. The Atlantic and N.W. Pacific basins are exceptions to this general rule as tropical cyclones in these areas have been monitored by aircraft reconnaissance since the mid-1940s. Additional ship-borne observations, including wave height, can be obtained from Comprehensive Ocean-Atmosphere Data Set (COADS; Elms, 1995).

In spite of these limited data, several studies of interdecadal tropical cyclone variability have been performed. For many years, individual tropical cyclones have been categorized by a correlation of their meteorological intensity to their impact on agriculture and housing (e.g., Saffir-Simpson scale, Philippine Typhoon Damage scale). Another simple but useful definition of extreme tropical cyclones is when particular thresholds are exceeded for a given value of maximum sustained winds. For example, tropical cyclones whose maximum sustained winds exceed 95 kts or 49 ms^{-1} comprise categories 3, 4, 5 of the Saffir-Simpson scale. These storms are categorized as "intense hurricanes" in the Atlantic Basin. Additional indices have been proposed and/or used in the Atlantic Basin and, to a more limited extent, in other basins: number of tropical storms (>=34 kt, 18 ms^{-1}); number of hurricanes (>=64 knots, 33 ms^{-1}); number of intense hurricanes; number of hurricane days.

The working group recommends that these indices be computed for all basins.

The frequency of intense hurricanes in the Atlantic and super typhoons in the N.W. Pacific over the last several decades (since the advent of aircraft reconnaissance) is not well documented because of a bias in the databases. From the mid-1940s through the late-1960s, the wind strength of these storms was overestimated in comparison to the minimum central pressure observed (Landsea, 1993; Black, 1993). There is an immediate need to reanalyze these data sets and remove this overestimation bias for these strongest storms. Intense hurricanes are of extreme interest, because even though they account for only 21% of all U.S. landfalling tropical cyclones, they cause over 82% of the normalized tropical cyclone damage (Pielke and Landsea, 1998).

One outstanding question concerns the extent to which landfalling tropical cyclones should be analyzed separately from all tropical cyclones. The advantage of the latter is that events are not as rare and thus can be analyzed with more certainty. Despite the smaller frequency, we can extend the reliable record further into the past by only using landfalling tropical cyclones (i.e., U.S. landfalling hurricanes provide a reliable record back to the turn of the 20th century (Neumann et al., 1993)).

Other data sets could be used to derive extremes indices for tropical cyclones if they were systematically acquired for that purpose. Hourly and daily synoptic reports could be quality controlled and merged with other tropical cyclone data. Damage assessments could also be acquired from the re-insurance industry. Regional hurricane centers charged with hurricane monitoring could regularly observe additional variables and regularly update their data sets. The working group advocates implementation of all of these data acquisition efforts.

The working group also suggests that each regional hurricane center tasked with a basin of responsibility include the following variables in their tropical cyclone data sets: radius of maximum wind; radius of 34 kt ($18 ms^{-1}$) wind; radius of 64 kt ($33 ms^{-1}$) wind; annotation of asymmetric and convective structure; and the date and time of landfalling tropical cyclones. The group recommends that regular reports of a tropical cyclone include the date, latitude, longitude, maximum sustained winds and gusts. If synoptic stations are encompassed by a tropical cyclone, acquisition of temperature, dew point, sea-level pressure, and precipitation data is also desirable. Following the landfall of a tropical cyclone (especially "intense" tropical cyclones), centers are encouraged to provide the following information: maximum wind swath, storm surge, rainfall distribution and damage/fatality reports. While it is important to implement these recommendations for future storms, it is also highly desirable to build historical databases of the same variables, wherever the information is available.

Based upon expanded data available, as described above, the following extremes indices could be developed: strongest cyclone; mean cyclone intensity; largest cyclone (34 kt radius); mean cyclone size; extreme track variation;

normalized/actual impacts (damages). The latter includes normalization for effects such as changes in number and value of structures vulnerable to the storms and inflation. Research should continue into better indices that better integrate and characterize the total impact of tropical storms such as the eddy kinetic energy of the storm.

In addition to broadening the number of parameters, the working group encourages hurricane centers to provide the climatological community with regular, annual updates of their databases, including both original and corrected values.

4. Extratropical Cyclones

Extratropical cyclones are baroclinic low pressure systems that occur throughout the middle latitudes. Their broad potential for causing property damage in the mid-latitudes, particularly in the form of winter storms, is well-documented. Here the main interest is in wind, as the other working groups will cover temperature and precipitation. However, in the absence of exhaustive metadata, wind measurements are deemed unreliable for analysis as they are particularly subject to numerous inconsistencies, including instrumentation/station relocations, instrumentation malfunction (anemometers need regular servicing and calibration because of their moving parts) and differences in measurement methodologies (e.g., frequency, duration, type (absolute gust, fastest mile, fastest minute)), and siting changes (such as growth of nearby trees that produce a sheltering effect). In place of direct wind measurements, derived wind measurements based on pressure gradients/geostrophy are encouraged.

Extratropical cyclones generate economically significant costs and benefits. Rainfall and snowfall, in particular, often bring much needed moisture, although excesses can cause flooding. Extratropical cyclones are major producers of disruptive snowstorms, frequently at surprisingly low latitude. Given increasing and high population densities and property values along the coasts, coastal storms are among the most significant events in terms of potential economic impacts and the vulnerability of property is increasing. The relatively high quality of land and near shore (shipping track) observations (from data sets like COADS) indicates that meaningful historical and secular climatologies of these events either can or have been developed for specific locations. Davis and Dolan (1993) used extratropical cyclone frequencies and intensities to calculate damaging wave climates for a small portion of the Atlantic Coast. This type of work can be generalized to much longer coastlines. Hayden (1982) amassed the cyclone record of eastern North America and the western Atlantic Ocean.

Data directly associated with extratropical cyclones generally consist of a temporal series of positions, known as tracks. In some cases, data regarding the intensity of a storm is also available. These are both barometrically-derived parameters. In the United States and eastern portions of the North Atlantic Ocean, the tracks of individual extratropical storms have been published since 1874. The National Climatic Data Center's Global Tropical and Extratropical Cyclone Climatic Atlas (GTECCA) (NCDC, 1997) provides northern hemispheric extratropical tracks for the period 1965-1995.

Several uses have been made of extratropical cyclone track data. One such study tallies the monthly passages of storms through 182 2.5^o latitude by 5° longitude grid cells from 55N, 125W to 22.5N, 55W from 1885 through 1996. The data set provides a record of storm frequency dating back into the late 19th century which can be used to study relations between storm tracks and climate variables (snowfall, temperature, precipitation, etc.) as well as both temporal and spatial changes in atmospheric circulation patterns over the course of the last century (existing upper air records only extend back to the late 1950s in the Northern Hemisphere). Knappenberger and Michaels (1993) demonstrated that a strong relationship between cyclone tracks and snow distributions can be developed and that this relationship displays considerable secular change. Michaels et al. (1990) have also demonstrated that cyclones tracks can be used to reconstruct upper air thickness records many decades before direct measurements of these values had been made. Unfortunately, the extratropical cyclone track data set does not include central pressure and thus cannot be directly used in studies of storm intensities, which limits its usefulness in circulation change studies. Some intensity proxies are available from wind records as well as from surface or sea-level pressure records (available back in to the nineteenth century in some locations) in the vicinity of extratropical storms.

Where high quality, homogeneous *in situ* observations of sea-level pressure are available, they can be used either alone or in conjunction with track data and serve as the basis for the creation of new extremes indices. In the United States, daily sea-level pressure maps of the region north of 20° N exist since 1899, and monthly means of these from NCAR have been evaluated extensively by Trenberth and Paolino (1980). The United Kingdom Meteorological Office (UKMO) has an alternative version of monthly mean maps dating from 1873 north of 15° N (see Jones, 1987). No analyses are available, however, across the Pacific Ocean between 1881 and 1898, and the UKMO and NCAR versions are essentially the same from 1899-1939 (Williams and van Loon, 1976). Changes in coverage of observations have impacted some areas and changes in procedures, especially corrections from surface to sea-level pressures in regions of high topography, have also had a substantial impact on the homogeneity of the record. Problems are especially apparent over the Arctic region before about 1931 where analysts widely assumed the existence of a high pressure system that was not in fact present (Madden, 1976; Trenberth and Paolino, 1980; Jones, 1987) and Jones has devised corrections for these analyses. For the NCAR data set, Trenberth and Paolino (1980) devised a number of corrections over the Himalayan-Tibetan Plateau complex that coincided with known changes in procedures to adjust for the main discontinuities, but these have been applied only to the monthly means. Detailed evaluations including comparisons with many station records led to the conclusion that the analyses were most reliable only after about 1924.

A number of regional pressure analyses have been performed in the southern hemisphere but the first extensive series of daily hemispheric maps were prepared leading up to and following the International Geophysical Year (IGY) (1957-58) when there was a major expansion in the observing network. A series of these maps were published by the South African Weather Bureau in *Notos* for 1951-62, and these formed the basis for many studies that culminated in the papers published in the *Southern Hemisphere Monograph*, see especially van Loon (1972). More recently, gridded hemispheric sea-level pressure maps have been made available by the World Meteorological Centre in Melbourne beginning in 1972, and these have been evaluated and used by Trenberth (1979, 1984) and Swanson and Trenberth (1981). Further evaluation and extensions of both sets of analyses has been carried out by Jones and Wigley (1988) for the Antarctic region and Jones (1991) for the region south of 15 S. In the southern hemisphere, spatial data gaps are a substantial problem in producing reliable analyses as the frequency of ship reports over the southern oceans is usually too few to allow reliable analyses to be constructed. In earlier years in the summer half year, whaling and sealing fishing vessels improved the data coverage. More recently, following the very positive experience during the Global Weather Experiment in 1979, drifting buoys have been used over the southern oceans to provide a basic network of surface observations.

Track and sea-level pressure can be used jointly in quantifying the frequency, track and duration of extratropical cyclones at varying thresholds of maximum intensity. Sea-level pressure extremes and six-hourly pressure changes at a station can be analyzed. In addition, regional indices can be based upon pressure differences between two or more stations. Geostrophic winds can be calculated from pressure gradients and compared with actual winds. Finally, large-scale indices can be based upon analyzed sea-level pressure fields, and can then be used to calculate circulation indices of zonality and/or meridionality (blocking).

Further new data sets that can be used in the analysis of extratropical cyclones are the recent reanalyses (e.g., the National Centers for Environmental Prediction (NCEP)/NCAR reanalyses or those from the European Centre for Medium Range Weather Forecasts (ECMWF)). The reanalyses can provide statistics of various kinds relevant to storms and storm tracks throughout the atmosphere. These include total and band-passed (e.g., high frequency 2 to 8 day period) statistics of variances and covariances involving pressure or geopotential height, temperature, specific humidity, horizontal velocity, vertical velocity (or omega), as well as model-derived fields of precipitation. Precipitation fields can be further broken up into large-scale and convective contributions, and evaluated and compared with observed precipitation. Reanalyses of global data are expected to extend back to about 1956, but the temporal extent of the database varies and may be inadequate in some periods for certain regions, especially in the southern hemisphere. As well as relating the variance and covariance statistics to mean fields, the reanalysis data can provide quantitative measures of the magnitude and spatial distribution of variability that arises from individual storms. These will likely facilitate analysis of interannual and longer-term variations of storm tracks.

The availability of these comprehensive new data sets allow evaluation of the utility of storm track data sets and the NCDC's Comprehensive Aerological Reference Data Set (CARDS; Eskridge et al., 1995) can also be useful in providing a more complete picture. The NCDC is also planning a comprehensive global surface data set of hourly and daily data. Such a data set would be very beneficial in studying extreme events. However, the major recommendation from this section concerns generation of local wind statistics, as given above.

5. Thunderstorms (and Associated Small-Scale Phenomena)

Thunderstorms and small-scale weather phenomena (SCWP) are primarily characterized by quasi-random temporal and spatial events. These events, in turn, have local and regional impacts, often with significant damage and sometimes even loss of life. Tornadoes and thunderstorms with related phenomena such as lightning, hailstorms, wind, dust storms, water spouts, downpours and cloud-bursts belong to this group. The damage arises from strong winds, blowing debris, lightning strikes and fires, flooding, and hail damage (which is more severe when the hail stones are larger). Characteristics of observed SCWP are estimated (frequency, intensity) or measured (amount, value). Corresponding data have hourly or daily resolution, from which monthly or annual series can be compiled. Occurrence/non-occurrence of the phenomena are regularly noted by meteorological stations in summary statistics. These summaries are generally available in digital form from national meteorological services for recent decades (in Europe, primarily for the period 1961-1990 and again after 1990). Qualitative estimates of intensity are less readily available and are often not digitized.

Unfortunately, areally consistent values of SCWP are inherently elusive. Statistics of relatively rare events are not stable at single stations and stations are too far apart to aggregate and obtain reliable area averages. Observation practices can be subjective and vary broadly between nations, and the metadata outlining these practices are often not readily available to researchers. Relevant metadata include information about the definition of phenomena in instructions for observers, changes of observers and in observation site relocation. Only high-quality stations with professional observers can produce valuable time series of SCWP, while observations by voluntary observers can be biased by some errors (e.g., observations during the night). Metadata are less critical with automatic recording instrumentation, such as recording rain gauges or anemographs (wind speed and direction). Of course, degradation or failure of such instrumentation is not uncommon in the absence of vigilant maintenance schedules.

In light of the spatial variability of SCWP, the density of surface meteorological observing stations is too coarse to assure that all such events are directly measured. As a result, simple application of traditional methods of relative homogeneity testing on series of SCWP characteristics, such as comparing incidences at neighboring stations, is not possible due to very low correlation coefficients. For example, in a relatively small area of a few thousand km^2, agreement in concurrent occurrence of the same phenomena by two or more stations is relatively very low (Brazdil and Vais, 1997). However, single stations can be compared with areal series of SCWP statistics where the correlation will probably be higher. In addition, agreement in occurrence of local maxima and minima in smoothed series as well as trends of the same signs for different SCWP (e.g., thunderstorms, hailstorms, cloud bursts) can be used for checking possible disharmony between elaborated series (Brazdil and Vais, 1997).

A more comprehensive way to check on the historical record is to make use of remote sensing measurements. These include ground based radar and satellite observations both of which provide much more comprehensive spatial coverage but often only for recent times. Such data do, however, through comparisons with conventional reports, allow the historical record to be evaluated for how many SCWP might be missed and thus the record might be calibrated. It is recommended that this kind of evaluation should take place.

If time series of SCWP characteristics for high-quality stations or areal means for climatologically uniform regions are compiled, their analysis is possible. Such analysis should be based on statistical characteristics and the probability of occurrence for different SCWP. Temporal changes in series might be studied by using smoothing and linear trend analyses. For explaining observed trends, a synoptic-climatological analysis of SCWP can be used. For example, in south Moravia, Czech Republic, a decreasing linear trend of thunderstorms, hailstorms and heavy rain in 1946-1995 was connected with a significant decrease in the occurrence of these phenomena during cyclonic situations, which is when 90% of these phenomena occur (Brazdil and Vais, 1997).

Additional simple analyses of SCWP variability can be recommended for single stations or local-scale networks of stations where at least thirty years of homogeneous station observations exist and metadata are present to confirm no changes in siting or observation procedure. Recommended analyses of the parameters include basic statistics (mean, variance, percentiles, etc.), probability of occurrence (repetition), fluctuation (smoothing, cyclicity, etc.), trends, and explanations of observed fluctuations. In addition, the summary statistics can be used in correlative analyses with temperature, hail and heavy precipitation.

From a practical point of view, impacts of SCWP are important. But such information must be taken from original records of single weather stations or from other sources (e.g., data of insurance agencies, reports in newspapers). Changnon

(1997) states that in the United States, only 5 to 10% of all thunderstorms and 10 to 25% of all hail produce damage. This is consistent with results for central Europe (Brazdil and Vais, 1997). When combined with damage reports from the re-insurance industry, indices based upon the number of days (or fraction of days) with damage due to thunderstorms, hail, excessive precipitation, lightning and wind could be compiled. However, because of increases and dispersion of population, more property is becoming vulnerable and thus before conclusions can be drawn about SWCP changes, allowance has to be made for these factors.

There is a dearth of papers devoted to climatological analyses of SCWP with emphasis on variability and impacts (but see Changnon, 1985; 1997) and this is likely a function of the climate research community's desire to work with more reliable, larger-scale, more quantitatively-based data sets. However, it is clear that verification of mesoscale and higher resolution modeling results will require more attention to SCWP data in the near future. By combining the statistics with the National Severe Storms Laboratory's Severe Weather Climatology, it may be possible to make quantitative, larger-scale assessments in the United States that may be more broadly embraced by climate researchers. The statistics could be used to verify other model results at the local-, meso- and regional-scale, respectively. Remotely sensed data, especially from radar, may provide useful spatial statistics for recent periods.

Model verification of SWCP is being performed over central Europe in advance of IPCC 2000, and emphasis will be placed on analyzing SCWP using high-quality stations and corresponding areal means. The working group recommends that high-quality SCWP stations be identified and that the same methodology be applied in other parts of the world (e.g., North America, other parts of Europe). A comparison of these results might help improve the currently weak understanding of variability of these phenomena. The working group also recommends that the longest possible series of SCWP should be digitized on the level of national meteorological services and prepared for international exchange. However, an awareness of the limitations of these data is essential if they are to be used properly.

Acknowledgements

The contributions of the following individuals to this report are gratefully appreciated: Dr. Rudolf Brazdil, Department of Geography, Masaryk University, Czech Republic; Richard Heim, NOAA/National Climatic Data Center, U.S.A.; Dr. Aida Jose, Phillipine Atmospheric, Geophysical and Astronomical Services Administration, Philippines; Dr. Paul Knappenberger, Department of Environmental Sciences, University of Virgina, U.S.A.; Dr. Chris Landsea, NOAA/AOML/Hurricane Research Division, U.S.A.; Neal Lott, NOAA/National Climatic Data Center, U.S.A.; Dr. Patrick Michaels, Department of Environmental Sciences, University of Virgina, U.S.A.; and Dr. Gudrun Rosenhagen, Deutscher Wetterdienst, Germany. In addition, the participation of the following individuals in Breakout Group A during the June 1997 Workshop on Indices and Indicators for Climate Extremes is duly noted: George H. Baldwin, Gerhard Berz, David Changnon, Stanley Chagnon, Graham Cook, Kenneth Davidson, Richard Davis, Henry Diaz, Raymond Fosse, Chuck Hakkarinen, Kenneth Kunkel, William Murray, Fielding Norton, Richard Smith and Nobou Yamazaki.

References

Black, P.G., 1993: Evolution of maximum wind estimates in typhoons. *Tropical Cyclone Disasters*, J. Lighthill, Z. Zhemin, G. Holland, and K. Emanuel, Eds. Beijing: Peking Univ. Press, 104-115.

Brazdil, R. and T. Vais, 1997: Thunderstorms and related weather extremes in south Moravia, Czech Republic, in 1946-1995: Data, results, impacts. *Preprints of the Workshop on Indices and Indicators for Climate Extremes*, NOAA/NCDC, Asheville NC USA, 3-6 June 1997, 4 pp.

Changnon, S.A., 1985: Secular variations in thunder-day frequencies in the twentieth century. *J. Geophys. Res.*, **90**, 6181-6194.

Changnon, D., 1997: Damaging storms in the United States: Selection of quality data and monitoring indices. *Preprints of the Workshop on Indices and Indicators for Climate Extremes*, NOAA/NCDC, Asheville NC USA, 3-6 June 1997, 24 pp.

Church, C., D. Burgess, C. Doswell, and R. Davies-Jones, 1993: The Tornado: Its Structure, Dynamics, Prediction and Hazards. In *Geophysical Monograph 79*. Washington: American Geophysical Union, pp. 459-466.

Davis, R.E. and R. Dolan, 1993: Nor'easters, *American Scientist*, **81**, 428-439.

Doswell, C.A. and D.W. Burgess, 1988: On some issues of United States tornado climatology. *Mon. Wea. Rev.*, **116**, 495-501.

Elms, J.D., 1995: COADS Project Report: Early data digitization and United States code history. *Proceedings of the International COADS Winds Workshop*, Kiel, Germany, 31 May-2 June 1994, U.S. Department of Commerce, NOAA/ERL/CDC and Institut fur Meereskunde, 29-36.

Eskridge, R.E., O.A. Alduchov, I.V. Chernykh, Z. Panmao, A.C. Polansky and S.R. Doty, 1995: A Comprehensive Aerological Reference Data Set (CARDS): Rough and Systematic Errors. *Bull. Amer. Met. Soc.*, **76**(10), 1759-1775.

Gray, W.M., 1984: Atlantic seasonal hurricane frequency: Part I. El Niño and 30 mb quasi-biennial oscillation influences. *Mon. Wea. Rev.*, **112**, 1649-1668.

Hayden, B.P., 1982: Secular variation in Atlantic Coast Extratropical Cyclones. *Mon. Wea. Rev.*, **109**, 159-167.

Hebert, P.J., J.D. Jarrell, and M. Mayfield, 1996: The deadliest, costliest, and most intense United States hurricanes of this century (and other frequently requested hurricane facts) *NOAA Tech. Memo.*, NWS TPC-1, Miami, Florida, 75-104.

Holland, G.J., 1993: Ready Reckoner - Chapter 9, *Global Guide to Tropical Cyclone Forecasting*, WMO/TC - No. 560., Report No. TCP-31, World Meteorological Organization, Geneva.

Hurrell, J. W., 1995: Decadal trends in the North Atlantic oscillation regional temperatures and precipitation. *Science*, **269**, 676-679.

Jones, P. D., 1987: The early twentieth century Arctic high -fact or fiction? *Climate Dyn.*, **1**, 63-75.

Jones, P. D., 1991: Southern Hemisphere sea level pressure data: An analysis and reconstructions back to 1951 and 1911. *Intl. J. Climatol.*, **11**, 585-607.

Jones, P. D., and T. M. L. Wigley, 1988: Antarctic gridded sea level pressure data: An analysis and reconstruction back to 1957. *J. Climate*, **1**, 1199-1220.

Jose, A.M. and V.C. Manalo, III, 1997: A Study of Potential Indicators/Indices of Tropical Cyclone Variability/Changes in the Phillipine Vicinity (1948-1996). *Preprints of the Workshop on Indices and Indicators for Climate Extremes*, NOAA/NCDC, Asheville, NC USA, 3-6 June 1997, 7 pp.

Knappenberger, P.C. and P.J. Michaels, 1993: Cyclone Tracks and Wintertime Climate in the Mid-Atlantic Region of the USA. *Intl. J. Climatol*, **13**, 509-531.

Krayer, W.R. and R.D. Marshall, 1992: Gust factors applied to hurricane winds. *Bull. Amer. Met. Soc.*, **73**, 613-617.

Lander, M., 1994: An exploratory analysis of the relationship between tropical storm formation in the Western North Pacific and ENSO. *Mon. Wea. Rev.*, **122**, 636-651.

Landsea, C.W., 1993; A climatology of intense (or major) Atlantic hurricanes. *Mon. Wea. Rev.*, **121**, 1703-1713.

Madden, R. A., 1976: Estimates of the natural variability of time-averaged sea-level pressure. *Mon. Wea. Rev.,* **104**, 942-952.

Michaels, P.J., D.E. Sappington, D.E. Stooksbury, and B.P. Hayden, 1990: Regional 500mb heights and U.S. 1000-500mb thickness prior to the radiosonde era. *Theor. and Appl. Clim.*, **42**, 149-154.

National Climatic Data Center (NCDC), 1997: *Products and Services Guide.* Asheville, NC: U.S. Department of Commerce, NOAA, 60 pp.

Nicholls, N., 1979: A possible method for predicting seasonal tropical cyclone activity in the Australian region. *Mon. Wea. Rev.*, **107**, 1221-1224.

Neumann, C.J., B.R. Jarvinen, C.J. McAdie, and J.D. Elms, 1993: *Tropical Cyclones of the North Atlantic Ocean, 1871-1992*, Prepared by the National Climatic Data Center, Asheville, NC, in cooperation with the National Hurricane Center, Coral Gables, FL, 193 pp.

Pielke, R.A., Jr., and C.W. Landsea, 1998: Normalized Atlantic hurricane damage, 1925-1995. *Wea. Forecasting*, in press.

Revell, C.G. and S.W. Goulter, 1986: South Pacific tropical cyclones and the Southern Oscillation. *Mon. Wea. Rev.*, **114**, 1138-1145.

Swanson, G. S., and K. E. Trenberth, 1981: Trends in the Southern Hemisphere tropospheric circulation. *Mon. Wea. Rev.*, **109**, 1879-1889.

Trenberth, K. E., and D. A. Paolino, 1980: The Northern Hemisphere sea level pressure data set: Trends errors and discontinuities. *Mon. Wea. Rev.*, **108**, 855-872.

Trenberth, K. E., 1979: Interannual variability of the 500 mb zonal mean flow in the Southern Hemisphere. *Mon. Wea. Rev.*, **107**, 1515-1524.

Trenberth, K. E., 1984: Interannual variability of the Southern Hemisphere circulation: Representativeness of the year of the Global Weather Experiment. *Mon. Wea. Rev.*, **112**, 108-123.

van Loon, H., 1972: Pressure in the Southern Hemisphere. In *Meteorology of the Southern Hemisphere*, C. W. Newton (Ed.), *Met. Monogr.*, **13**, Amer. Meteor. Soc., 59-86.

Williams, J., and H. van Loon, 1976: An examination of the Northern Hemisphere sea level pressure data set. *Mon. Wea. Rev.*, **104**, 1354-1361.

(Received 5 November 1997; in revised form 11 August 1998)

WORKSHOP ON INDICES AND INDICATORS FOR CLIMATE EXTREMES: ASHEVILLE, NC, USA, 3-6 JUNE 1997 BREAKOUT GROUP B: PRECIPITATION

NEVILLE NICHOLLS[1] AND WILLIAM MURRAY[2]

[1] *Bureau of Meteorology Research Centre, BMRC, Melbourne, Australia*
[2] *NOAA Office of Global Programs, Silver Spring, Maryland, USA*

Abstract. A uniform, international reference system of precipitation indices would greatly facilitate assessment of changes in global precipitation patterns, intensities and extremes. However, national/ regional differences in precipitation monitoring standards, data quality control procedures, and product development practices complicate efforts to develop such a system. This report represents the results of Working Group B's concerted effort to examine in detail the problems associated with the development of the needed indicators and the Group's recommendations to address the identified issues. The Group concluded that a successful strategy must define a minimum set of indices/indicators based on higher quality data that would represent a global base set. The Group identified a set of indices for this purpose. This base data set should be complemented where possible by countries/ regions having the data sets and processing resources to do more.

1. Introduction

Heavy precipitation, on a variety of spatial and temporal scales, causes considerable damage and loss of life worldwide, each year. However, mechanisms to monitor, understand, and predict precipitation extremes are inadequate, largely because of the lack of high-quality data and the absence of a concerted, global effort to apply consistent analyses to the available data. There is also an absence of relatively simple, uniform indices and indicators of precipitation in large parts of the world that can be used to help answer basic questions such as the IPCC question of whether climate is getting more extreme - these answers can't be provided if each country or area has its own definitions.

Working Group B addressed the needs for such indices, the availability of precipitation data from countries around the globe, and efforts that should be made to integrate these data in a common data base for international use and assessment of climate extremes. The Working Group identified a number of indices and indicators of precipitation extremes which should be applied consistently to available global data. The Group also made recommendations regarding aggregation of these indices into indicators of extremes on large spatial scales, on the development of comprehensive data sets to enhance globally-consistent analysis, and on methods for quality control and the removal of inhomogeneities in data for monitoring precipitation extremes. Working group discussions are summarized in the following sections.

Climatic Change **42**: 23–29, 1999.
© 1999 *Kluwer Academic Publishers. Printed in the Netherlands.*

2. Indices For Monitoring Precipitation Extremes

Calculation of the following indices would provide a comprehensive description of the variation of important precipitation extremes. The Group recognized that other indices may provide further information, but felt that the following set would enable useful conclusions to be reached regarding the variation and distribution of extreme precipitation. Most of the recommended indices relate to daily rainfall. It was pointed out that hourly extreme precipitation indices will be more difficult to calculate in many countries, because of the lack of digitized data.

2.1. INDICES TO BE CALCULATED FROM DAILY DATA

• Number of dry days; hail days; snow days. The definition for each of these indicators would correspond to the definition from the country of interest.
• Frequency of exceeding specified thresholds - Number of days with precipitation exceeding 90th, 95th, 99th percentiles. Do this for calendar year, seasons, or other period as appropriate (e.g., wet season in monsoonal regions). Use only days with precipitation to calculate thresholds. Calculate thresholds by simple counting and/ or fitting gamma distribution. Calculate thresholds on 1961-90 reference period. Where appropriate repeat analysis separately for snow.
• Variations in magnitude of thresholds - Calculate 90th, 95th, 99th percentiles for each year separately, by fitting days with precipitation to gamma distribution. Plot as time series and smooth and calculate trends.
• Simple daily intensity index (total precipitation divided by number of days with precipitation).
• Maximum length of dry spell - number of contiguous days with precipitation below threshold (1st percentile calculated on only days with precipitation; use 1961-90 reference period).
• Using decadal moving 30-year window (eg., 1941-70, 1951-80, 1961-90) compare frequency distribution of daily rainfalls. Examine variations in extremes of distribution.
• Percentage of annual (or seasonal as appropriate) precipitation falling on days with rainfall above 90th, 95th, 99th percentiles.

2.2. INDICES CALCULATED WITH HOURLY DATA

• Number of hours with precipitation.
• Frequency of exceeding thresholds, as for daily data (90th, 95th, 99th percentiles, calculated from 1961-90 reference period either by counting or fitting gamma distribution to hours with precipitation).
• Variations in magnitude of thresholds, as for daily data.

2.3. INDICES CALCULATED WITH MONTHLY DATA

• Percentage of country/region with precipitation in lowest or highest 5% (severe drought/wet period) or 10% (serious drought/wet period) in a year/season.

2.4. A MINIMUM SET OF INDICATORS

The above list would provide a comprehensive description of precipitation extremes on a variety of time scales. The detail provided by the calculation of all these indicators may lead to some confusion. For simplicity of interpretation (especially in comparative analyses), the Group proposed the following high-priority list of indicators. These are all relatively easy to calculate and are easily interpreted.
• Variations in magnitude of 95th percentile - Calculate for each year (or season) separately, by fitting days with precipitation to gamma distribution.
• Percentage of annual (or seasonal as appropriate) precipitation falling on days with rainfall above 95th percentile.
• Percentage of country/region with precipitation in lowest or highest 5% (severe drought/wet period) in a year/season.

2.5. RECOMMENDATIONS FOR FITTING DISTRIBUTIONS TO DETERMINE THRESHOLDS

It is recommended that thresholds (eg., 95th percentile) be determined by fitting the gamma distribution to the data. Thresholds can be estimated simply by counting the observed data, but fitting a distribution is preferred for the following reasons.
• Estimates of extreme quantiles are more accurate (ie., smaller standard error) if a distribution is fit, compared with calculating quantiles by counting.
• Distribution-fitting approach copes better with missing values or with data sets with very few non-zero values.
• More sophisticated analyses are possible by fitting distributions (eg., examine seasonal or spatial variations in terms of the parameters of the distribution).
• Techniques are available for checking whether the distribution provides a good fit to the data (eg., Q-Q plots, formal tests of fit such as Kolmogorov-Smirnov).

The two-parameter gamma distribution is used very widely for rainfall data and is well-understood. The group consensus was that it should be used unless analysis of data indicates that it is not appropriate. It was recognized that widespread use of the same distribution would facilitate comparative analysis between regions.

The preferred method of fit is maximum likelihood, because of its power and generality. Alternatives include ordinary method of moments or probability-weighted moments.

2.6. SPATIAL AGGREGATION

The indices listed above generally are calculated at individual stations, without aggregation to large areas (except for the percentage of country/region with monthly rainfall above or below thresholds). An approach to aggregate the extremes indices calculated on daily data is needed, to provide consolidated information (regarding the extremes in a catchment for instance). It may be desirable to calculate spatial average extreme indices for a country, a regional grouping of countries, major catchments, or major economic or agricultural zones. Two approaches are possible, depending on the amount and quality of the data available.
• If there are sufficient daily data available it may be feasible to prepare objective daily rainfall analyses of the region of interest, through the historical period of interest. The analyses can then be used to calculate daily regional average rainfalls. The extremes indices noted above can then be applied to these regional average rainfalls.
• If there are insufficient daily data to allow a daily historical rainfall analysis, then the extreme indices will need to be calculated at individual stations. Some form of averaging of the values of these indices, across the region, may then be feasible.
• If only a few stations are available no spatial average will be able to be derived.

3. Data Issues

3.1. MISSING DATA

Gap-filling of daily data for analysis of extremes may be possible, if only a small number of days are missing in any year. Fitting a gamma distribution to the available data may provide a means of doing this. It will be necessary to test this approach on data from the station or region, to determine the severity of biases likely to arise from this gap-filling.

3.2. ACCUMULATED DAYS

In some countries daily precipitation values are at times accumulated over two or more days (eg., at weekends). Indiscriminate inclusion of such multi-day values in an extreme index for daily rainfall will introduce biases (eg., resulting in exaggerated large "daily" rainfalls). If the frequency of such accumulations changes with time this will lead to artificial trends in analyses of the extremes. The extent of such biases needs to be explored in regions where such accumulations are common. This can be done by recalculating the extreme indices for longer accumulation periods (eg., 3 days or 7 days) or by excluding accumulated data completely (eg., excluding Monday rainfalls where rainfall is accumulated over the weekend).

3.3. HOMOGENIZING PRECIPITATION DATA

In many countries precipitation gauges have improved with time, leading to apparently higher precipitation. This bias can produce artificial trends in analyses of changes in precipitation extremes. This bias is likely to be most severe for the analysis of light rainfalls (eg., counts of number of dry days) or snowfall. Heavy precipitation amounts, especially in areas without snow, are less likely to be affected.

3.4. DATA ACCESS

Ideally, consistent analysis of extreme precipitation should be undertaken for all areas of the globe where appropriate data are available. Consistency of analyses would be more likely if the available data were provided to a single center which could then undertake the analysis. The World Data Center A for Meteorology has agreed to archive all data provided for this analysis.

However, some countries may wish to participate in the examination of extreme precipitation, but be unable to provide their data to a central location for archiving and analysis. In such cases it may be necessary to provide appropriate software and scientific support to these countries, to assist in the analysis of extreme precipitation.

In some other cases it may be more practicable for countries or organizations to retain the responsibility for archiving data and providing them to other researchers. In these cases a WWW page could provide links to such data sets.

3.5. DATA QUALITY

A number of members of the group mentioned the importance of basing indices and indicators of precipitation extremes on higher quality data and the need for avoiding the inclusion of "junk" values when dealing with the precipitation data.

4. Analyses

4.1. RESULTS OF ANALYSES

A WWW site will be needed to consolidate the results of these analyses of precipitation data. A comprehensive description of the analysis methods should form part of this site. As well, the station and spatially-aggregated results should be provided in a consistent format to allow comparison between stations, countries, and regions.

4.2. REMOTE-SENSED DATA

The above description of analyses for examining precipitation extremes relies solely on conventional precipitation data. Improved satellite and radar precipitation estimates are likely to become available in the near future. Some well-instrumented sites already combine high-density conventional data with radar and satellite observations. These sites can provide much higher quality information about the spatial variations of precipitation extremes. It would be useful to commence using data from a few such "flagship" sites to provide more spatial information regarding variations in extremes.

4.3. PRESENTATION OF ANALYSES

Most of the analyses described above will be carried out at individual stations. The station results will then be aggregated over a larger area (alternatively, station rainfalls may be analyzed, then the spatial average rainfall subjected to the analysis of extremes). It is recommended that, as well as providing the spatial average information, that spatial analyses of the station information be provided. This will provide a more illuminating exposition of the variations in extremes, and will allow users to note relationships between orographic features and the behavior of the extremes.

5. Summary

The following points summarize the Group discussions:

1. It is both possible and desirable to develop a set of indices and indicators that offers insight into precipitation extremes. To be most useful, the set selected needs to be feasible globally; that is, it must be relatively simple and straightforward since some countries are quite limited in terms of both the data sets available and the sophistication of processing and analysis capabilities that can be employed. Issues that the Group considered during discussions, in addition to selection of the recommended list of indicators, included the need to spatially aggregate data, problems associated with missing data, needs for homogenizing data, as well as data availability, access, and quality.

2. The Group recommended adoption of a minimum set of indicators that are relatively easy to calculate and interpret. Three high priority indicators were proposed: 1) variations in the magnitude of the 95th percentile value for days with precipitation; 2) percentage of annual (or seasonal) precipitation falling on days with rainfall above the 95th percentile; and 3) percentage of a specific country or region with precipitation in the lowest and highest 5% in a year or season to identify severe droughts and wet periods.

3. The above described minimum set of indices/indicators should represent a global base set. Group and plenary discussions emphasized that this base should be complemented where possible by countries/regions which have the data sets and processing resources to do more.

4. A WWW site will be needed to describe analysis methods used, consolidate the results, and disseminate the information. Station and spatially-aggregated results should be provided in a consistent format to allow comparison between stations, countries, and regions. Conventional precipitation data will be used for the recommended analyses of precipitation extremes, but improved satellite and radar precipitation estimates are likely to become available in the near future. The use of remotely sensed data to complement the conventional data, at least at selected sites, promises to provide additional useful information, particularly with regard to variations of precipitation extremes.

(Received 5 November 1997; in revised form 1 September 1998)

WORKSHOP ON INDICES AND INDICATORS FOR CLIMATE EXTREMES, ASHEVILLE, NC, USA, 3-6 JUNE 1997

BREAKOUT GROUP C: TEMPERATURE INDICES FOR CLIMATE EXTREMES

C.K. FOLLAND[1] (Chair), C. MILLER[2] (Rapporteur), D. BADER[3], M. CROWE[4], P. JONES[5], N. PLUMMER[6], M. RICHMAN[7], D.E. PARKER[1], J. ROGERS[8] and P SCHOLEFIELD[9]

[1] Hadley Centre, Meteorological Office, Bracknell, Berkshire, RG12 2SY, U.K.
[2] NOAA/NESDIS E/EI, Silver Spring, MD 20910, U.S.A.
[3] Pacific Northwest National Laboratory, Washington, DC, 20024, U.S.A.
[4] NOAA/NCDC, Asheville, NC, 28801, U.S.A.
[5] Climatic Research Unit, University of East Anglia, Norwich, NR4 7TJ, U.K.
[6] National Climate Centre, Bureau of Meteorology, Melbourne, Victoria 3001, Australia
[7] CIMMS, University of Oklahoma, Norman, OK, 73019, U.S.A.
[8] Department of Geography, Ohio State University, Columbus, OH, 43210, U.S.A.
[9] World Climate Data and Monitoring Programme, WMO, Geneva 2, Switzerland

Abstract. The "temperature" breakout group at the 1997 Asheville Workshop on Indicators and Indices for Climate Extremes reviewed and developed the rationale for a choice of temperature indices for monitoring changes in climate extremes, and the supporting data required. A set of basic and supplementary key indices was drawn up. The key indices are meant to be easy to interpret, be relevant to the practical concerns of policy makers and others in the public sector and provide potential inputs into the Third Assessment Review of the Intergovernmental Panel on Climate Change that is expected to report in 2001. The indices are expressed in various ways to facilitate spatial and temporal trend detection and impact analysis. There is flexibility in the number and the form of the indices identified and the choice for any particular application is subject to further analysis and prioritization. The success of this endeavor will depend on original work being done to further develop the indices and on the cooperation of organizations globally to provide the data necessary for the development and the implementation of the indices. This paper summarizes the group's recommendations.

1. Introduction

1.1. RATIONALE BEHIND THE CHOICE OF TEMPERATURE INDICES

This is one of three broad issue papers on climate extremes indices designed to address the needs of policymakers and their scientific advisors in the forthcoming Third Assessment Report of the Intergovernmental Panel on Climate Change, and beyond. The three key issues are storms, precipitation and temperature, the latter the subject of this paper. The emphasis is on indicators of meteorological extremes that will be of value in the context of natural and anthropogenic climate changes

Climatic Change **42**: 31–43, 1999.

and their detection, and societal responses to these. Easterling and Kates (1995) provide a discussion of the attributes of climate indicators that are likely to be of value for monitoring the impacts of climate. One key attribute is that indices should be well defined and another is that they should give warning of trends likely to have societal impact if they were to persist. Thus the indices should "lead" impacts in the sense that good stock market indices give warning of financial impacts if the recent trends in their values were to persist.

Extremes in temperature, particularly through occurrences of prolonged hot and cold events (usually hot summers and cold winters), can have significant socioeconomic impacts. In the United Kingdom, examples are the cold winter of 1947 and the hot dry summer of 1976 which were associated with substantial economic costs (Burroughs, 1997) and recently, a very cold spell in late December 1995 in Scotland where intense frost by day and by night did great damage to buildings over just a few days. Yet the relatively poor availability and quality of daily temperature data and lack of consistency in analyses have made it difficult to interpret the results of many past studies on changes in temperature extremes. However, GCM simulations with enhanced greenhouse gases suggest that some significant changes in temperature extremes are to be expected over the next century (Kattenberg et al., 1996). This is not surprising; even a moderate change in the mean can give a relatively large change in the probability of observing the rarer extremes.

It is important to realise that monitoring changes in temperature indices has to be accompanied by the monitoring of changes in atmospheric circulation indices like the North Atlantic Oscillation or Southern Oscillation. Discussion of circulation indices is beyond the scope of this paper but as regional temperature indices are developed, it will be necessary to relate their changes to those of regional atmospheric circulation. Jones et al., 1998, this volume, discuss examples of indices of the latter.

Other desirable characteristics of indices for monitoring climate extremes include:

a. A good signal-to-noise ratio for the detection of a trend or a change from one period to another. "Signal" refers to the size of the trend or change being estimated; "noise" refers to the real physical noise in the atmospheric system and not to non-physical measurement error.

b. Relevance to economic activity and other aspects of human society and ecological systems.

c. Sensitivity to likely anthropogenically-induced or natural variations in climate.

d. It should usually be possible to define them meaningfully over continental to global scale regions, though not necessarily everywhere.

e. They should be calculable from available (or potentially available) observational and model data.

The precise method of calculating the temperature indices discussed here needs significant work and testing, especially the daily indices. For example, how much missing data can be tolerated or how sensitive is an index to different ways of calculating probability distributions? Issues of data homogeneity need to be investigated. It is recommended that this research be done by a consortium of scientists, e.g., the authors of this paper, working with the newly reconstituted WMO Commission for Climatology/CLIVAR Working Group (WG) on Climate Change Detection.

We propose that a subset of the indices be highlighted as being particularly suitable for presentation to policy makers ("key indices"). They would be readily understandable, with minimum explanation. Their choice would be skewed to being of particular importance to the global warming debate and so would be especially suitable for use in the Third Assessment of the Intergovernmental Panel on Climate Change. In this paper we suggest several candidate key indices, but we recognize that additional indices may be needed.

1.2. DATA TO BE USED FOR INDEX CALCULATIONS

A better system of data exchange than currently exists is required if many of the indices defined below are to be kept up-to-date or developed with good spatial coverage. A very important issue is the efficient and timely implementation of the international exchange of data, especially daily data. In addition to relevant existing Global Climate Observing System (GCOS) initiatives, it is recommended that a Task Group comprising a key representative of each WMO Regional Association or continental area be appointed. These persons should be chosen for their ability to facilitate the flow of data. For Europe and Australasia, the WMO Regional Association framework could be used, as it is currently active and effective there, though other solutions may be necessary in some other regions.

It is recommended that, initially, a subset of about 50% of the proposed 1,000 stations in the GCOS Surface Network (GSN) (Peterson et al., 1997) be chosen for test calculations of selected indices. They would be the stations for which a substantial amount of daily data was readily available in the next one to two years and for which near real-time receipt of good data was likely. To encourage the continuity of selected stations, administrative and other key personnel in each country should be informed of the ongoing results of the project in a timely way via the WMO WG on Climate Change Detection and the WMO/GCOS Office.

2. Basic Indices and Supplementary Key Indices

2.1. DAILY PERCENTILES OF MAXIMUM, MINIMUM AND MEAN TEMPERATURE - BASIC INDEX 1

Emerging evidence for an anthropogenically induced warming signal (Santer et al., 1996; Tett et al., 1996; Folland et al., 1998) adds urgency to the importance of monitoring maximum and minimum temperature, as well as the mean. One of the expectations from greenhouse warming is a somewhat greater increase in minimum relative to maximum temperature. In addition, the impacts of changing maximum and minimum temperature are not the same and their future changes may be influenced by changing weather patterns. Thus some weather patterns give relatively larger changes in maximum than minimum temperatures; occasionally, these may even be in opposite directions (Karl et al., 1996).

Estimation of daily temperature percentiles requires a consistent method of calculation worldwide. We recommend that each station's daily maximum, minimum and mean temperature anomalies be fitted in principle for each calendar day of the year to a suitable distribution. We provisionally recommend the modified two parameter gamma distribution, using procedures described by Horton et al. (1998) and used in a companion paper in this volume by Jones et al. (1998). The parameters should be estimated by a maximum likelihood approach. In contrast to a standard two-parameter distribution, the modification, which uses an extra parameter, allows positive and negative temperature anomalies to be fitted.

An important issue is the method of calculating the daily mean climatology. Harmonic methods can have limited accuracy as the climatic mean may vary in unexpected ways through the annual cycle and these variations should be retained. Thus, a consistent multi-week fluctuation in the calendar daily mean of Central England temperature is seen for many 30-year periods since the late eighteenth century, including 1961-90, between late January (warm peak) and early February (pronounced cold trough, Parker et al., 1992). This relates to a climatological tendency to a strong westerly atmospheric circulation in late January and more blocked conditions in early February. So we provisionally recommend that raw calendar daily averages, calculated for 1961-90, be low-pass binomially filtered through the calendar year with a filter of half power in the period range of five to ten days. Tests should be carried out to confirm the best choice of method and filtering period.

In practice the gamma fitting procedure is liable to lead to somewhat unstable parameter estimates between successive calendar days. One way to avoid this is to pool anomaly data from the climatological period being used (normally 1961-90), centered on the given day, for several proximate calendar days. The additional days should be far enough apart to be effectively independent, but close enough to represent accurately the distribution for the given day. The gamma distribution is fitted to this greater number of data and allocated to the given day. The process is repeated, stepping forward by one calendar day.

This method has been successfully implemented in Jones et al. (1998) for Central England temperatures. Similar principles should be used for creating all daily temperature indices.

Additionally, plots of the sample data points against the theoretical probability density function should be visually examined for a range of climates to further ensure that the modeled data fit the sample before concluding that this procedure is always adequate. These steps will help to minimize potential sources of error, especially in situations where the number of sample points is considerably less than the number of outcome states one wishes to estimate. If the fit is poor, the gamma distribution will be rejected and others tested for suitable fit. Since the data will also be subject to various sampling errors, this procedure requires skilled judgment. In addition, stringent quality control is needed.

The expected outcome will be a smooth fit over the full range of values which will allow the selection of percentiles in both tails of the distribution, e.g. for values of 2, 5, 10, 90, 95 and 98%. Daily, monthly, seasonal, and annual indices of temperature extremes can be defined as the percentage of days per month, three-month season (December-February, etc.), summer and winter half-year, and full calendar year, exceeding the higher thresholds or not exceeding the lower thresholds. A further index can be derived to represent combined warm and cold extremes. The indices can be tabulated and plotted at each location on a global map. The gamma distribution method can be used, with the help of an appropriate gridding method, to calculate time series of the percentage of a given area and time interval (e.g., continent, globe for a month) that exceeds a given high temperature percentile or is below a given low temperature percentile.

Such maps and time series will enable analyses to be made of the stability of the climatology of extremes of maximum, minimum and mean temperature. These maps can also provide useful diagnostics for model simulations of recent climate change and variability. Moreover, end users, such as the insurance industry, can utilize the changing incidence of extreme values (e.g., 2nd or 5th percentiles of the minimum temperature) to assess the likelihood of trends that may adversely or beneficially impact their business, e.g., through changing impacts on crops and frozen pipes.

2.2. MONTHLY MEAN GRIDDED TEMPERATURE DATA - BASIC INDEX 2

Percentiles of gridded monthly, seasonal, and annual anomalies will be estimated in the same way as for daily station data; example analyses have recently appeared in WMO (1998). Although such data are unlikely to be highly skewed (the result of the central limit theorem operating on the more highly skewed daily data), the sample size is low. Therefore, maximum likelihood fitting methods are again recommended to give the best estimates. In addition, five-year mean temperature percentiles will be calculated, provisionally based on the longer period 1891-1990. Sea surface temperatures, as well as land surface air temperatures, will be included, though, in principle, it is better to use air temperatures over the oceans. However the two oceanic parameters are often similar when expressed as anomalies and the sea surface temperature data are often of higher quality and more spatially complete. This is the reason why sea surface temperature is currently used for monitoring global and regional temperatures. Better analyses of night marine air temperature are expected in the next few years and their percentiles should be compared with those of sea surface temperature.

Maps of the percentiles will be created, along with time series of the percentage of the available area of the globe (or a large region) that exceeds a high percentile or is below a low percentile. This can be done for the land and ocean separately, and their combination. The statistical significance of trends in these percentage areas can be estimated, with due allowance for serial correlation.

For land, the above percentiles are most readily derived from station temperature time series. To facilitate comparisons with changes in mean temperatures, and in the diurnal temperature range changes discussed below, an appropriate method of aggregating station data would be to average their series into 5° x 5° boxes, initially as simple averages of anomalies. The ocean temperatures are already available in this form (Parker et al., 1995). However, some experiments are needed with a weighting procedure for the land data to reduce the irregularities that might arise from the varying numbers of stations in both time and space. This problem is already allowed for in the individual ship data upon which the oceanic analyses are based.

2.3. DIURNAL TEMPERATURE RANGE - BASIC INDEX 3

Variations in the diurnal temperature range over land (DTR), defined as the difference between average monthly maximum and minimum temperatures (e.g., Karl et al., 1993; Easterling et al, 1997) have provided important insights on recent climate changes. These analyses should be extended to encompass more regions and longer periods. The analyses should be gridded, enabling the patterns of change by season to be illustrated. As discussed in the Second IPCC Scientific Assessment (Nicholls et al., 1996), the currently observed reductions in diurnal range that have accompanied the recent warming over many continental areas would have significant impacts if they continued. Consequently, DTR has become an important parameter.

Currently, the most useful base period for DTR calculations is 1951-80, as used in the Second IPCC Scientific Assessment. Calculations are often made of the difference between 1981-90 and 1951-80 or of the trend over 1951-90, two possible forms of a DTR index. However, it would be an advantage to change the base period to 1961-90 as soon as practicable to be consistent with calculations using the mean temperature. The main problem is the shortness of the record that can be analyzed on a quasi-global scale. In the future, it is likely that data on diurnal temperature range will become more readily available before 1950. Also, there are significant homogeneity issues, as the DTR may be affected by instrumental changes (Parker, 1994). In fact, rather large inhomogeneities are possible if biases in maximum and minimum temperatures develop, or change, in opposite directions (Quayle et al., 1991).

2.4. INTER-PERIOD TEMPERATURE DIFFERENCES - BASIC INDEX 4

A measure of the variability of temperature is given by the mean absolute value of differences in anomalies between two adjacent periods of time. We call these inter-period differences, e.g. the mean absolute difference of anomalies between two consecutive five-day periods over a decade for a specific season. Inter-period differences can provide additional information not reflected in other indices. While absolute extremes may not be changing significantly (as suggested by Basic Index 1 or 2), an increase (decrease) in inter-daily or inter- ten-day variability may still be suggestive of a more (less) extreme climate from a practical perspective. Conversely, the impacts of a trend towards a more extreme climate, as suggested by Basic Index 1, could be exacerbated by reduced inter-period variability that leads to longer spells of extremes.

2.4.1 Calculation of inter-period temperature indices

Karl et al. (1995) made the first attempt to calculate inter-period temperature differences over large areas. In statistical terms these are the first differences of the temperature series. We suggest that inter-period differences for one, five and ten-day periods are likely to be particularly useful, along with interannual differences in the seasonal mean. The one-day time scale simply represents day-to-day variability and so, for a particular season, the absolute differences between consecutive daily temperature anomalies are averaged. First differences are much less prone to confounding effects of high and low frequency variability than can result from using the standard deviation as an index (Karl et al., 1995). Inter-period differences are an efficient high pass filter and so have a clearer meaning (one of the ideal attributes of a climate index). Anomalies are calculated relative to calendar daily mean values as described for Basic Index 1. We recommend that the differences for the five-day and ten-day time scales (within given seasons) be calculated in the same way, using non-overlapping intervals, when investigating trends in such indices. Plummer (1996) calculated overlapping running differences where averages of the daily temperature anomalies were calculated for days one to ten, days two to eleven, days three to twelve, etc. Then the absolute differences between these consecutive average values were computed. The average of these absolute differences gave a quantitative measure of variability for a specific season but these were not readily related to inter-daily differences since they were so small. Our non-overlapping method is more suitable for a running mean index over several years (e.g., ten years), which seems appropriate in the current context. Experiments with overlapping differences will be made to test the stability of this scheme. Interannual seasonal variability is simply calculated by taking the absolute value of the difference of the seasonal mean temperature anomaly between one year and the next for a specific season. For the year as a whole, first differences between annual means can be calculated. However, as Plummer (1996) points out, an annual index based on the mean of the four inter-seasonal absolute differences can be more illuminating and can give very different results, at least for individual years. It is recommended that this be explored further.

Once "gridded," changes in time of the various indices can be displayed in map form (e.g., as trends), thus showing patterns of increases and decreases.

2.5. FROST/FREEZE INDICES - BASIC INDEX 5

2.5.1. *Frost/freeze severity index*
Spells of freezing conditions can have major economic effects, as mentioned above. An area particularly sensitive to such spells is the citrus growing area of Florida where recent spells of winter extreme minimum temperatures have done much damage (Burroughs, 1997). A frost/freeze severity index would use daily minimum air temperatures at individual stations. More than one index is possible, and the best indices need further investigation. A basic index would portray the percentage of time, for a particular month, that daily minimum air temperatures are less than, or equal to, $0°C$. To simulate ground frost conditions the index could also be computed with minimum temperature thresholds in the range $2-5°C$, though the choice of optimal values requires further study. Grass minimum temperature data would clearly be better but, although widely measured, they are not readily available.

2.5.2. *Frost/freeze duration index*
Again only a preliminary definition is given. A basic index could be calculated to depict the dates of the first and last minimum air temperatures for a frost season that were below $0°C$. The frost seasons are defined as 1 August to 31 July in the Northern Hemisphere and 1 February to 31 January in the Southern Hemisphere. As in sub-section 2.5.1, the minimum temperature threshold could be increased to values in the range $2-5°C$ to account for ground frost conditions.

2.6. HEAT WAVE DURATION INDEX (HWDI) AND COLD WAVE DURATION INDEX (CWDI) - BASIC INDEX 6

Cold waves and heat waves typically occur on synoptic time scales (2-8 days) and are often associated with the passage of large high pressure systems in the extratropics. While these systems may be stationary or quasi-stationary, winter anticyclones can often be quite mobile (Rogers and Rohli, 1991). Heat and cold waves have hitherto been typically defined (mostly for the USA) in terms of the occurrence of extreme high or low air temperature values that persist for a period of days. (Cold waves have also been measured by the size of the temperature drop during their onset). Considering the first type of measure, use of threshold air temperature values is only suitable at a specific location or for a homogeneous region. A general index must use percentiles. Percentiles are needed rather than absolute temperature thresholds because the latter would be impossible to extend globally and analyses would show marked heterogeneities across areas with significant changes in elevation. At a given location, the indices will have a zero value if critical percentiles are never exceeded in a month or season. Multiple occurrences of cold or warm waves, with corresponding percentile values, may occur in a given month or season. An appropriate heat or cold wave duration index should have the property of frequently being numerically similar at individual locations over a sizable region.

Despite the above remarks, it is recognized that for particular applications in specific regions thresholds based on absolute values may be the most useful. So we can expect that a number of such indices will emerge related to particular physical systems including individual types of crop etc. So the generalized measures we develop can be used as a background to many specialized absolute threshold indices that may emerge. In fact, the frost/freeze severity index described above can be regarded as such an index, though of unusually general application.

The definition of the generalized heat wave and cold wave duration indices we consider here can be expressed in more than one way. We present alternatives that may require a good deal of experimentation.

We define HWDI in two parts. Firstly, we note the percentage of consecutive days in a month or season for which the maximum air temperature at a location is equal to, or above, a specific percentile. Secondly, we note the number of separate occurrences of such consecutive days. This allows us to distinguish between, say, a single twenty-day spell of such temperatures from five four-day spells. Note that the minimum hot or cold "spell" recorded is two days long with this definition. This minimum could be lengthened, depending on the application, but for a general index we suggest this choice. It is also suggested that choices of threshold be made from the 90th, 85th or 80th percentiles. A variation on the index would be to combine the time of persistence of maximum temperatures above an x percentile mark over n days with that of minimum temperatures that simultaneously remain above the y percentile mark.

Correspondingly, a CWDI is defined as the percentage of consecutive days in a month or season in which the minimum air temperature at a location is equal to or below the 10th, 15th or 20th percentile. The additional definitions for the HWDI would apply here as well. Both the HWDI and CWDI must be calculated with care for events extending over two adjacent months or seasons where some experiments are clearly needed. In addition, it would be useful to plot the indices hemispherically or globally in percentile form.

These indices will provide evidence of interannual and interdecadal variability for a month or season in the tendency for heat or cold waves to occur, and their changing duration. Heat waves in summer and cold waves in winter will have the most obvious impacts on human and environmental systems. It should be noted that little is known about how these indices behave, especially spatially, so considerable exploratory work will be essential.

2.7. STRESS INDICES - BASIC INDEX 7

2.7.1. Heat stress index
We briefly considered a "heat stress" index that would be a function of temperature and dew point. Several such indices exist. If such an index is worthwhile, it raises for the GSN the complex issue of humidity measurements and the standards that should be maintained, e.g., for manual versus automatic measurements where unexpected problems can arise (Folland, 1977). Because surface humidity data are likely to become important within a purely climate change detection context, as well as for impacts, it is recommended that the WMO Working Group on Climate Change Detection and GCOS give this parameter much more consideration than it has had hitherto.

2.7.2. Wind chill index
This could also be considered as another type of "cross-cutting" index involving two parameters. This type of index exists and would be a function of temperature and surface wind speed. However, it is difficult to calculate, except locally, because of the problem of a representative wind speed.

2.8. KEY TEMPERATURE INDICES

These indices have been selected principally because GCM simulations suggest that significant changes may be expected with increasing greenhouse gases and the indices recommended here can be readily interpreted and used by the non-scientific community. We recommend that the following three indices be considered as key indices to be presented to policy makers and decision makers. They might have priority, among the temperature indices, for inclusion in the Third Assessment Report of the Intergovernmental Panel on Climate Change.

2.8.1 *Key index 1 - Seasonal and annual mean extremes*

This index is based on one of the products of Basic Index 2. We recommend that a pair of time series, drawn on the same diagram, be created of the percentage of the area available for analysis that is warmer than the 90th percentile and colder than the 10th percentile for annual and seasonal data. These indices should be calculated for the globe and selected regions of interest for sufficiently long periods to show trends. It may be desirable to re-express the percentiles as return periods when presenting the information since, in our experience, the latter are readily usable by policy makers.

2.8.2. *Key index 2 - Diurnal temperature range*

We recommend that a suitable key index is the change in the global annual mean diurnal temperature range between the most recently available 20-year period and 1951-70. This is a concise way of presenting the information. It also separates the most recent period of warming of the global mean that began in the middle 1970s from the previous period of little global mean temperature change. An interim alternative is to use the difference between 1981-90 and 1951-80 until such time as the gap in data between about 1990 and the present (regularly telecommunicated) maximum and minimum temperature data is filled.

2.8.3. *Key index 3 - Frost severity index*

The frost severity index (see Basic Index 5) will be calculated for complete calendar years over selected, mainly extratropical, regions of continental size, and plotted on a single graph.

3. Conclusions

The proposed indices will be of substantial value, not only for climate monitoring and for policy makers, but also for research into natural and anthropogenic climatic variations. For example, the propensity to extremes may be a function of the phase of the El Niño/Southern Oscillation even where the mean value is unaffected. Extremes are also likely to undergo natural interdecadal fluctuations, partly due to fluctuations in the global ocean circulation, the evidence for which is rapidly increasing. They are also likely to depend on the time- and space-varying forcings from greenhouse gases, tropospheric aerosols, (natural) volcanic aerosols in the stratosphere, and solar influences.

Acknowledgments

The following contributed to discussions that led to the above report: M.Beniston, E. Førland, J. Gavin, R. Heino, R. Lee, C. Pfister, S. Rosner, A. Sun, A. Tank, J. Wibig, M. Changery, D. Easterling, K. Gallo, R. Heim and T. Peterson. C.K. Folland and D.E. Parker were supported by UK Public Meteorological Service Contract MSG-2/97.

References

Burroughs, W.J.: 1997, *Does the weather really matter? The social implications of climate change*, Cambridge University Press, Cambridge, UK, 230 pp.

Easterling, D.R., Horton, B., Jones, P.D., Peterson, T.C., Karl, T.R., Parker, D.E., Salinger, M.J., Razuvayev, V., Plummer, N., Jamason, P. and Folland,C.K.: 1997, 'Maximum and minimum temperature trends for the globe', *Science*, **277**, 364-367.

Easterling, W.E. and Kates,R.W.: 1995, 'Indexes of leading climate indicators for impact assessment', *Clim.Change*, **31**, 623-648.

Folland, C.K.: 1977, 'The psychrometer coefficient of the wet bulb thermometers used in the Meteorological Office Large Thermometer Screen', *Meteorological Office Scient.Pap.* **38**, 37pp.

Folland, C.K., Sexton, D.M.H., Karoly, D.J., Johnson, C.E., Rowell, D.P. and Parker,D.E.: 1998, 'Influences of anthropogenic and oceanic forcing on recent climate change', *Geophys.Res.Lett.,***25**, 353-356.

Horton, B., Folland, C.K., and Parker, D.E.: 1998, 'Monitoring percentile values of gridded worldwide monthly to five-year surface mean temperatures', *Clim.Change* (submitted).

Jones, P.D., Horton, E.B., Folland, C.K., Hulme M., Parker, D.E. and Basnett, T.A.: 1998, 'The use of indices to identify changes in climatic extremes', *Clim.Change* (in press).

Karl, T.R., Jones, P.D., Knight, R.W., Kukla, G., Plummer, N., Razuvayev, V., Gallo, K.P., Lindseay, J., Charlson, R.J. and Peterson, T.C.: 1993, 'A new perspective on recent global warming: Asymmetric trends of daily maximum and minimum temperature', *Bull.Am.Meteorol.Soc.,***74**, 1007-1023.

Karl, T.R, Knight, R.W., and Plummer, N.:1995, 'Trends in high frequency climate variability in the twentieth century', *Nature*, **377**, 217-220.

Karl, T.R., Knight, R.W. Easterling, D.R., and Quayle, R.G.: 1996, 'Indices of climate change for the United States', *Bull.Am.Meteorol. Soc.*, **77**, 279-292.

Kattenberg, A., Giorgi, F., Grassl, H., Meehl, G.A., Mitchell, J.F.B., Stouffer, R.J., Tokioka, T., Waver, A.J. and Wigley, T.M.L.: 1996, 'Climate models - projections of future climate', in Houghton, J.T., Meiro Filho, L.G., Callander, B.A., Harris, N., Kattenberg, A. and Maskell, K. (eds.), *Climate Change 1995: Intergovernmental Panel on Climate Change*, Cambridge University Press, Cambridge, pp.285-357.

Nicholls, N., Gruza, G.V., Jouzel, J., Karl, T.R., Ogallo, L.A. and Parker, D.E.: 1996, 'Observed climate variability and change', in Houghton, J.T., Meiro Filho, L.G., Callander, B.A., Harris, N., Kattenberg, A. and Maskell, K. (eds.), *Climate Change 1995: Intergovernmental Panel on Climate Change*, Cambridge University Press, Cambridge. pp.133-192.

Parker, D.E.: 1994, 'Effects of changing exposure of thermometers at land stations', *Int.J.Climatol.*, **14**, 1-31.

Parker D.E., Folland, C.K. and Jackson, M.: 1995, 'Marine surface temperature: observed variations and data requirements', *Climate Change*, **31**, 559-600, and in: *Long-term Climate Monitoring by the Global Climate Observing System*, Ed: T. Karl, pp.429-470, Kluwer, Dordrecht.

Parker, D.E., Legg, T.P., and Folland, C.K.: 1992, 'A new, daily Central England temperature series 1772-1991', *Int.J.Climatol.*, **12**, 317-342.

Peterson, T., Daan, H., and Jones, P.: 1997, 'Initial selection of a GCOS Surface Network', *GCOS-34, WMO/TD* No.799, World Met.Org., Geneva, 14pp.

Plummer, N.: 1996, 'Temperature variability and extremes over Australia: Part 1 - recent observed changes', *Aust.Met.Mag*, **45**, 233-250.

Quayle, R.G., Easterling, D.R., Karl, T.R. and Hughes, P.M.: 1991, 'Effects of recent thermometer changes in the cooperative station network', *Bull.Am.Meteorol.Soc.*, **72**, 1718-1723.

Rogers, J.C. and Rohli, R.V.: 1991, 'Florida citrus freezes and polar anticyclones in the Great Plains', *J.Climate*, **4**, 1103-1113.

Santer, B D., Taylor, K.E., Wigley, T.M.L., Johns, T.C., Jones, P.D., Karoly, D.J., Mitchell, J.F.B., Oort, A.H., Penner, J.E., Ramaswamy, V., Schwarzkopf, M.D., Stouffer, R.J., and Tett, S.: 1996, 'A search for human influences on the thermal structure of the atmosphere', *Nature*, **382**, 39-46.

Tett, S.F.B., Mitchell, J.F.B., Parker, D.E. and Allen, M.R.: 1996, 'Human influence on the atmospheric vertical temperature structure: Detection and observations', *Science*, **96**, 1170-1173.

WMO: 1998, 'The Global Climate System Review- climate system monitoring December 1993-May 1996', Ed: J.M.Nicholls, World Climate Data and Monitoring Programme, WMO No 856, 95 pp.

(Received 5 November 1997; in revised form 27 August 1998)

GLOBAL CLIMATE CHANGE:
WHY U.S. INSURERS CARE

FRANKLIN W. NUTTER

Reinsurance Association of America, 1301 Pennsylvania Avenue, NW, Washington, DC 20004

Abstract. The number of natural catastrophes in this decade is four times greater than in the 1960s; economic losses are eight times greater; and insured losses are 15 times greater. The insurance industry's financial interest is inter-dependent with climate and weather. Natural events drive the demand for insurance coverage and can threaten the viability of an insurer if it is over-exposed in high risk areas. Early in the 1990s, the industry began to recognize that historical data were potentially misleading with respect to future natural catastrophe exposure. The U.S. insurance industry is pursuing a variety of new approaches including: the use of catastrophe computer models to integrate the natural knowledge about extreme events taken from the sciences into the actuarial sciences. The evaluation of building codes and building code enforcement in every community in the country enhanced its support for hazard mitigation.

1. Introduction

The scope of losses paid by the insurance industry as a result of natural catastrophes in recent years is well documented. The seven largest economic losses from natural catastrophes have all occurred in the last five years. Five of these were wind and water related: Hurricanes Andrew, Muriel and Hugo; the Mid-West floods of 1993; and the European storms of 1992.

Munich Reinsurance reports that the number of natural catastrophes in this decade is four times greater than in the 1960s; economic losses are eight times greater; and insured losses are 15 times greater than the 1960s, even after adjusting for inflation. Since 1989, the U.S. insurance industry has paid more than 75 billion dollars in losses from natural catastrophes, a figure which exceeds by 50 percent all such losses paid in the preceding 40 years. Notwithstanding these recent losses, until the early 1990s insurers charged little in premiums for catastrophe-related coverages. More alarming than the losses that have been paid is the potential for a mega-catastrophe in the U.S. Some estimates for insured losses are as high as 50 to 60 billion dollars from a single hurricane.

The causes behind the dramatic rise in catastrophe losses are several:

- Population growth in high risk areas, particularly the East Coast, Long Island and Florida.;
- Dramatic increases in insured values (estimates are that by 2000 the insured property value in the state of Florida alone will exceed one trillion dollars);

Climatic Change **42**: 45–49, 1999.

- The insurance industry's own expansion of coverage during the '60s, '70s, and '80s, which had the effect of increasing potential insured damage; deductibles were lowered and full replacement cost added to homeowners' policies in that period.;
- Government policy which either endorsed weak building codes or failed to enforce existing building codes, thereby giving rise to greater damage; for example, estimates suggest losses in Hurricane Andrew were 30 percent greater than if existing building codes had been enforced; and
- Climate change and the incidence of more frequent extreme events. Whether due to natural variability or enhanced by man's activities, most scientists believe we are beginning a period of increased catastrophe level weather and climate variability.

2. Why Does the Insurance Industry Care?

The insurance industry's financial interest is inter-dependent with climate and weather. It is the risk of natural events which drives the demand for insurance coverage and yet, if not properly managed, can threaten the viability of an insurer if it is over-exposed in high risk areas. An insurance company thrives or dies from its ability to make estimates of the economic consequences of future events.

Most insurance coverages are priced based upon historical data which is then trended forward using adjustments for inflation and other economic factors. This was true of insurance coverage for natural disasters until early in the 1990s when, as a result of Hurricane Andrew, a paradigm shift occurred. At this point, the industry began to recognize that, due to unanticipated climate variability, historical data were potentially misleading with respect to future natural catastrophe exposure. Insurers began to understand that in a variable and changing climate that at times is not well understood, a company's financial viability could be threatened. Thus, there is a real dilemma: we now know that climate is not and will not be linear (i.e., one disaster after another) or exactly cyclical (i.e., similar disasters occurring in a climate cycle). Even if climate is approximately cyclical, what if the industry does not fully understand the recurrence period of catastrophic natural events? And, what if, due to human intervention, climate is now fundamentally changed, causing changes in these weather patterns and cycles and leading to more extreme events?

3. Industry Reaction

In the context of this paradigm shift, and being somewhat uncertain about the status of scientific assessment of climate change, the industry has reacted in three stages. The first stage was the classic reaction of business in the face of uncertainty -- in several high risk areas insurers sought to freeze in place their existing catastrophe exposure. Some insurers even sought to withdraw from insurance markets in heavily populated or extreme coastal areas. Insurers have also sought to raise insurance premiums to match the risk exposure in those areas prone to major or extreme climate events. Those rates are, of course, subject to review and approval by insurance commissioners who have shown some scepticism about assumptions regarding future extreme events.

A second stage of reaction by the industry was the recognition that government, both state and Federal, has an interest in disaster financing as well as stable insurance markets. Thus, the industry and government at both the state and Federal levels have engaged in an ongoing debate about the proper role of government in financing catastrophe risk. In three states -- Florida, Hawaii, and California -- state funds were created to provide insurance or reinsurance to help stabilize insurance markets. While the debate has continued at the Federal level, no action yet has been taken to address the proper role, if any, for the Federal government in catastrophe risk financing.

Within the industry, the debate has centered around the issue of whether government should address the solvency of the industry by providing a financial safety net under the industry, or directly engage itself in providing catastrophe insurance to the public. The underlying issue in this debate is whether natural events such as hurricanes, floods and earthquakes are insurable, as most insurers believe, or whether they are uninsurable, as a few insurers argue. Those who believe extreme natural events are uninsurable promote the greater intervention of government into the private insurance sector. The National Flood Insurance Program is an example of this approach.

Lacking in the first and second phases of the industry's recognition of a paradigm shift in natural catastrophes has been a broad assessment of science related to natural events. The U.S. insurance industry has not historically been science-based in its natural catastrophe or strategic evaluation. Although a number of notable European insurers and reinsurers have shown greater interest in understanding the *causes* of climate change, including the impact of global warming, U.S. insurers have been more focused on the *effect* of natural disasters. Thus, the U.S. industry has been more attentive to approaches to mitigate the effect of natural catastrophes and other extreme events; while some European insurers have called upon their governments to reduce the human factors they believe contribute to global warming.

4. Emerging Industry Approach

As a third phase, the U.S. insurance industry is pursuing a variety of new approaches to address problems associated with natural catastrophes. The most significant is improving the financing of catastrophe exposures through retained earnings, higher limits of reinsurance, corporate restructuring and the use of new, innovative capital market products. The initiative most related to science and scientific assessment of global climate change, is the use of catastrophe computer models to integrate the natural knowledge about extreme events taken from the sciences into the actuarial sciences. These catastrophe models incorporate scientific assumptions about climate trends and the probability of future extreme events, then produce estimated prospective costs associated with natural catastrophes. They assist an insurer with an analysis of its potential exposure and are used to support rates filed for approval with insurance departments. It is the classic example of using insurance to translate scientific analysis and data into the economic consequences of people's behavior, i.e., where they live and the value of properties in those areas. The result of the use of catastrophe models is the application of risk based premiums for insured property. The use of the models remains somewhat controversial with insurance departments because the assumptions were all considered proprietary.

The second component of the industry's current approach is to broaden the options policyholders have for insurance coverage. By giving people optional deductibles and coverages, insureds can select insurance commensurate with the level of premium with which they are comfortable.

The third component of the current agenda is the industry's evaluation of building codes and building code enforcement in every community in the country. In this way, the industry will integrate into the rating system the quality of mitigation that each community applies. We know from past disaster assessments that window shutters, tie-downs, and hurricane straps are cost-effective mitigation techniques. Policyholders in communities with high quality building codes and appropriate enforcement will benefit through their insurance rates.

The industry is also supporting scientific research. In 1996 and 1997, the insurance industry lobbied in support of increased government funding of research and monitoring of climate and climate change through NASA's Mission to Planet Earth, and in support of increased funding for the National Hurricane Center. Some U.S. insurers are supporting private climate research through the Risk Prediction Initiative at the Bermuda Biological Station. Funding scientific research of climate with a view toward integrating this information into the insurance community will improve the industry's understanding of global climate change and the potential consequences of extreme events.

Lastly, through the Institute for Business & Home Safety, the U.S. industry has greatly enhanced its support for hazard mitigation by conducting research on building design and building materials. By recognizing communities with the best hazard mitigation approaches, the Institute is bringing increased attention to the appropriate steps to take to reduce natural hazard damages. IBHS has also funded some climate related scientific research (an analysis of El Niño being the most recent example).

5. Conclusion

No financial services business is more dependent on the vagaries of climate and weather than property and casualty insurers. The industry is at great risk if it does not understand global climate variability and the frequency of extreme events. It must recognize that it must do more than be a pass-through mechanism for the costs associated with natural disasters. Understanding global climate change and integrating that information into the insurance system is an essential part of addressing climate extremes and conveying information to governments and the public about the economic consequences of human activity in the face of changing global climate.

(Received 5 November 1997; in revised form 7 August 1998)

INTERACTIONS BETWEEN THE ATMOSPHERIC SCIENCES AND INSURERS IN THE UNITED STATES

STANLEY A. CHANGNON

Changnon Climatologist, Mahomet, IL 61853 U.S.A.

E. RAY FOSSE

Insurance Consultant, Goreville, IL 62939 U.S.A.

EUGENE L. LECOMTE

Institute for Business and Home Safety, Boston, MA 02108 U.S.A.

Abstract. Insurance and insurers are directly affected and concerned about climate fluctuations in the United States. Growing losses in the 1990s awakened many in the insurance business to the enormity of weather-related problems they faced and a need for better information about climate and its fluctuations including a new potential problem, climate change. This paper presents the views of experts from the crop insurance sector, the property-casualty insurance sector, and from the atmospheric sciences sector based on atmospheric research accomplished for the insurance industry for decades. The paper addresses how climatologists and insurers can and need to work together to effectively bring understanding and wise consideration of climate conditions and their future fluctuations and extremes to insurers. Considerable climatological analyses have been employed in the past by the crop insurance industry, but less by property insurers. Insurers can adjust to a change in climate but to do so will require clear evidence, which does not exist now for the United States, and an understanding as well as wide acceptance of the on-going change by regulators and the buying public. Atmospheric scientists can help insurers to mitigate weather losses, to assess risks, to measure critical perils, to educate about risks, and to learn about critical issues like climate change and long-range forecasts.

1. Introduction

One of the major private sector components most directly and immediately effected by weather and climate extremes is the insurance industry. The industry can be grouped into four major components: the crop insurance sector, the property-casualty sector, the life-health sector, and the reinsurance sector, which serves the other sectors. In labeling the "insurance industry," please understand the industry in the U.S. is not monolithic -there are 4,000 property-casualty companies, 2,000 life-health companies, 50 crop insurance firms, and hundreds of companies handling reinsurance, each operating independently and competitively. Insurers are regulated by the states with general operational and performance standards set by state regulators (these standards differ between states) who also approve the rates sought by individual insurers. The regulators thus control how insurers function to assure their solvency through approval of rates that are reasonable, adequate, and not discriminatory.

Climatic Change **42**: 51–67, 1999.
© 1999 *Kluwer Academic Publishers. Printed in the Netherlands.*

Some insurers use climate data and information effectively, others do not. All are concerned about the potential for climate change and how to respond.How atmospheric scientists and insurance leaders interact to serve this key sector is a terribly important issue. The goal of both groups is the provision of data and information to insurers about current weather conditions including extremes. Extremes of interest to insurers fall into two broad classes: (1) weather extremes such as storms and cold/heat waves, and (2) climate extremes such as droughts, multi-year cold or wet periods, etc.

Many atmospheric scientists are working on various facets of the climate change issue. Those who have worked in the arena of impacts of weather and climate realize that one of the key issues concerning all weather-sensitive individuals and institutions is the extremes of weather and climate. Mean annual temperatures may change 1 to 5 degrees F and create challenges for society, but shifts in the extremes, either to more frequent events or to extremes of greater intensity, are seen as a major problem facing insurers.

This paper first presents a section describing the crop insurance sector including its unique data which is potentially useful in defining future changes in hail, and the industry's interests in interactions with climate specialists. The second section describes the interests, activities, and concerns of the property-casualty insurers in the climate change issue. The third section of the paper describes how the atmospheric sciences community has been interacting with insurers and defines the needs for future interactions on climate issues including climate change.

2. The Crop Insurance Industry

2.1. INTRODUCTION

Climates conducive to the production of crops also include events which cause damage to crops. One sector of the insurance industry specializes in providing coverage against such crop loss or damage. While the economic consequences of weather damage to crops are rarely of national significance, individual farms, communities, and a few states regularly suffer severe financial hardships as a result of weather events affecting crop production. There are two basic types of weather damage to growing crops: storm events of hours/days duration, and climate events lasting months/years, such as droughts. Yield reductions due to these events occur every year, but the major events and large losses are periodic, such as the drought of 1988, the floods of 1993, and extreme hail-loss years like 1961 and 1992.

In the process of providing the crop insurance coverage, organizations sponsored by most firms have gathered data and used statistical data methods for the promulgation of premium rates for nearly a century. They also have engaged in analyzing weather records accumulated by federal and state agencies. In more recent years they have also availed themselves of the expertise of climatologists for insights into the use of various available data to address the problem of developing equitable rate structures and to consider issues like climate change (Fosse and Changnon, 1993). In these activities, considerable analyses of hail insurance statistics have been accomplished, leading to a growing awareness of the value of crop insurance historical data in the study of climate. Following is a brief description of the dimensions of the crop insurance business, the nature of the insurance provided, selected statistics resulting from insurance data, and needs of the industry for atmospheric sciences help.

2.2. TYPES OF CROP INSURANCE

The U.S. has two major forms of crop insurance, hail insurance and Multiple Peril Crop Insurance (MPCI). Early difficulties in underwriting hail insurance led the companies to form organizations aiming to collect hail insurance statistics for rate-making purposes and to develop and improve upon procedures for assessment of loss of crops caused by hail. The first such organization dates to about 1915. The current services organization for crop insurance is the National Crop Insurance Services, Inc. (NCIS). Among the first actions of these early organizations was the formulation of statistical plans, i.e., the order and scope of data collected and recorded primarily for use in rate-making. Consequently, recordings of loss over the years have been consistent, yielding a database for many insurance applications and research. While useful data under the present plan begin with 1924, the preponderance of data used in rate-making today begins with 1948.

Currently, there are some 50 individual companies, or groups of companies, or management organizations conducting crop insurance business in the U.S. There are a few single-state operations and several regional and country-wide operations. Liability assumed on hail insurance reached an estimated $16 billion in 1996, with total premiums written of about $630 million. Multiple Peril Crop Insurance premium volume exceeded $1 billion on 1996.

The crop insurance companies are but a small portion of the approximately 4,000 U.S. property and casualty companies, and all crop insurance is only about 7 percent of the approximately $50 billion annual premium volume written for the basic property lines of insurance. While reinsurance is a vital resource in the conduct of crop insurance business, the statistics recorded by NCIS are for direct writings, irrespective of reinsurance cessions and recoveries.

2.3. THE NATURE OF CROP INSURANCE RECORDS

For both hail insurance and MPCI, the coverage is site specific to the level of township (36 square miles or 94 km^2) in most states, and to the county-level in a few states. When an application for insurance is completed, it provides the legal location of each crop insured.

The first step in the assessment of loss is verification of location, since premium rates and other insuring conditions may vary by township or county. Records of insurance written and loses paid are maintained by year, by crop, by kind of policy, and by location. The record includes the amount of insurance coverage (liability in $), the premium ($) collected, amount of loss ($) paid, cause of loss, and date of loss.

These data are processed with various automated error checking procedures. For example, the NCIS maintains a master location file, against which location descriptions on the applications are checked (for possible location and for township within county). The accuracy of the records is of great importance since their principal propose is for rate-making that must be defended before state regulatory bodies.

There are limitations on the use of these crop insurance data for purposes other than rate-making. For example, not all regions of a state have significant crop production, and hence there is a dearth of crop loss statistics for such regions. Secondly, crop production in most areas is a seasonal enterprise. Thus, there may be extreme weather events in non-growing season periods (and hence have no impact on growing crops) and hence no record of these events is made. The amount of liability (insurance coverage) assumed by farmers also fluctuates with time, another factor that affects loss statistics. Finally, there is great variability in the effect of damaging weather, and especially hail, on the kind of crop and the stage of crop growth at the time of the event.

Even so, there is value in the crop insurance statistics. They are highly useful for measuring the impact of weather extremes. They are retained at a detail level and may be extracted in various combinations and summary forms. The currently advanced knowledge about hail in the U.S., for example, has been facilitated considerably by research of crop-hail insurance data.

The industry computes, for each area (township, county, state, region, and nation) the Loss Cost. It is the ratio of losses paid to the liability on the acreage suffering the loss, expressed in dollars per $100 liability. For example, a loss cost of $2.40 means that for the area and liability affected, the amount of $2.40 was paid for losses for each $100 of liability in force. Loss cost is the best value for comparing loss results among areas, crops, and years. Loss ratio is the ratio of losses paid to the amount of premium written. For example, if there is a loss ratio of 60 percent for a given area, there was, for the period being considered, losses paid equal to 60 percent of the premium for the subject area, crop, and year.

Tables 1 and 2 present summaries of the 1996 annual insurance statistics for the top ten states, based on the amount of liability, for the crop-hail and the MPCI coverage. Values in table 1 reveal that high losses occurred in four states: Kansas, Montana, Nebraska, and North Carolina. The crop-hail losses of $389.8

Table 1. The top ten states based on amount of liability for crop-hail insurance in 1996, and their premiums and losses. Based on data supplied by the National Crop Insurance Services.

State	Liability ($000)	Premium ($000)	Losses ($000)	Loss cost, $
Iowa	1,603,000	43,770	14,580	0.91
Illinois	1,530.000	15,659	3,442	0.22
Minnesota	1,146,000	48,296	14,394	1.26
North Dakota	1,020,000	67,424	29,477	2.89
Nebraska	900,000	60,063	46,471	5.16
Washington	706,000	9,399	2,443	0.35
Idaho	619,000	14,165	2,577	0.42
Kansas	598,633	36,313	38,830	6.49
North Carol.	518,750	28,398	111,674	21.53
Montana	468,400	39,815	33,419	7.13
U.S. Totals	13,019,750	493,651	389,786	2.99

million in 1996 (Table 1) rank as the highest one-year value since national records began in 1948. However, the 1996 loss cost value, representing a time normalized expression of loss, was $2.99, a value that ranked as the eleventh highest since 1948. The 1996 loss ratio (losses divided by premiums) was 78 percent, a high ranking value when compared to the long-term average of 67 percent. Insurance companies lose money when the ratio is 70 percent or higher, revealing that 1996 was not a profit-making year.

The MPCI values (Table 2) reveal very high crop losses in 1996 occurred in Texas and Kansas where losses exceeded premiums with loss ratios of 1.60 in Texas and 1.57 in Kansas. The national MPCI values for 1996 shown in table 2 reveal that the loss ratio (losses divided by premiums) was 77 percent, a relatively low ratio for this type of insurance. The average loss ratio for 1981-1985 was 144 percent, for 1986-1990 it was 148 percent, and for 1991-1995 it was 123 percent. The only lower annual value than that in 1996 occurred in 1994 which had a loss ratio of 62 percent. However, the $1.408 billion in insured crop losses in 1996 (table 2) ranks as the third highest annual value since 1990, with 1993 having losses of $1.601 billion and 1995 with $1.510 billion. The national totals of crop insurance in tables 1 and 2 reveal that $40 billion in liability were taken on U.S. crops in 1996, with losses totaling $1.8 billion.

These records could be employed in determining regional indicators and indices of climate extremes. A measurable change in climate could have an enormous impact on the usefulness of the cumulative data for crop insurance purposes.

Table 2. The top ten states based on amount of liability for Multiple Peril Crop Insurance in 1996, and the premiums and losses for 1996. Based on data from the Federal Crop Insurance Corporation.

State	Liability ($000)	Premium ($000)	Indemnity losses ($000)
Iowa	3,553,512	168,546	41,048
Minnesota	2,282,570	143,158	35,420
Illinois	2,216,880	100,579	56,823
Nebraska	2,011,435	109,035	45,939
Texas	1,704,615	237,301	380,172
California	1,536,979	77,677	27,853
North Dakota	1,332,908	122,533	57,500
Kansas	1,252,205	96,110	151,285
Indiana	933,367	43,980	38,905
South Dakota	907,322	82,532	54,574
U.S. Totals	26,608,124	1,821,479	1,408,514

2.4. INTERACTIONS OF CROP INSURANCE INDUSTRY AND ATMOSPHERIC SCIENTISTS

There is a considerable history of interaction between the crop insurers and segments of the scientific community. Since 1960 the Illinois State Water Survey, a state agency which includes atmospheric scientists, has conducted research in behalf of and in cooperation with the hail insurance company organizations.

As an example of the value derived from such activities, a specific project affecting hail rate-making in Illinois is described. In 1960 the data base for rating in Illinois was limited because extensive hail insurance coverage did not exist until 1948. Weather Bureau hail-day frequency data were compared with the annual insurance experience (loss costs) for 12 years (1948-1959), revealing a good association. Resulting correlation coefficients were used with hail-day frequencies for years during 1900-1947 to estimate annual loss costs for these years (Changnon, 1960). These extended the period of record used in promulgating hail insurance rates in Illinois.

Another illustration involved the successful development of statistical techniques, utilizing annual hail insurance loss costs, to determine future hail loss trends on a state-by-state basis (Neill et al., 1979). In addition to the results of such services, there has been a more important result—an understanding relationship between the scientific and business entities necessary for effective communication and interaction. Future productive interactions will occur only if such understanding and appreciation exist.

What of future interactions? A critical issue is climate change. Within the crop insurers there is limited attention to the subject, mostly because there is yet to be an effective consensus within the scientific community about its occurrence and no evidence of a change in crop-damaging weather (Fosse, 1996). Yet, no other financial group could be more seriously affected by climate change than insurers (Changnon et al., 1997). The use of historical data in rating crop insurance relies upon the assumption of representativeness of the variability found in the annual occurrences of the insured weather elements. The industry is capable of making adjustments which identifiable changes would require. So, what are the problems with climate change that atmospheric scientists can help with?

First, the identification of indicators which would signal climate change with the force of confidence sufficient to provoke action is needed. Primary would be a change in rates that is acceptable to the industry and to government regulators. Secondly, performance of educational functions necessary for the adoption of sensible public policy concerning climate change. To be more specific about the second problem, insurers, and certainly the crop insurance segment, will need an understanding public and regulators. This is essential so that modifications in rating and underwriting, as indicated by the effects of climate change, will be accepted. The financial services provided by insurance addresses obvious social needs, but if it is to be provided by profit-oriented institutions, there must be an environment in which the risks may be managed efficiently and economically effectively. Arbitrary decisions influenced more by political considerations than by respect for facts ill serve the public interest.

3. The Challenges of Climate for the Property-Casualty Industry

3.1. INTEREST AND INVOLVEMENT IN THE CLIMATE CHANGE ISSUE

A question frequently asked, especially in the aftermath of Hurricane Andrew, an event which shook the financial stability of many property-casualty insurers, is: Why aren't property-casualty insurers more interested in climate change?

The answer is: They are interested! Further, they have for several decades been vigorously working to eliminate the potential for weather-related damage from occurring. Property-casualty insurers know that adverse weather cannot be eliminated; they know that mother nature cannot be turned back; and they know that hurricanes, tornadoes, severe wind storms, hailstorms, snow and rain storms, will occur. The real question therefore becomes, "What can be done to lessen the number of deaths, injuries, property damage, and economic loss which these natural hazard events cause?"

Before answering the foregoing question, a brief explanation is offered regarding why insurers are avoiding involvement in the scientific discussions pertaining to climate change. The decision not to engage in the debate over scientific issues is based on the fact that insurers are not technically equipped to enter into those types of discussions or the controversial causation issues swirling about those subjects. Underwriters, however, acknowledge that they must become more knowledgeable about the atmospheric sciences. Further, insurers acknowledge that they and the other stakeholders must more effectively come to grips with the consequences of weather-related events.

The insured property losses of 1991-1994 were defined as "disastrous years for the property insurance industry and the $21 billion in insured losses in 1992 traumatized the industry" (Roth, 1996). The property insurance industry does not keep industry-wide weather loss data like the crop industry does, but it does estimate the losses caused by weather events that produce $5 million or more in property losses, events defined as catastrophes. The 1996 catastrophes caused losses amounting $8.2 billion, almost five times the total insured crop losses for 1996. The 1991-1994 losses were created by 145 catastrophes and amounted to $33.8 billion, more than the total of the previous 20 years. However, adjustment of the catastrophic losses during the 1950s and early 1960s for inflation, changing insurance coverage, population growth, and demographic shifts, reveals the earlier weather losses approximated those of the 1990s (Changnon, et al., 1997). The huge losses from weather catastrophes of the 1990s led many companies to seek rate increases and some to withdraw coverage in high loss areas (Roth, 1996). Other industry actions are described in the following section.

3.2. ACTIONS TO ADDRESS CLIMATE FLUCTUATIONS AND CLIMATE CHANGE

In the early 1980s before Hurricanes Hugo and Andrew, many insurers had become concerned about the expanding demographics in the Sunbelt regions of the United States. They recognized the growth in the exposures, i.e., the value of residential and commercial structures in these regions. They were also concerned about the adequacy and enforcement of building codes and construction practices. Further, they foresaw the need for changes in structural design and for use of materials which would provide greater resistivity to wind. Also, they recognized the problems associated with an inadequate, obsolete, and a decaying infrastructure. Additionally, most insurers envisioned the need for land use considerations and control measures as a vital link in making America a safer place. Finally, insures saw the need to inform all stakeholders (policy and decision makers, insurers, businesses, emergency planners, lenders, designers, builders and the general public) to gain a better understanding of the full breadth of the weather-related issues, as well as the specific problems which each faced. Insurers have learned from years of experience of loss reduction work that they cannot do it alone. All stakeholders must be involved and must set their parochial considerations aside and work as one.

A major action to address weather hazards and to deal with the possibility of climate change was the formation of the Insurance Institute for Property Loss Reduction in 1994, subsequently renamed in 1997 as the Institute for Business and

Home Safety (IBHS). Insurers recognized the potential of an increasing number of natural hazard events and foresaw the escalating cost of future catastrophes due to the burgeoning population and the rapidly expanding exposures. They concluded that for the financial well being of property insurers, as well as society as a whole, a primary need was to eliminate as many losses as possible. Thus, the property insurers' focus was fixed on taking steps to assure that losses were reduced.

By working to reach out to the public, IBHS seeks to make all stakeholders aware of natural hazards, understand the associated risks, know how to reduce these risks, and take rational actions to mitigate the level of risk to which they are exposed. Major components of the IBHS program are public relations, education, and the articulation of response and recover issues. IBHS encourages responsible decisions about the density, type, and location of structures, and creates incentives to reduce development in high risk areas subject to natural hazards. By seeking stronger building codes, IBHS assists in ensuring that all new structures will be designed, engineered, and constructed using up-to-date techniques and materials. IBHS believes that codes, code enforcement, and land use control measures will, when coupled with an effective education program, bring a safer future for all Americans, while concurrently helping to ensure the availability and affordability of property insurance. By promoting the retrofitting of existing structures, IBHS participates in the reduction of potential deaths, injuries and property damage.

Finally, IBHS has undertaken development of an insured "paid loss" database for catastrophic events. Analysis will address specific weather damages to different classes of structures and types of materials. The database will provide valuable information and contribute to the reduction or elimination of losses and understanding of weather effects.

IBHS is convinced that the impact of catastrophic weather hazards, which may increase in a changed climate, will be significantly reduced. Regardless of the cause of the natural hazard events, i.e., caused by natural cyclical happenings or by human-induced climate change, these actions form an appropriate response and one which will safeguard society.

The insurers do have needs which the atmospheric sciences can provide. In considering these items, recall that historically the focus of the property insurance business has been on the fire peril (Roth, 1996). Also, note that most insurer's "statistical plans" have included no data regarding weather information. This situation is changing, and with available climatological data, as well as the paid loss data and other information being compiled in the IBHS Data Base, insurers will have the opportunity to address weather-related perils with the vigor previously reserved for the fire peril.

Climatological data, weather observations, and weather/climate forecasts are desired by property insurance interests to address five needs.

1. To provide a sound basis for an insurer's decisions to increase or reduce insurance so as to control its (the insurer's) exposure and assure solvency.

2. To calculate risk probabilities by location or area—this will be of considerable value, both as respects the foregoing item as well as in the determination of an insurer's premium needs.

3. To assist architects, loss control specialists, engineers, code designers and officials, and building contractors in the design and construction of safer, loss free structures.

4. To develop more accurate future loss estimates—"modelers" need accurate extended forecasts.

5. To assist in logistical planning, i.e., to determine the number of loss adjusters required and the locations for their most effective deployment—accurate short-range forecasts and climate statistics are required.

As the climatological data becomes more effectively used, as better weather observations are made, as weather and climate forecasts become more accurate (as shown by the ENSO-based predictions in 1997-1998), and as the skills of insurance practitioners are honed, greater use will be made of these products. In the interim, it behooves all stakeholders to work together in the problem-solving process.

4. Perspectives of the Atmospheric Sciences on Interactions with Insurers

4.1. PAST SERVICES AND INTERACTIONS WITH INSURERS

If one is to be effective in interactions with any potential "users" of atmospheric information, whether they are insurers (a company or an association of firms) or any other weather-sensitive group, one has to learn about the industry--its functions, its terminology, and its problems. This allows the weather scientist to appropriately guide how a problem is addressed and how the material is presented. Obviously, this atmospheric-insurance dialogue is a two-way street; the insurance user also has to become sufficiently aware of atmospheric issues like climate data limitations to intelligently utilize climate information.

Activities that an Illinois group has conducted for and with insurers since 1959 are summarized in Table 3. Review of these activities, and those of others who have worked extensively with the insurers, reveal they fall within four broad areas of application.

1. Perform climatological analyses such as defining historical fluctuations, probabilities of conditions, and prediction using insurance data and standard

Table 3. Types of research and assistance performed for insurers since 1959 by Illinois-based weather scientists.

- Assessing historical climate data to develop indices of past storm losses (Changnon, 1960, 1967, 1995).
- Performing detailed field studies of severe storms to provide data on the physical dimensions of damages (Changnon, 1962, 1964, 1966, 1970).
- Developing instruments and techniques to sense weather-damaged crops, including remote sensing of damage and hail-bearing storms (Changnon and Barron, 1971; Towery and Changnon, 1970; Changnon, 1992; Towery et al., 1976a).
- Utilizing historical climate data as a basis or proxy for developing rates, both geographically and by magnitude, say for rain insurance (Changnon and Changnon, 1989) and hail insurance (Fosse and Changnon, 1981; Changnon, 1984; Changnon, 1996).
- Assessing the viability of weather modification as a means to lessen storm damage (Changnon, 1977, 1978a), and the potential impact of weather modification on insurance industry (Friedman, 1976; Changnon et al., 1978).
- Assessing the issue of climate change (Changnon, 1988; Changnon and Huff, 1991; Fosse and Changnon, 1993).
- Studying insurance-based measures of past losses for measuring climate variability and the potential of climate change (Changnon, 1992; Changnon et al., 1996).
- Defining how atmospheric scientists interact with industry (Changnon et al., 1988; Changnon and Changnon, 1990; Changnon et al., 1996).
- Investigating statistical-based predictably of future trends in crop-loss data (Neill, et al., 1979).
- Developing crop yield-weather models for use in defining crop insurance risks (Changnon and Neill, 1968; Changnon and Sonka, 1981; Garcia et al., 1987).
- Analyzing loss data to define sizes of areas to use in developing rates for small areas and regions (Changnon and Stout, 1967), and to measure the spatial density of losses per unit area for designing coverage limitations (Changnon, 1968).
- Defining the characteristics of damage-producing weather conditions and how they create damage (Changnon 1970b, 1971, 1973, 1978b,; Towery et al., 1976b).
- Presenting talks at insurance conferences and workshops.

---

climatic data for assessing risk, rate making, decisions about use of reinsurance, and sales planning.

2. Develop and report on new techniques and data that have applications to a variety of insurance operations.

3. Provide expertise and advice on various issues including state of the science.

4. Work with insurers to collect data needed to better define risk or damage.

Much experience has been with the crop insurance sector (Changnon, 1978c; Changnon and Fosse, 1981). However, in recent years there have increasing efforts to assist the property insurers, done for two reasons: (1) to assist in developing appropriate risk measures for assessing rates and coverage in high risk areas, and (2) to interpret the recent extremes and their frequency of occurrence (Changnon et al., 1996). Their records of property loss catastrophes, defined as events that cause > $5 million in losses, have been useful in assessing the fluctuation of major damaging events (Changnon and Changnon, 1992). In 1997, the definition of a catastrophe was shifted to losses of $25 million or more. Work with the reinsurance industry has been primarily as an advisor to provide expertise on issues like severe storms and climate change.

Property insurance firms in the United States have not had a weather loss data gathering system comparable to that which has existed since 1948 in the crop-hail insurance industry. Hence, assessing the weather risks for property has often involved the use of climatic data and indicators derived therefrom (Changnon, 1989; Changnon, 1988). One major insurance firm for many years had a staff with meteorological expertise, and this firm made numerous studies of historical data and collected field loss data to assess storm risks for the company (Friedman and Shortell, 1967; Collins and Howe, 1964). However, most firms have not had the database nor the interest until recently to effectively deal with weather risks.

Hurricane Andrew provided a wake-up call for the property insurance industry as to the importance of weather-related perils (Roth, 1996). Until the 1990s the property industry was focused on its historical peril-fire. Most insurers' statistical systems had done little to accommodate weather loss information; hence, there was no data base.

Putting weather information into perspective with other factors influencing insurance is an important part of an effective interaction with the insurers. The huge weather-induced losses in the 1990s were related to several non-weather factors. Roth (1996) identified three reasons for corporate problems that developed in the 1990s. First, there had been prolonged favorable loss experiences before 1990, causing rate levels to be set lower than needed, especially in the southeastern U.S. Second, a significant demographic change in population, with increases of 75% from 1970 to 1990 in coastal areas, went unnoticed by many insurers along with a significant increase in the values of property. Third was a growing lack of enforcement of building codes and shoddy construction workmanship ($4 billion of Hurricane Andrew's $15 billion in insured losses were estimated to be due to these construction factors).

4.2. AREAS NEEDING ATTENTION BY ATMOSPHERIC SCIENTISTS

The 172 property catastrophes (each causing losses > $5 million) and near record crop losses during 1991-1995 produced a series of major impacts in all sectors of the insurance (Changnon et al., 1997). Insured property losses reached $48 billion, crop-hail insured losses were $1.6 billion, and other insured weather-related crop losses were $4.5 billion. A few property insurance firms went broke. These losses led to an amazing array of responses. Property insurers raised rates, reduced coverage in high risk locations, and sought greater financial stability. Problems with crop insurance led to a change in federal crop insurance laws.

Table 4. Examples of recent weather-insurance collaborations to address the needs of insurers.

- Assessing the incidence of thresholds of sub-freezing temperatures in the southern U.S. where frozen pipes are a major problem.
- Defining the rainfall regions of the U.S., based on hourly rain rates, to revise inadequate rain insurance rates.
- Defining the damaging characteristics of hailstones needed for assessing roofing materials as input for improving building codes.
- Conducting workshops for insurance leaders on topical issues like climate change and variability, and on climate predictions, and involving insurance leaders in the planning for the U.S. Weather Research Program.
- Assessing the regional risk of damage to property from hail, lightning, tornadoes, winds and winter storms.
- Developing means to measure peak winds within hurricane damaged areas to have more representative data needed to assess the type of property damage.
- Measuring weather-damaged areas rapidly after the event with new sensors like NEXRAD systems to provide guidance for field assessments and in-season monitoring of loss.
- Evaluating the potential damage by winds and developing a PC program so insurers can use this wind information in underwriting and in advising property owners about risk.
- Assessing hail frequency and estimating intensity from climatic records for many decades before insurance records began.

Reinsurance firms raised rates, sought and secured new sources of funds, and some withdrew from the marketplace. Property insurers and reinsurers became much more aware of weather problems than ever before, and this has created opportunities for closer interactions with atmospheric scientists. A multi-sector assessment of the impacts and responses of insurers to the severe weather conditions of the 1990s has shown there is a great need for closer interactions between the atmospheric sciences community and insurers (Changnon et al., 1997).

Table 4 presents a list of several recent and ongoing collaborations between atmospheric scientists and insurers, illustrating the dimensions of the work. These projects and recent insurance meetings involving atmospheric scientists reveal there are two topics of concern to the industry: long-term forecasts, and global climate change. An area of deep interest, as might be expected, is in long-range forecasts, and in particular those with that address the likelihood of future extremes (Sonka et al., 1992). Climatologically based probabilities and models may be the best information available at this time, but this type of information is not being accessed by most insurance interests. However, the current situation in which several "weather experts" make proclamations about next summer's weather conditions with predictions that vary considerably (wet versus dry, hot versus cool), is one that has deeply hurt the credibility of the atmospheric sciences with insurers. There is need to have reputable persons/institutions explain this situation and direct interested users to what is available and reliable.

The global climate change issue has great relevance to the weather insurance industry; a shift in climate to more or fewer storms, or to more intense events, would create major problems. Insurance rates and their justification are traditionally based on historical data, not on future expectations. However, for a variety of planning purposes, insurers can use reliable information about the onset of a change in climate, and about its dimensions as they relate to crop and property damaging weather.

5. Conclusions:
Guidelines for Effective Interaction with Insurance Interests

First, atmospheric scientists have to invest time and seek opportunities to learn about the insurance business including its operations and functions. Those who seek to work effectively with insurers must develop an appreciation of the forces affecting insurers, in addition to weather. Economic pressures of all types exist in the market place and sometimes lead to what appear to be scientifically unrealistic decisions in the insurance field. Then, there are governmental regulations that affect decisions about how insurance companies can plan and adjust to changing conditions. For example, a major problem for U.S. insurers, in adjusting to a potentially changing climate, either to more or fewer extremes, is the fact that future rates are based *solely* on past records, not on future expectations. Insurers rely on past experience and cannot establish rates on the basis of predicted shifts in weather. However, there are other economic actions and adjustments a company could pursue if it had knowledge of a shift in future weather conditions. A major recommendation for interactions with insurers is to deal with things one has the expertise to handle. Candid and unbiased information about controversial issues like climate change, long-range predictions, and weather modification capabilities is essential.

Second, atmospheric scientists need to prepare products that can be used rather easily and without many adjustments by the industry -they should be user friendly. Some products will not be used, no matter how carefully developed and presented. Insurance decision makers generally understand how to use a new product, but at times can be reluctant to change how they make decisions. Performing work for an individual firm that operates in a competitive world can also mean restricting a product to that firm's sole use.

Third, to be effective atmospheric scientists need to understand the areas where the industry has needs for atmospheric data and information. Several examples have been offered in this paper and they are summarized below. The industry can use atmospheric sciences help:

- to mitigate weather losses.
- to assess risks in space and time.
- to measure critical perils and their damages.
- to educate the public, government, and builders about risks.
- to learn about major atmospheric issues and advances in the field that are applicable to insurance interests.

The weather extremes of the 1990s served a useful purpose -they awakened insurers to the need to more wisely address weather hazards. The effect of the weather extremes have created an opportunity for the atmospheric sciences.

Acknowledgments

We appreciate the use of data supplied by the National Crop Insurance Services of Overland Park, KS. Portions of the atmospheric research herein were supported by the Electric Power Research Institute, but the results presented do not reflect the views of EPRI.

References

Changnon, S.A.: 1960, *Relations in Illinois Between Annual Hail Loss Cost Insurance Data and Climatological Hail Data.* Research Report 3, Crop-Hail Insurance Association, Chicago, 18pp.

Changnon, S.A.: 1962, *Hailstorms on May 17, 1962.* Research Report 16, Crop-Hail Insurance Actuarial Association, Chicago, 15pp.

Changnon, S.A.: 1964, Surface Features of Two Intersecting Hailstorms. *Nubila*, 6, 74-86.

Changnon, S.A.: 1966, *Disastrous Hailstorms on June 19-20, 1964 in Illinois.* Research Report 31, Crop-hail Insurance Actuarial Association, Chicago, 37pp.

Changnon, S.A.: 1967, Method of Evaluating Substation Records of Hail and Thunder. *Monthly Weather Rev.*, 95, 209-212.

Changnon, S.A., and G. Stout: 1967, Crop-hail Intensities in Central and Northwest United States. *J. Applied Meteorology*, 6, 542-548.

Changnon, S.A.: 1968, Effect of Sampling Density on the Areal Extent of Damaging Hail. *Journal of Applied Meteorology*, 7, 518-521.

Changnon, S.A. and J.C. Neill: 1968, Mesoscale Study of Corn-Weather Responses on Cash-Grain Farms. *J. Appl. Meteoro.*, 7, 94-100.

Changnon, S.A.: 1970a, Major Hailstorms Retrace Tri-State Tornado Track in Illinois. *Transactions Illinois Academy of Sciences*, 63, 34-41.

Changnon, S.A.: 1970b, Hailstreaks. *J. Atmospheric Sciences*, 27, 109-125.

Changnon, S.A.: 1971, Hailfall Characteristics Related to Crop Damage. *J. Applied Meteorology*, 10, 270-274.

Changnon, S.A. and N. Barron: 1971, Quantification of Crop-Hail Losses by Aerial Photography. *J. Applied Meteorology*, 10, 86-96.

Changnon, S.A.: 1973, Hail Sensing and Small Scale Variability of Windblown Hail. *J. Weather Modification*, 5, 30-42.

Changnon, S.A.: 1977, On the Status of Hail Suppression. *Bull. Amer.Meteoro.Soc.*, 58, 20-28.

Changnon, S.A.: 1978a, Weather Modification in a Socioeconomic Context: Its Proper Setting.

Weather Modification Technology and Law, Amer. Assoc. Advancement of Science,Washington, D.C., 71-79.

Changnon, S.A.: 1978b, Heavy Falls of Rain and Hail Leading to Roof Collapse. *J.Structural Division, ASCE*, 104, 198-200.

Changnon, S.A.: 1978c, The Climatology of Hail in North America. Monograph 38, American Meteorological Society, *Review of Hail Science and Hail Suppression*, 107-128.

Changnon, S.A., Farhar, B., and E. Swanson: 1978, Hail Suppression and Society. *Science*, 200, 387-394.

Changnon, S.A.: 1981, *Weather-Crop Yield-Crop Practice Relationships*. Contract Report 264, Illinois State Water Survey, Champaign, IL, 29pp.

Changnon, S.A., and E.R. Fosse: 1981, Impacts and Use of Climatological Information in the Hail Insurance Industry. *Proc.Climate and Risk Conference*, Mitre Corp., 28pp.

Changnon, S.A.: 1984, Temporal and Spatial Variations in Hail in the Upper Great Plains and Midwest. *J. Applied Meteorology*, 23, 1531-1541.

Changnon, S.A.: 1988, Climatography of Thunder Events in the Conterminous U.S. Part 1: Temporal Aspects. *J. Climate,* 1, 389-398.

Changnon, S.A.: 1989, Temporal Features in Thunder Days in the United States. *Climatic Change,*15, 455-477.

Changnon, S.A., Sonka, S., and S. Hofing: 1988, Assessing Climate Information Use in Agribusiness. Part 1: Actual and Potential Use and Impediments to Use. *J. Climate*, 1, 757-765.

Changnon, S.A., and J.M. Changnon: 1989, Developing Rainfall Insurance Rates for the Contiguous United States. *J. Applied Meteoro.*, 28, 1185-1196.

Changnon, S.A., and J.M. Changnon: 1990, Use of Climatological data in Weather Insurance. *J.Climate*, 3, 568-576.

Changnon, S.A., and F.A. Huff: 1991, Potential Effects of A Changed Climate on Rainfall Frequencies in the Midwest. *Water Resources Bull.*, 27, 1-7.

Changnon, S.A.: 1992, Use of NEXRAD Data in Assessing Climate Change. *Proc.Conference on Managing Water Resources During Global Climate Change*, Am. Water Res. Assoc., Reno, 71-74.

Changnon, S.A., and J.M. Changnon: 1992, Temporal Fluctuations in Weather Disasters: 1950-1989. Climatic Change, 22, 191-208.

Changnon, S.A.: 1995, *Temporal Fluctuations of Hail in Illinois*. Miscellaneous Publication 167, Illinois State Water Survey, Champaign, 18pp.

Changnon, S.A., Changnon, D., Fosse, E., Hoganson, D., Roth, R., and J. Totsch: 1997, Effects of Recent Weather Extremes on the Insurance Industry: Major Implications for the Atmospheric Sciences. *Bull. Amer. Meteoro. Soc.,* 78, 425-435.

Changnon, S.A.: 1996, *Climatology of Hail Risk in the United States*. CRR-40, Changnon Climatologist, Mahomet, IL, 76pp.

Collins, G.F., and G.M. Howe: 1964, *Weather and Extended Coverage*. TRC Service Corp., Hartford, CT, 51pp.

Fosse, E.R.: 1996, Impacts and Responses to the 1991-1994 Weather: The Crop-Insurance Industry. Chapter 4 in *Impacts and Responses of the Weather Insurance Industry to Recent Weather Extremes*, Changnon Climatologist, Mahomet, IL, 75-100.

Fosse, E.R., and S.A. Changnon: 1993, Potential Impacts of Shifts in Climate on the Crop Insurance Industry. *Bull.Amer. Meteoro. Soc.*, 74, 1703-1708.

Friedman, D., and P Shortell: 1967, *Prospective Weather Hazard Rating in the Midwest with Special Reference to Kansas and Missouri*. The Travelers Co., Hartford, CT, 103pp.

Friedman, D.G.: 1976, *Hail Suppression's Impact on the Property Insurance Industry*. TASH Report 11, Illinois State Water Survey, Champaign, IL, 69pp.

Garcia, P., Offutt, S., Pinar, M., and S.A. Changnon: 1987, Corn Yield Behavior: Effects of Technological Advances and Weather Conditions. *J. Clim. Appl.Meteoro.,*26, 1014-1024.

Garcia, P., Offutt, S., Pinar, M., and S.A. Changnon: 1987, Corn Yield Behavior: Effects of Technological Advances and Weather Conditions. *J. Clim. Appl.Meteoro.,*26, 1014-1024.

Neill, J., Changnon, S., and P. Reddy: 1979, *Prediction of Fluctuations in Crop-Hail Loss Data.* CHIAA Research Report 44, Chicago, IL, 90pp.

Roth, R.J.: 1996, The Property-Casualty Insurance Industry and the Weather of 1991-1994. Chapter 5 in *Impacts and Responses of the Weather Insurance Industry to Recent Weather Extremes*, Changnon Climatologist, Mahomet, IL, 101-132.

Sonka S., Changnon, S.A., and S. Hofing: 1992, How Agribusiness Uses Climate Predictions: Implications for Climate Research and Provision of Predictions. *Bull.Amer. Meteoro. Soc.,* 73, 1999-2008.

Towery, N.G., and S.A. Changnon: 1970, Characteristics of Hail-Producing Radar Echoes in Illinois. *Mo. Wea. Rev.*, 98, 346-353.

Towery, N.G., Changnon, S.A., and G. Morgan: 1976a, A Review of Hail-Measuring Instruments. *Bull.Amer.Meteoro. Soc.*, 57, 1132-1140.

Towery, N.G., Morgan, G., and S.A. Changnon: 1976b, Examples of the Wind Factor in Crop-Hail Damage. *J. Applied Meteorology*, 15, 1117-1120.

(Received 5 November 1997; in revised form 19 November 1998)

DAMAGING WEATHER CONDITIONS IN THE UNITED STATES: A SELECTION OF DATA QUALITY AND MONITORING ISSUES

BRIAN D. SKINNER[1], DAVID CHANGNON[2], MICHAEL B. RICHMAN[1], and PETER J. LAMB[1]

[1] *Cooperative Institute for Mesoscale Meteorological Studies (CIMMS), University of Oklahoma, Norman, OK, 73019-0628*
[2] *Department of Geography, Northern Illinois University, DeKalb, IL, 60115-2854*

Abstract. The limitations of observational data available for the study of damaging weather conditions (e.g., storms and extreme temperature events) are discussed. Crop and property insurance loss records are advocated as a potential supplement to traditional weather observations, as they integrate specific information about the spatial dimension of damaging weather conditions and the cost of damage they cause. Insurance loss data may also be analyzed in combination with meteorological data sets to derive indicator variables for the detection of damaging weather events.

Two sets of insurance data are described. One record provides adjusted property losses associated with "catastrophic" weather events since 1949, and the other is an index of the amount of crop-hail losses per year since 1948. Additionally, an example of the benefits of the combination of insurance and meteorological data is presented through a selection of results from a recent study of freezing temperatures in the southeastern United States and associated insurance claims related to pipe bursting.

If insurance data are to be applied in the future in similar studies of damaging weather conditions, it is essential that the insurance industry continues to collect and adjust loss data and periodically confirm that adjustment factors are temporally consistent.

1. Introduction

Some of the most interesting and important aspects of weather are extreme temperature events and various storms. These weather conditions cause loss of life, economic devastation, and serious environmental impacts that far outweigh the effects of most other weather extremes. Unfortunately, any assessment of the frequency and occurrence of such events, through indices for monitoring fluctuations and shifts in climate, rests on the availability of suitable data and their interpretation. A meaningful assessment requires information (e.g., magnitude, intensity, and areal scale of the conditions that cause damage) which is often not available.

Since the early 1890s, the United States has maintained a fairly dense Cooperative Observer Network of locations where potentially damaging weather conditions are routinely recorded. At its maximum coverage, this network included approximately 10,000 sites, and currently includes about 8,000. However, only about 1,000 stations have existed continuously since the

Climatic Change **42**: 69–87, 1999.
© 1999 *Kluwer Academic Publishers. Printed in the Netherlands.*

network's onset. Observations include, among others, the amounts of rain and snow, the incidence of thunderstorms, hail, and freezing rain, and the occurrence of extreme temperatures. At a much less dense scale, the incidence of daily pressures and wind speeds have also been recorded since the 1890s at about 200 First-Order observation stations. Thus, there are 100+ years of various indicators of potentially damaging weather conditions that can be used to assess historical fluctuations of climate. However, the extraction of damaging weather conditions from historical weather records is complex. Certain factors must be considered when defining indices based on observed weather variables, in order that they be useful as direct and indirect measures of damaging conditions.

First, many of the available observations in weather records at a point location do not directly measure the occurrence of a *damaging* storm. For example, it is known that damage is caused by only 5–10% of all thunderstorms (Court and Griffiths, 1986), 10–15% of all cases of freezing rain (Changnon, 1969), 10–25% of all hailfalls (Changnon et al., 1978), and 30–40% of all tornadoes (Fujita, 1981; Changnon, 1982). Damage is thus not guaranteed by the occurrence of a 6-inch (15.3 cm) snowfall, a 2-inch (5.1 cm) rainfall in 24 hours, a thunderstorm, freezing rain, a day with hail, or a particular temperature. It only indicates that a potentially damaging event occurred -- an indirect measure. Furthermore, most point records for weather phenomena like thunderstorms, freezing rain, and hail, do not quantify the event as to its local intensity (e.g., size of hailstones or thickness of ice). Similarly, point measures of pressure and wind speeds, as indicators of cyclonic-scale storms, also suffer limitations. The particular pressure reading which is associated with damaging systems may be unknown, and even if it were, the network may not be sufficiently dense to sense these identifying thresholds. Use of threshold wind speeds as an indicator of local damage have been defined, but the historical wind records are not suitable for temporal analyses, since there have been a series of changes in instruments, heights of sensors, and locations of measurements that collectively impair the records for these purposes.

Second, there is the frequent absence of a spatial dimension, based on intensity and areal extent, which must be inferred. Some studies have been able to show a relationship between the areal coverage of a weather condition on a given day and the likelihood of damaging weather conditions. For example, the areal extent of high frequencies of hail days (measured at standard weather stations) during the growing season relates well to the typical crop-hail losses experienced in that area (Changnon and Changnon, 1997a). Studies of damaging winter storms in the Midwest, based on weather station reports, found a good relationship between the areal extent of freezing rain and the incidence of damage (Changnon, 1969).

The third factor affecting the incidence of damaging weather conditions is the nature of the "target". For example, an intense hailstorm on May 15 will cause greater damage to a maturing wheat field in Texas than to an emerging corn crop in Illinois. The damage caused by this same storm affecting the

Chicago or Denver urban areas would be still greater.

Thus, analysis of damaging weather conditions requires data on the point intensity, areal scale, and target (land use and season) involved. Unfortunately, the U.S. historical records of damaging weather conditions kept since 1955 in *Storm Data* are inconsistent and incomplete in these important respects, and are not adjusted for changing socioeconomic conditions that partly determine the value of the impacts. For these reasons, crop and property insurance loss records have great utility as indicators of damaging weather conditions. Insurance loss data, after adjustments for temporal trends in liability and other societal changes, are excellent measures of damaging weather conditions and are being explored for use in climate change studies (Changnon et al., 1997; Changnon and Changnon, 1997b). They integrate information about damaging weather conditions according to their type, physical characteristics, and the amount of damage they cause, providing a logical method for assessment of extreme weather events. These and other insurance data, known to be associated with particular weather phenomenon, may also be *combined* with existing meteorological data sets to derive indicator variables for the detection of damaging weather events. We offer here a description of two important sets of data on weather-related insurance claims and losses, and one example of how insurance data and climatological data may be used in combination.

2. Insurance-Related Data Bases and Indices

The weather insurance industry has developed and used, over the past 47 years, the two sets of data described below that integrate weather event types with the losses they caused (Changnon and Changnon, 1990). One record provides adjusted property losses resulting from "catastrophic" events, and the other is an index of the crop-hail losses per year. It is important to acknowledge that incorporation of damage amounts brings another measure of uncertainty into the record of past events -- the ever-changing sensitivity of society to storms, in terms of differences in the cost of repairs, shifts in dollar value, variations in local and regional population densities, and changes in construction methods. Fortunately, the insurance industry has developed processes for adjusting records for individual storm losses to these changing socioeconomic conditions.

2.1. PROPERTY "CATASTROPHE" DATA SET

Beginning in 1949, the property insurance industry identified all weather events causing a total of $1 million or more in losses in the affected area (local to multi-state regions) and labeled these as "catastrophes." The minimum value of $1 million was kept constant until 1983, when the level was increased to $5 million in response to the changing dollar value. The Property Claims Services of the American Insurance Services Group, Inc., has gathered the readily available

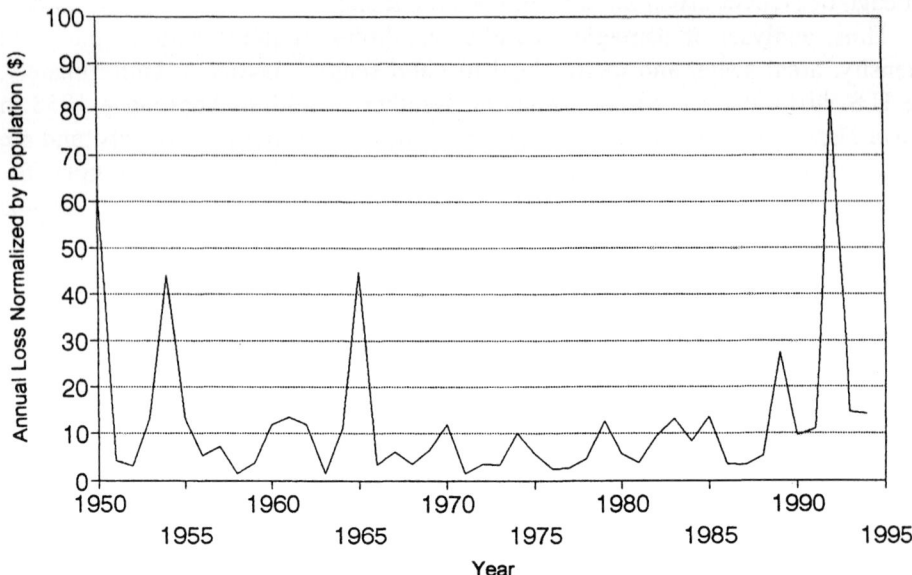

Figure 1. Normalization of adjusted catastrophe losses by U.S. population for 1950-1994, based on annual values of losses due to catastrophes causing greater than $35 million.

direct loss data using a consistent procedure throughout the period since 1949, and made estimated assessments of total catastrophe losses for the insurance industry. These estimates have reflected "judgement of the total net insurance payments for real and personal property included under 1) fire and extended coverages, 2) time-element coverages, 3) inland marine fixed property coverages, and 4) comprehensive to automobile coverages" (Property Claims Services, 1995).

One major insurance company adjusted each catastrophe record for the above set of changing conditions so that losses from any given past year could be compared with other years in the period 1949–1994. Three adjustment factors were used, which were applied according to the date, type of weather phenomenon, and the geographical location of the damage. The first adjustment factor employed integrated changes of property values and cost of repairs over time. The second adjustment factor addressed the relative growth in the size of the fixed property market in the area(s) affected. The third adjustment factor represented an estimate of the relative change over time in the share of the total property market that was insured against weather perils. The primary data available in the adjusted data set include the annual number of catastrophes, the annual total adjusted losses from all catastrophes, and the annual catastrophe intensities (total losses divided by the total number of catastrophes).

Annual adjusted losses associated with catastrophes causing greater than $35 million in damage were normalized by U.S. population (Figure 1). A minimum value of $35 million was selected because it represents the adjusted value for a

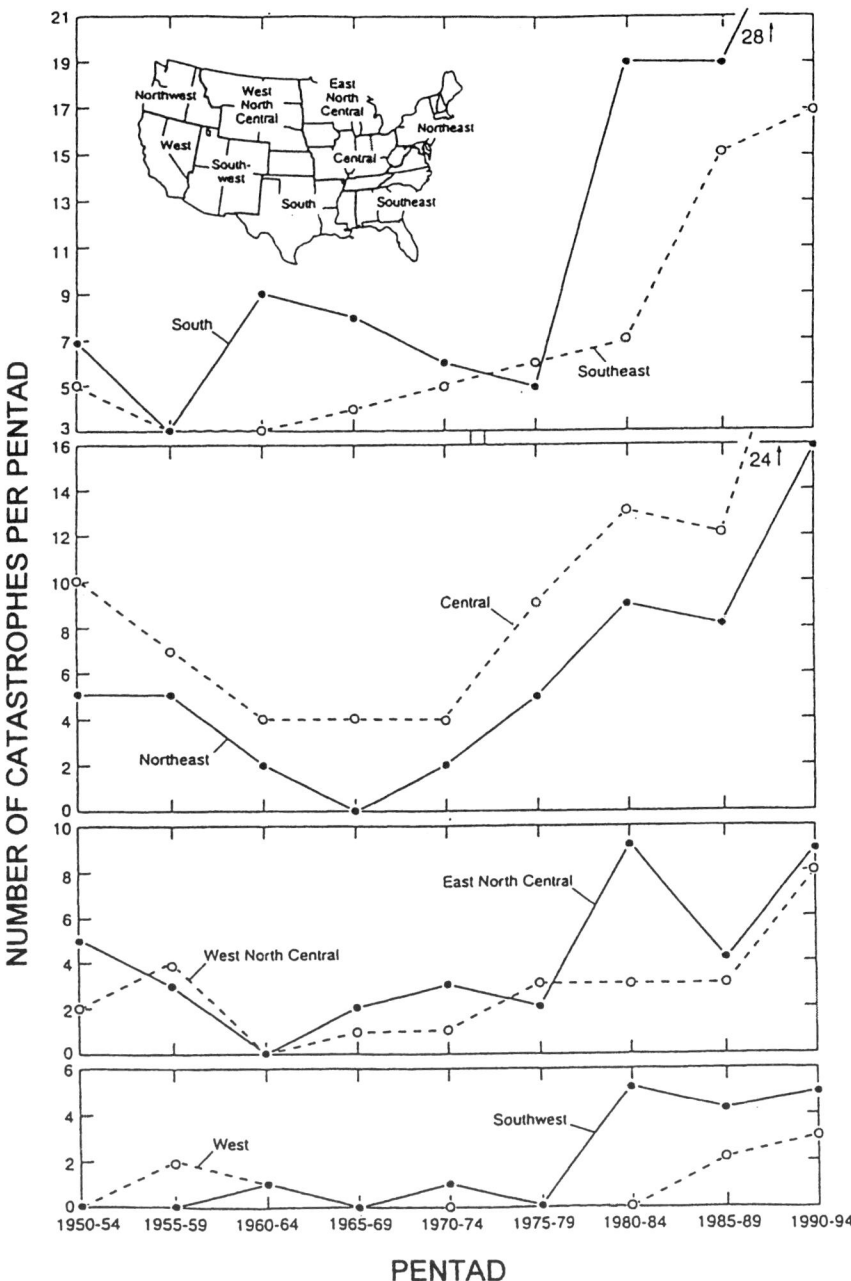

Figure 2. Regional frequencies of catastrophes causing greater than $100 million damage for 1950–1994 (from Changnon and Changnon 1997b).

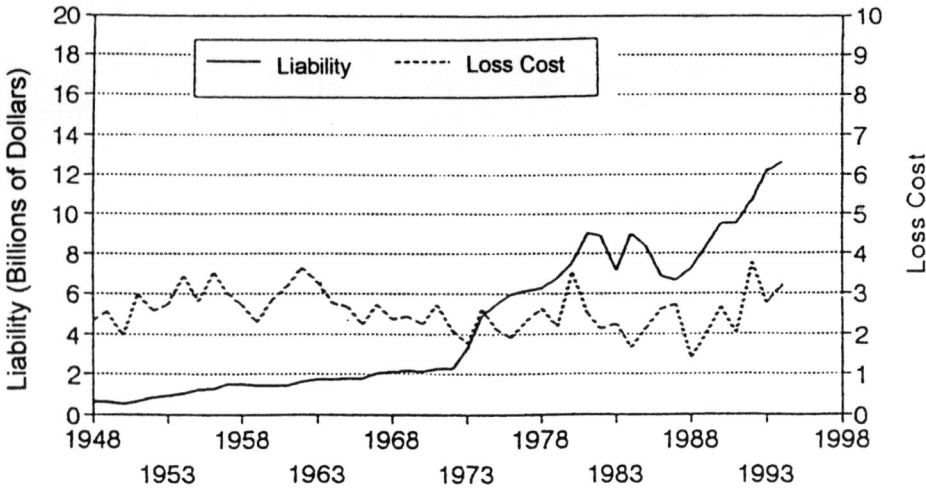

Figure 3. Annual values of loss cost and liability for the United States, 1948–1994.

$1 million loss that occurred in 1950. Results indicate essentially no trend, with spikes in five years when major hurricanes hit the U.S. (Changnon and Changnon, 1997b). Time series of losses associated with catastrophes causing greater than $100 million in damage in the affected area were constructed for specific regions within the U.S. and are shown in Figure 2. These graphs indicate that in the South and West, which have experienced increases in population due recent migration patterns, the number of catastrophes has generally also increased over time, an indication of the influence of changing "targets" for damaging weather conditions.

2.2. CROP-HAIL "LOSS COST" DATA SET

The crop-hail insurance industry began collecting data on hail loss to crops in 1948 (Crop-Hail Insurance Actuarial Association, 1978). The National Crop Insurance Services (NCIS) and its predecessor agency, the Crop-Hail Insurance Actuarial Association (CHIAA), have accumulated daily and annual crop-hail loss data since that time. These data cover most of the crop-hail insurance written in the United States. Annual liability, premiums, and losses were recorded and are published (with county-level resolution) for each crop, for individual states, and for the nation (National Crop Insurance Services, 1995).

The insurance data have been adjusted for temporal changes in liability (coverage), dollar values, and other factors by using the "loss cost." The annual loss cost value is calculated as the ratio of annual losses to annual liability, expressed in loss dollars per $100 dollars of total liability for the county concerned. Losses are defined as the dollar amount paid for losses incurred under the policy conditions, and the liability is the dollar amount of insurance placed on the insured crop (Changnon and Changnon, 1997a). For example, a loss cost of

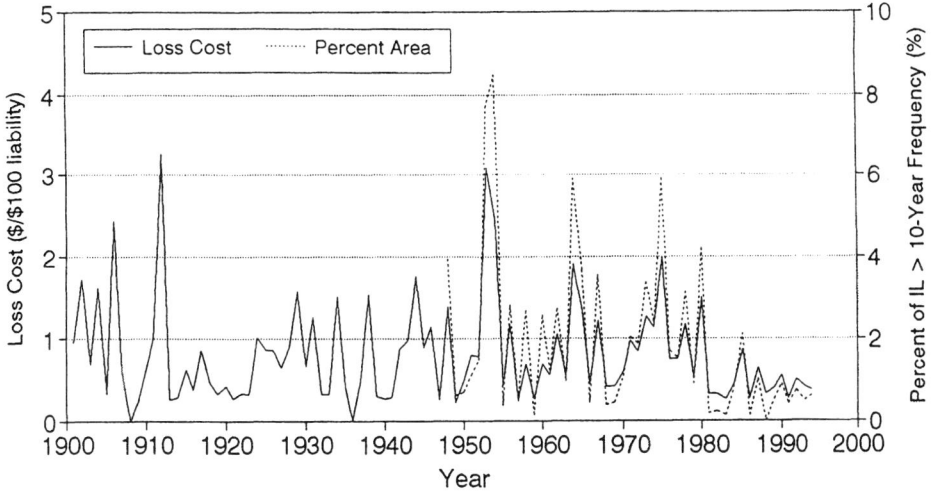

Figure 4. Illinois annual crop-hail loss cost, 1901–1947 calculated and 1948–1994 observed, and the 1948–1994 percent area of Illinois with extensive insurance coverage which experiences more than a 10-year hail-day frequency.

$3.37 means that for the area and liability affected, the amount of $3.37 was paid for losses for each $100 of total liability in force in each spatial unit.

Loss costs may be examined at the county, state, or national level (Figure 3). For example, Changnon and Changnon (1997a) identified strong relationships between the annual percent area of a state that experienced a one-in-ten year frequency in hail days and that state's annual loss cost (Figure 4). This relationship, based on data from 1948–1994, was then used to determine the loss costs for years before 1948.

3. Case Study: Extreme Freezing Temperatures and Property Losses due to Pipe Bursting

An example of the benefits of the use of insurance data in applied climatological research is found in a study of the occurrence of extreme freezing temperatures in the southeastern United States and their relationship with insurance claims and losses due to pipe bursting (Skinner, 1997; Skinner et al., 1998). Spurred by an increase in recent years in the number of pipe-bursting claims, the Institute for Business and Home Safety (IBHS), a national consortium of 70 insurance groups with over 300 insurance companies, sponsored this comprehensive investigation. The intent was to provide new information for application within the insurance

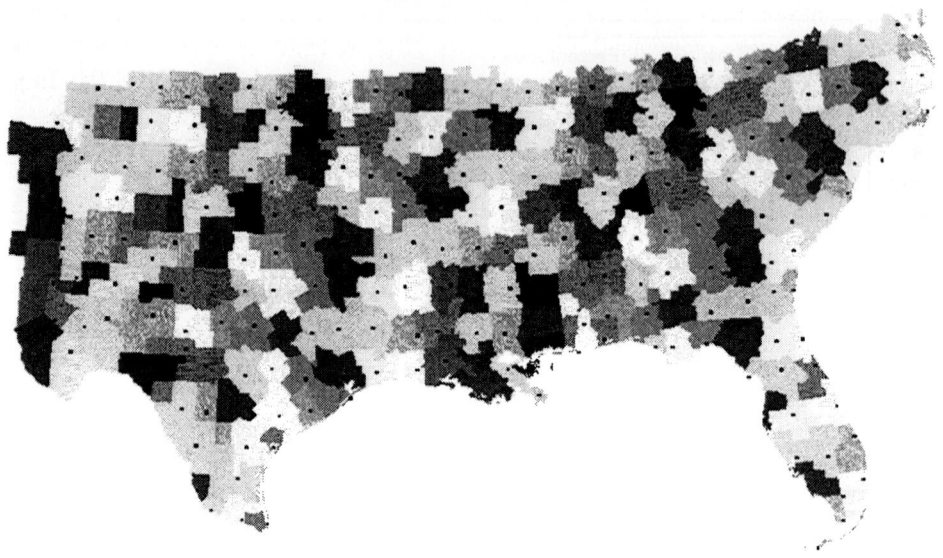

Figure 5. Insurance districts for combined temperature and insurance data set. Temperature data set locations are plotted within corresponding insurance districts of one or more counties.

industry to help reduce losses from pipe-freezing events. In addition, some of the results also provide new information about climate change in the Southeast in terms of the frequency of extreme freezing temperatures.

3.1. DATA

Temperature data used in the investigation consisted of daily maximum and minimum station observations taken as a subset of a larger data set that was developed over the last 15 years (e.g., Richman and Lamb, 1985; Richman et al., 1991; Gong and Richman, 1995). This subset spans the cool season (October–April) during 1950–1992 at 241 locations in 16 south and southeastern states shown in Figure 5.

In cooperation with IBHS, two individual sets of claim and loss information for damages related to pipe-bursting events were obtained from one member company for use in the investigation. These data sets are excellent examples of the types of proprietary data which may be employed in conjunction with meteorological data in future applications. Each data set contained records organized at a county resolution, with information about the total number of claims and corresponding total amount of building and contents losses. The larger of the two data sets included individual daily records per county for the time period 1986–1995, which included the catastrophic pipe-freezing event of December 1989. The second data set contained county records exclusively from the similarly catastrophic pipe-freezing event of December 1983.

To combine the meteorological and insurance data sets, a geographical information system was used to calculate the centroid of each county and determine the temperature station closest to the centroid location. By this method, groups of counties were assigned to the appropriate nearest temperature station to form the so-called *insurance districts* shown in Figure 5. Claim records for the cool-season months (October–April) for time periods for which there were both temperature and claim data (December 1983 and January 1986 through December 1992) were retained, and limited to only those records for which there was a daily minimum temperature less than or equal to 0°C.

3.2. DEVELOPMENT OF INDICATOR THRESHOLDS FOR PIPE BURSTING

A large number of initial calculations and analyses were performed to outline the relationship between the occurrence of freezing temperatures and the occurrence of claims. One important, basic finding was that the correlation of the time series of total freeze occurrence (number of days) in a given time period with the time series of associated claim occurrence (total number of claims) increases as successively colder temperature thresholds were employed. However, with each colder threshold, there was a decrease in the number of temperature stations at which such temperatures were observed. The method used to determine temperature indicator thresholds for pipe bursting had to accommodate this trade-off between reaching higher correlation values and still retaining the largest number of insurance districts possible, and identified the one temperature threshold which optimized both properties.

Correlation-curves were established for each location, which are characteristic shapes or profiles of the relationship between claim occurrence and freezing temperatures. To produce each correlation-curve, time series of claim occurrence and freeze occurrence (each composed of 99 half-month time periods) were correlated, using a series of temperature thresholds from 0°C to –20°C, at 1°C intervals. Undefined correlations (due to lack of occurrence of temperatures below the threshold value) were assigned a value of zero. Results were then represented graphically, by plotting the correlation values as a function of the temperature threshold. An explanatory schematic of the method used to create correlation-curves for each location is given in Figure 6.

VARIMAX-rotated Principal Component Analysis (VPCA) was then applied to group sites with different "types" of correlation-curve similarities into natural subregions of the domain (Richman and Lamb, 1985; Richman, 1986). Four regions were ultimately derived from the VPCA, which were numbered Regions 1–4 from north to south and are shown in Figure 7. These delineated regions represent areas of similar pipe-bursting claim-response to the particular freezing temperature thresholds also indicated in Figure 7.

A three-step optimization technique was applied to determine the specific indicator thresholds for pipe bursting in each of the four regions. First, regional-average correlation-curves were created by averaging individual correlation

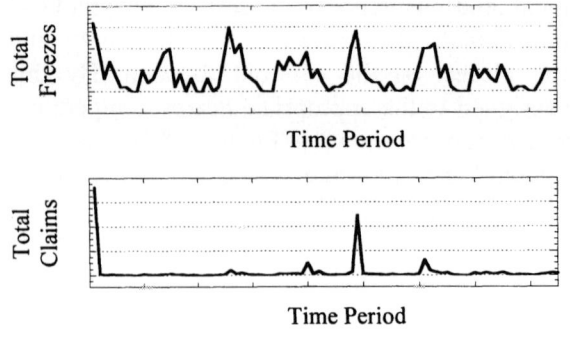

Analyses were based on the correlation between time series of claim occurrence (lower panel) and occurrence of freezing temperatures (upper panel). Example time series from Insurance District 121 near Dallas, TX, are shown at left. Each time series consisted of values from 99 half-month time periods from December 1983 and January 1986 through December 1992.

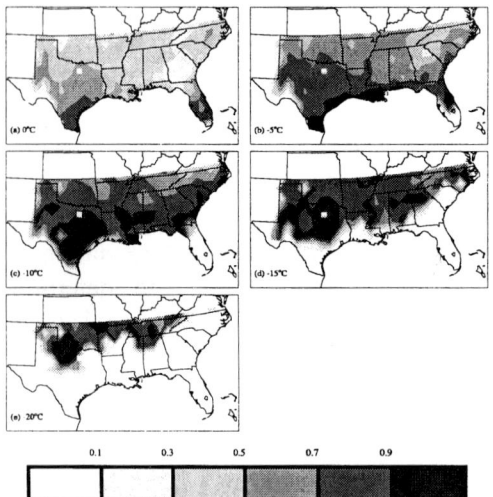

Correlations were obtained for each station using a series of temperature thresholds, from 0°C to –20°C at 1°C intervals. Results for particular temperature thresholds may be contoured as in the figure at left. Undefined correlations, due to lack of freeze occurrence, are given the value zero. In the development of correlation-curves, values corresponding to individual locations are displayed on one graph. For example, values for Insurance District 121 are identified on the panels at left by a square white bullet. These values were then displayed again as numerical values on the correlation-curve below.

Thus, for Insurance District 121, the correlation between claim occurrence and freeze occurrence at 0°C equals approximately 0.6, which is shown at right and is the same as the value in the region of the square bullet in plot (a) above.

The completed correlation-curve is a characteristic profile of the claim-response to freezing temperatures at different levels. For Insurance District 121, there is an increase in correlation value until a temperature threshold of –16°C is reached, when the value becomes undefined (due to an absence of such cold temperatures in the historical record) and is set equal to zero. This curve can then be compared to curves from other locations to find regions of similar response (see text).

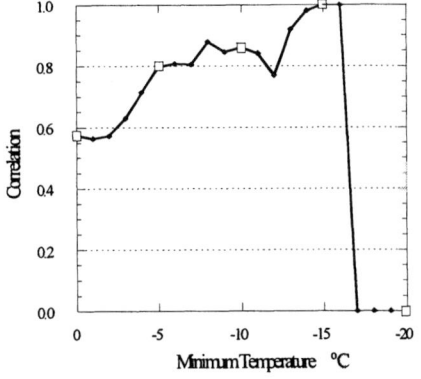

Figure 6. Explanatory schematic of method used to develop correlation-curves for individual insurance districts

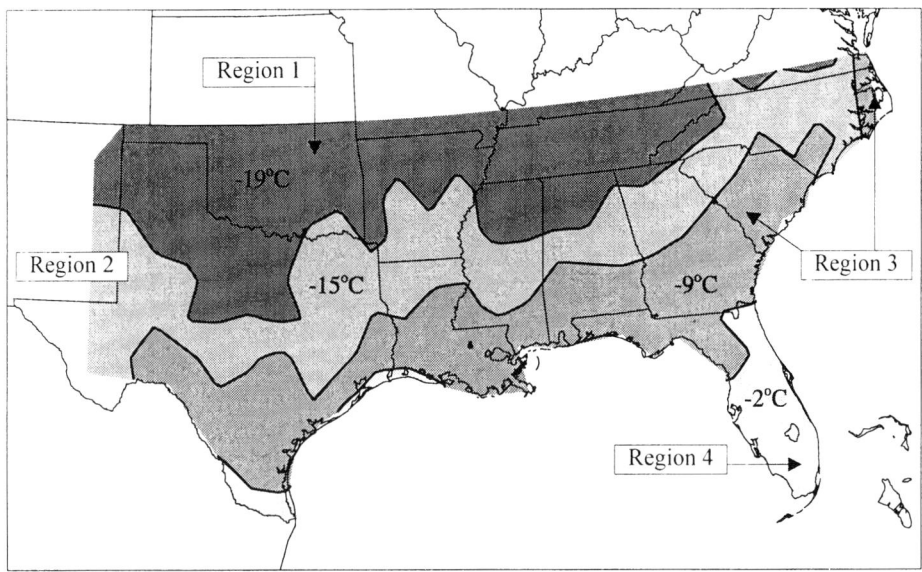

Figure 7. Regional boundaries, derived from correlation-curve PCA, of areas of similar claim-response to particular freezing temperatures. Indicator temperature thresholds for pipe bursting in each region are shown.

values from insurance districts within each region to produce a single curve. This procedure considered only those insurance districts with defined correlations at each threshold. The resulting curves are illustrated in Figure 8a, and may be compared to assess differences in the relationship between temperature and claims in each region. Corresponding regional curves of the fraction of insurance districts which have defined correlation values at each temperature threshold, were created in the second step of the optimization technique. Results are shown in Figure 8b, in which the differences between the regional curves can again be compared. Each of the four regions exhibits 100% station coverage at $0°C$, but this decreases rapidly as lower temperature thresholds are reached. In the final and key procedure of the optimization technique, each regional correlation-curve value was weighted by its corresponding regional fraction of insurance districts (threshold by threshold) to produce the *threshold-index* curves shown in Figure 8c. The temperature corresponding to the maximum (peak) value of the threshold-index curve in each region represented the optimal temperature for use in the present context, and is taken to be the regional indicator temperature threshold. Thus, Regions 1–4 were each assigned respective indicator thresholds of $-19°C$, $-15°C$, $-9°C$, and $-2°C$, temperatures which yield the best relationship between claims and freezes while also considering the station coverage involved. These best indicator temperatures for pipe bursting are displayed in Figure 7, along with the previously delineated regional boundaries.

Linear interpolation (with some introduced boundary conditions) was then

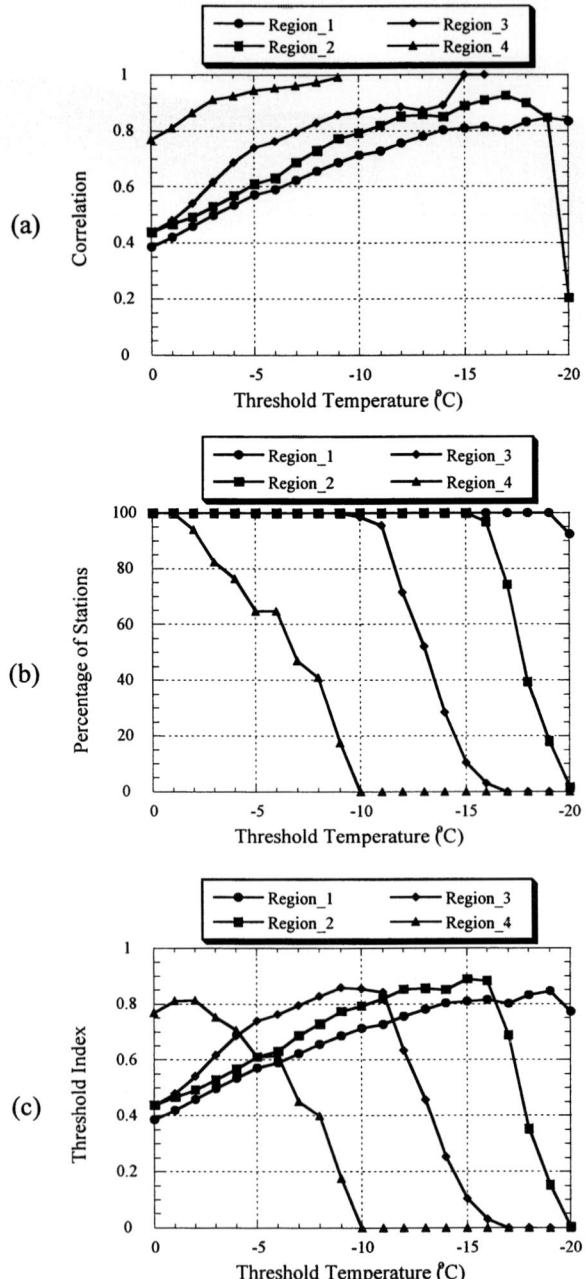

Figure 8. (a) Regional-average correlation curves for Regions 1–4 in Figure 7. Average correlation values at each temperature were calculated from only the insurance districts with defined correlations (see text). (b) Percentage of insurance districts within each Region 1–4 which reported claims associated with minimum temperatures less than or equal to each threshold. (c) Threshold index curves for Regions 1–4 corresponding to temperatures from 0°C to –20°C.

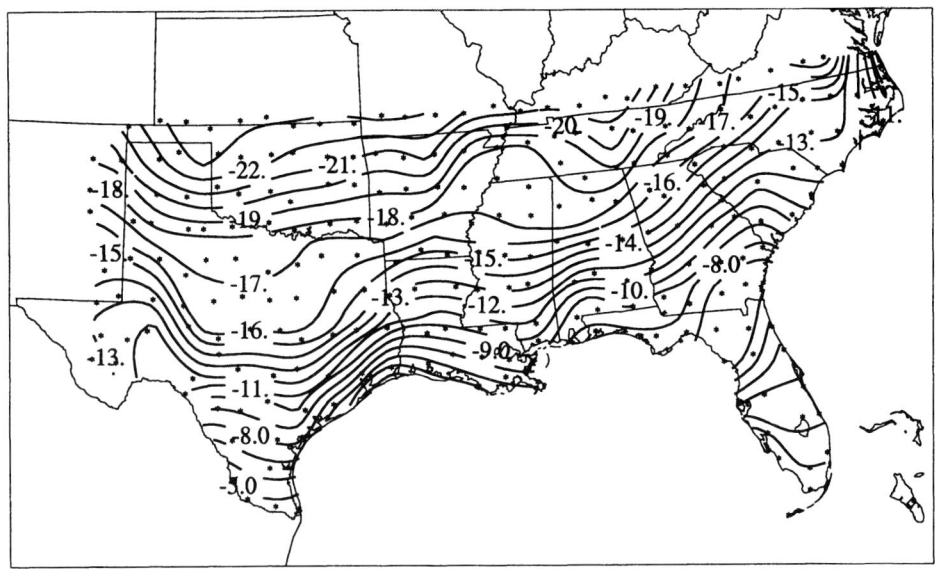

Figure 9. Contours of indicator threshold temperatures resulting from linear interpolation of Figure 7. Asterisks are station locations for the temperature data set, consistent with Figure 5.

applied to the step-function-like pattern in Figure 7 to produce indicator threshold temperatures for individual insurance districts. The resulting pattern is shown in Figure 9. A crucial question was whether the indicator thresholds for each region implicit in Figure 9 do represent the threshold beyond which the number of claims per freeze event rises significantly. Scatter plots (not shown) of the relative departure of the minimum temperature from the indicator threshold versus the associated number of claims indicated that, although small numbers of claims occur for all temperatures, very large numbers of claims are generally confined to temperatures near or below the indicator threshold. Thus, the derived indicator thresholds serve as a method of estimating which freezes are likely to produce property damage due to pipe bursting, and which are not. Furthermore, the identification of these critical temperature thresholds for pipe-bursting damages allows for the development of mitigation efforts specific to individual locations.

3.3. APPLICATION OF INDICATOR THRESHOLDS

The occurrence of a minimum temperature below the indicator threshold in an insurance district was referred to as a *severe freeze*. The entire 43-year temperature record from 1950–1992 was used to determine the average monthly occurrence of severe freezes in each insurance district. Results (not shown) indicated that the occurrence of severe freezes is somewhat rare, and is limited to

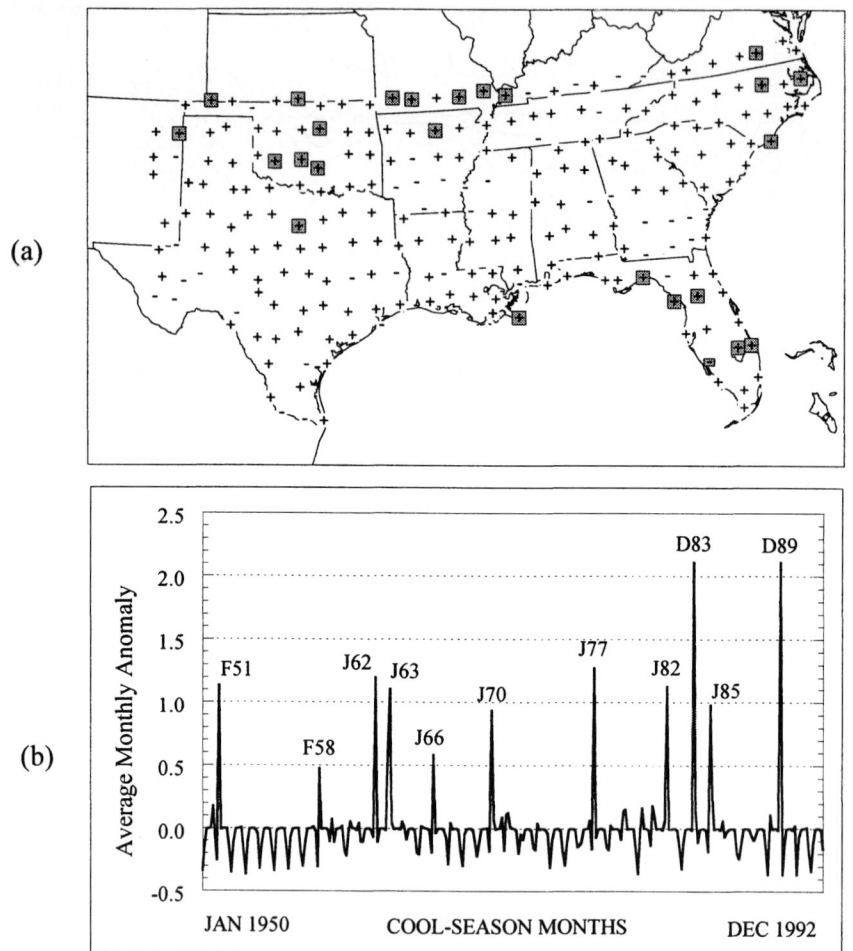

Figure 10. (a) Signs of linear regression slope coefficients for time series of severe-freeze occurrence for the period 1950–1992. Positive and negative slopes are indicated by '+' and '–' signs, respectively. Insurance districts exhibiting statistically significant trends at the 95% confidence level are shaded. (b) Time series of domain-wide average monthly anomaly in severe-freeze occurrence for cool-season months in the period 1950–1992. Labeled peaks indicate month and year.

the traditional "winter" months of December, January, and February. In general, severe-freeze occurrence in these months averages about 0.1 to 0.5 freezes per month, meaning a severe freeze once every 2 to 10 years.

To establish whether trends in severe-freeze occurrence existed during the period 1950–1992, linear and parabolic regression analyses were performed. Overall trends for all cool-season months were considered in one analysis, using a 301-month-long time series that was constructed by removing long-term means

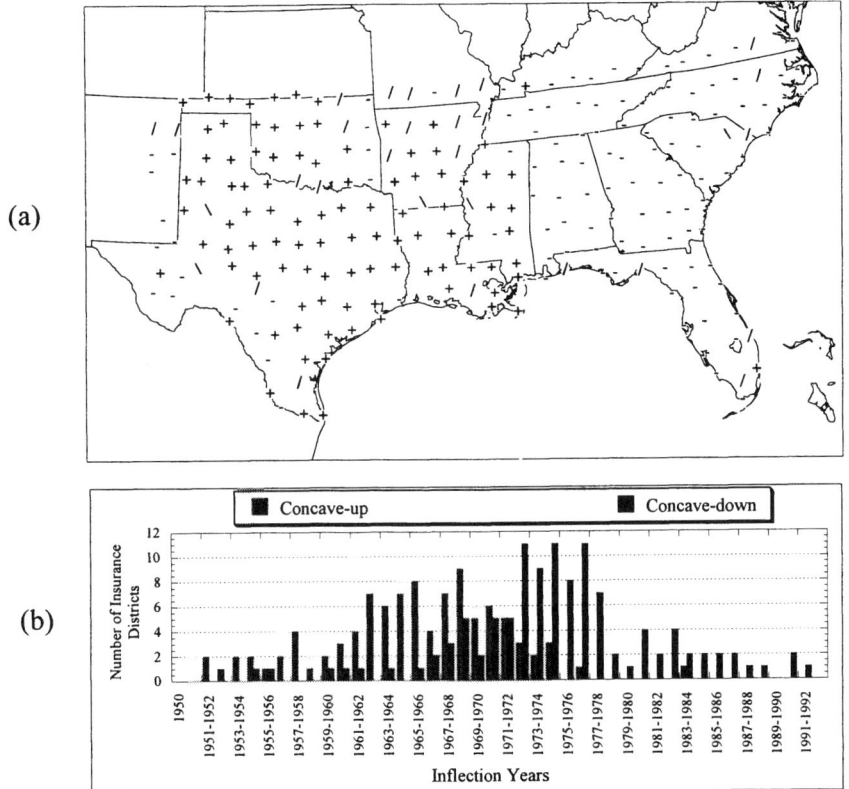

Figure 11. (a) Trends resulting from the parabolic regression of the time series of severe-freeze occurrence for the period 1950–1992. Concave-up (i.e., "mid"-period maximum), concave-down, continuously increasing, and continuously decreasing trends are denoted by '+', '–', '/', and '\' signs, respectively. (b) Histogram of the inflection years corresponding to insurance districts with concave-down and concave-up trends. The total number of districts exhibiting an inflection point within each cool season is indicated.

from each of the seven calendar monthly time series (43 years each), and chaining the resulting monthly anomalies in successive order (ONDJFMA–ONDJFMA...). Least-squares linear regression was applied and the significance of slope values for individual insurance districts was assessed using a standard t-test at the 95% significance level. Results in Figure 10a indicate a distinct dominance of positive trends over the region, but with the number of locally significant trends achieving a field significance (Livezey and Chen, 1983) at only the 82% level. However, the time series of domain-wide average monthly anomaly in number of severe freezes (Figure 10b) does reveal an increase in the intensity of events in the latter part of the time period, with the Decembers of 1983 and 1989 being particularly affected.

Parabolic regression of the same overall time series of severe-freeze occurrence was completed to afford a more detailed assessment of temporal trends (Figure 11). Concave-down trends dominate across the eastern half of the study region, and along the extreme western border, while an area of concave-up trends is located across much of Texas, Oklahoma, Louisiana, Arkansas, and Mississippi. These east-west differences in the trends of severe-freeze occurrence are in contrast to the more spatially uniform results of the linear regression. A histogram indicating the years for which the inflection points of the concave-up and concave-down trends occurred (Figure 11b) reveals that the distribution of inflection years for districts with concave-up trends centers around 1966, whereas the distribution for those with concave-down trends centers around 1975. Future research will address how these time periods of changes in geographical patterns of freezing temperature occurrence relate to the variability of the global climate system (e.g., upper-tropospheric flow patterns, sea-air interaction).

A final application of the indicator thresholds was the calculation of the general probability of daily severe-freeze occurrence during the winter season for each insurance district. These probabilities translate the earlier results about the frequency of severe-freeze occurrence into a format that may be more applicable to determining risk. Individual daily minimum temperature observations for the period 1950–1992 were assembled for each insurance district, using data from the "winter" months (Dec.–Feb.), which formed an approximately normal distribution. The mean and standard deviation of each distribution were calculated, and probabilities of the occurrence of temperatures below the indicator threshold were determined. The probability of a severe-freeze occurrence on any given day ranges from less than 0.2% in areas of Oklahoma, northern Texas, and southern Florida, to greater than 3% in southern Georgia, southern Alabama, and northern Florida. The average probability for the entire Southeast equals about 0.9%.

The above regression and probability analyses provide valuable information to the insurance industry about the frequency of severe freezing temperatures and trends in their occurrence. These results may be translated into information about the risk of incurring pipe-bursting damages within individual insurance districts.

4. Summary and Conclusions

Damaging weather conditions in the United States occur in all seasons, with spatial scales that vary from less than a square mile to large areas of the country. As a result, weather observations from the Cooperative Observer Network or First-Order stations are not adequate for detecting all occurrences of smaller-scale phenomena such as wind gusts, hail, and tornadoes. Records of routine manual or remotely-sensed observations can be used to determine the frequency and sometimes the intensity of related *larger-scale* weather phenomena, but there is not necessarily any indication/confirmation of the occurrence of a damaging smaller-scale event. These observations also may give no indication of the area affected by a particular event, and do not usually provide information about the affected location.

The property catastrophe and crop-hail loss-cost adjusted data bases represent excellent alternative indices for monitoring climate because they integrate both the type and physical characteristics of severe weather events and the amount of damage they cause. Furthermore, the weather insurance industry has developed processes for adjusting individual storm losses included in these data bases to changing socioeconomic conditions. Still, important concerns exist with these types of indices if they are to be used for long-term climate monitoring. The insurance industry must continue collecting and adjusting loss data, and those data should be examined periodically to confirm that adjustment factors are temporally consistent.

We have demonstrated that these types of insurance data bases may also be used in conjunction with existing large-scale weather observations to *derive* indicators for climate variability. Such indicators, as in the case of the pipe freezing study, augment standard weather observations by determining which extreme events produce damage (those meeting thresholds or criteria) and which do not. These combinations of insurance and weather data prove beneficial to both the meteorological community and the insurance industry.

Acknowledgments

The insurance data-set research included in Section 2 was funded in part by UCAR award S95-53410. We extend our thanks to Stanley A. Changnon for his comments and suggestions regarding the text.

The case study described in Section 3 was supported by the Institute for Business & Home Safety (IBHS), Boston. The assistance of Richard J. Roth, Sr., Eugene Lecomte, and the late Paul Cogswell was instrumental to the success of that investigation, and is greatly appreciated.

References

Changnon, D. and Changnon, S. A.: 1997a, 'Surrogate Data to Estimate Crop-Hail Loss', *J. Applied Meteor.* **36**, 1202-1210.

Changnon, D. and Changnon, S. A.: 1997b, 'Evaluation of Weather Catastrophe Data for Use in Climate Change Investigations', *Climatic Change* **38**, 435-445.

Changnon, S. A.: 1969, *Climatology of Severe Winter Storms in Illinois*, Bulletin #53, Illinois State Water Survey, Urbana, IL, pp. 1-45.

Changnon, S. A.: 1982, 'User Beware: The Upward Trend in Tornado Frequencies', *Weatherwise* **35**, 64-69.

Changnon, S. A., Farhar, B., and Swanson, E. R.: 1978, 'Hail Suppression and Society', *Science* **200**, 387-394.

Changnon, S. A., Changnon, D., Fosse, E. R., Hoganson, D. C., Roth, R. J., and Totsch, J. M.: 1997, 'Effects of Recent Weather Extremes on the Insurance Industry: Major Implications for the Atmospheric Sciences', *Bull. Amer. Meteor. Soc.* **78**, 425-435.

Changnon, S. A., and Changnon, J. M.: 1990, 'Use of Climatological Data in Weather Insurance', *J. Climate* **3**, 569-576.

Court, A., and Griffiths, J. F.: 1986, 'Thunderstorm Climatology', in Kessler, E. (ed.), *Thunderstorm Morphology and Dynamics*, University of Oklahoma Press, Norman, OK, pp. 9-40.

Crop-Hail Insurance Actuarial Association: 1978, *Crop-Hail Insurance Statistics*, Chicago, pp. 1-72.

Fujita, T. T.: 1981, 'Tornadoes and Downbursts in the Context of Generalized Planetary Scales', *J. Atmos. Sciences* **38**, 1511-1534.

Gong, X., and Richman, M. B.: 1995, 'On the Application of Cluster Analysis to Growing Season Precipitation Data in North America East of the Rockies', *J. Climate* **8**, 897-931.

Gordon, J. R.: 1995a, *An Investigation into Freezing and Bursting Water Pipes in Residential Construction: Phase 2 Report*, School of Architecture, Building Research Council, University of Illinois at Urbana-Champaign, pp. 1-49.

Gordon, J. R.: 1994, *An Investigation Into Freezing and Bursting Water Pipes in Residential Construction: Phase 1 Report*, School of Architecture, Building Research Council, University of Illinois at Urbana-Champaign, pp. 1-63.

Gordon, J. R.: 1995b, *An Investigation into Freezing and Bursting Water Pipes in Residential Construction: Loss Prevention Strategies, Recommendations*, School of Architecture, Building Research Council, University of Illinois at Urbana-Champaign, pp. 1-14.

Gordon, J. R.: 1995c, *On Establishing a Temperature Alert Threshold for Freezing Water Pipes*, School of Architecture, Building Research Council, University of Illinois at Urbana-Champaign, pp. 1-4.

Livezey, R. E., and Chen, W. Y.: 1983, 'Statistical Field Significance and Its Determination by Monte Carlo Techniques', *Mon. Wea. Rev.* **111**, 46-59.

National Crop Insurance Services: 1995, *National Crop Insurance Services 1994*, Overland Park, KS, pp. 1-25.

Property Claims Services: 1995, *Property Claims Services Report. Catastrophe Record of 1994*, American Insurance Service Group, pp. 1-3.

Richman, M. B.: 1986, 'Rotation of Principal Components', *J. Climatology* **6**, 293-335.

Richman, M. B., and Lamb, P. J.: 1985, 'Climatic Pattern Analysis of Three- and Seven-Day Summer Rainfall in the Central United States: Some Methodological Considerations and a Regionalization', *J. Clim. Appl. Meteor.* **24**, 1325-1343.

Richman, M. B., Lamb, P. J., and Angel, J. R.: 1991, 'Multivariate Statistical Analysis of Precipitation Associated with North American Winter Storm Tracks', *Preprints of Fifth Conference on Climate Variations*, American Meteorological Society, Boston, 47-51.

Skinner, B. D.: 1997, 'An Investigation Into the Occurrence of Freezing Temperatures in the

Southeastern United States and Their Relationship with Insurance Claims and Losses Due to Pipe Bursting', M. S. Thesis, The University of Oklahoma, pp. 1-320.

Skinner, B. D., Richman M. B., and Lamb, P. J.: 1998, 'An Investigation Into the Occurrence of Freezing Temperatures in the Southeastern United States and Their Relationship with Insurance Claims and Losses Due to Pipe Bursting', *Preprints of Ninth Symposium on Global Change Studies*, American Meteorological Society, Boston, 121-125.

(Received 5 November 1997; in revised form 21 September 1998)

ATLANTIC BASIN HURRICANES: INDICES OF CLIMATIC CHANGES

CHRISTOPHER W. LANDSEA
NOAA/AOML/Hurricane Research Division, 4301 Rickenbacker Causeway, Miami, Florida 33149,
U.S.A. Email: landsea@aoml.noaa.gov

ROGER A. PIELKE, JR.
NCAR/Environmental and Societal Impacts Group, Boulder, Colorado 80307, U.S.A.

ALBERTO M. MESTAS-NUÑEZ
CIMAS/University of Miami, Miami, Florida 33149, U.S.A.

JOHN A. KNAFF
NOAA/Cooperative Institute for Research in the Atmosphere, Fort Collins, Colorado 80523, U.S.A.

Abstract. Accurate records of basinwide Atlantic and U.S. landfalling hurricanes extend back to the mid 1940s and the turn of the century, respectively, as a result of aircraft reconnaissance and instrumented weather stations along the U.S. coasts. Such long–term records are not exceeded elsewhere in the tropics. The Atlantic hurricanes, U.S. landfalling hurricanes and U.S. normalized damage time series are examined for interannual trends and multidecadal variability. It is found that only weak linear trends can be ascribed to the hurricane activity and that multidecadal variability is more characteristic of the region. Various environmental factors including Caribbean sea level pressures and 200mb zonal winds, the stratospheric Quasi–Biennial Oscillation, the El Niño–Southern Oscillation, African West Sahel rainfall and Atlantic sea surface temperatures, are analyzed for interannual links to the Atlantic hurricane activity. All show significant, concurrent relationships to the frequency, intensity and duration of Atlantic hurricanes. Additionally, variations in the El Niño–Southern Oscillation are significantly linked to changes in U.S. tropical cyclone–caused damages. Finally, much of the multidecadal hurricane activity can be linked to the Atlantic Multidecadal Mode – an empirical orthogonal function pattern derived from a global sea surface temperature record. Such linkages may allow for prediction of Atlantic hurricane activity on a multidecadal basis. These results are placed into the context of climate change and natural hazards policy.

1. Introduction

The United Nation's Intergovernmental Panel on Climate Change (IPCC) has speculated that climate change due to increasing amounts of anthropogenic "greenhouse" gases may result in increased tropical sea surface temperatures (SSTs) and increased tropical rainfall associated with a slightly stronger intertropical convergence zone (ITCZ) (Houghton et al., 1990, 1992, 1996). Because tropical cyclones extract latent and sensible heat from the warm tropical oceans and release the heat in its upper tropospheric outflow to fuel

Climatic Change **42**: 89–129, 1999.
© 1999 *Kluwer Academic Publishers. Printed in the Netherlands.*

the storm's spin up, early work of the IPCC expressed concern that warmer SSTs will lead to more frequent and intense hurricanes, typhoons and severe tropical cyclones. These concerns prompted the IPCC (Houghton et al. 1990) to suggest in 1990 that:

> There is some evidence from model simulations and empirical consid-erations that the frequency per year, intensity and area of occurrence of tropical disturbances may increase [in a doubled carbon dioxide world], though it is not yet compelling.

However, any changes in tropical cyclone activity are intrinsically also tied to large–scale changes in the tropical atmosphere. As a result, SSTs by themselves cannot be considered without corresponding information regard-ing the moisture and stability in the tropical troposphere. What has been identified in the current climate as being necessary for genesis and main-tenance for tropical cyclones (e.g. SSTs of at least 26.5°C – Gray 1968) would change in an enhanced CO_2 world because of possible changes in the moisture or stability. It is quite reasonable that an increase in tropical and subtropical SSTs would be also accompanied by an increase in the SST threshold value needed for cyclogenesis because of compensating changes in the tropospheric moist static stability (Emanuel 1995).

In addition to the thermodynamic variables, changes in the tropical dy-namics also play a large role in determining changes in tropical cyclone activity. For example, if the vertical wind shear over the tropical North At-lantic moderately increased during the hurricane season in an increased CO_2 world – as what is typically seen during El Niño–Southern Oscillation warm phases (El Niño events), then we would most likely see a significant decrease in tropical cyclone activity. This is due to the Atlantic basin having a marginal climatology for tropical cyclone activity because of its sensitivity to changes in vertical wind shear and lack of an oceanic monsoon trough (Gray et al. 1993). In other less marginal tropical cyclone basins, changes in the vertical shear profile typically result in alterations in the preferred location of devel-opment (e.g. Nicholls 1979, Chan 1985, Revell and Goulter 1986, and Lander 1994).

These complications along with conflicting global circulation modeling (GCM) runs compelled the 1995 IPCC (Houghton et al. 1996) to express greater uncertainty about the nature of tropical cyclones in an enhanced CO_2 environment:

> The formation of tropical cyclones depends not only on sea surface tem-perature (SST), but also on a number of atmospheric factors. Although some models now represent tropical storms with some realism for present day climate, the state of the science does not allow assessment of future changes.

Most recently, Henderson–Sellers et al. (1998) addressed a few of the tropical cyclone–greenhouse warming problems. The first is that "there is no evidence to suggest any major changes in the area or global location of tropical cyclone genesis in greenhouse conditions." This conclusion is based upon Holland's (1997) thermodynamic tropical cyclone model which does show that in a greenhouse–warmed climate there is an upward alteration in the minimum SST from 26.5 to 28°C needed for tropical cyclogenesis. The additional conclusion from Henderson–Sellers et al. (1998) suggests "an increase in [maximum potential intensity] MPI of 10%–20% [in central pressure or 5%–10% in maximum sustained winds] for a doubled CO_2 climate but the known omissions (ocean spray, momentum restriction, and possibly also surface to 300 hPa lapse rate changes) all act to reduce these increases." This second finding is also based upon the thermodynamic models of Emanuel (1986) and Holland (1997), which also appears to corroborate similar findings for Northwest Pacific typhoons from a "downscaled" GCM to mesoscale model approach by Knutson et al. (1998). Henderson–Sellers et al. (1998) do not provide guidance for possible changes in tropical cyclone frequency, mean intensity, or area of occurrence.

Because of the enormous impacts that tropical cyclones have today (e.g. in 1995 total mainland U.S. hurricane damages averaged the order of $5 billion annually – Pielke and Landsea 1998), it is essential that detailed studies be made of observed tropical cyclone activity. An understanding of such activity plays an important role in both public and private policy decisions. Additionally, tropical cyclone activity has rather large interannual and interdecadal variations which are extremely important for their own sake and which could turn out to have a greater impact relative to changes forced by greenhouse warming, as suggested by Lighthill et al. (1994). A reliable assessment of what the future holds for tropical cyclone activity would also have significant policy utility.

This paper documents the long–term variations in tropical cyclone activity of the Atlantic basin (e.g. the North Atlantic Ocean, the Gulf of Mexico and the Caribbean Sea) from instrumental records and corresponding climatic fluctuations responsible for variability, on annual and longer timescales. Section II provides definitions of various tropical cyclone indices and acknowledges the datasets utilized. Section III reviews previous work on the physical mechanisms of interannual through multidecadal timescales that influence tropical cyclones in the Atlantic basin. Section IV analyzes indices of Atlantic basin tropical cyclone activity for both all–basin and landfalling cyclones and their quantitative relationship to various climatic forcing. Section V is a discussion of Atlantic hurricanes as measured through these indices and their relevance to policy issues. The final section summarizes and discusses our key findings.

2. Tropical Cyclone Definitions and Datasets

"Tropical cyclone" is the generic term for a non–frontal, synoptic scale, "warm–core" low-pressure system that develops over tropical or sub–tropical waters with organized convection and a well–defined closed cyclonic surface wind circulation. It derives its energy primarily from latent and sensible heat flux from the ocean which is enhanced by strong winds and lowered surface pressure. These energy sources are tapped through condensation in convective clouds concentrated near the cyclone's center (Holland 1993).

The tropical cyclone designation is a broad term under which various strength systems in the Atlantic basin are divided into:

- **Tropical Storm:** Maximum sustained (1 min mean) surface (10 m) wind speed 18 to 32 m s^{-1}

- **Hurricane:** Wind speed at least 33 m s^{-1}

- **Intense (or Major) Hurricane**: Wind speed at least 50 m s^{-1}. These are the category 3, 4 or 5 hurricanes on the Saffir–Simpson (Simpson 1974) Hurricane Scale (Table I)

The Atlantic basin is usually active during the months of June through November, comprising the traditional "hurricane season" (Neumann et al. 1993). However, the large majority of intense hurricanes occur in just the three months of August, September and October (Landsea 1993).

Tropical storms and hurricanes here are collectively referred to as **named storms** [in deference to the fact that since 1950 all tropical cyclones that were of at least tropical storm force were given a name for identification (Neumann et al., 1993), although some cyclones were determined to be of tropical storm strength after the fact and thus lack a formal name].

When considering the variations in named storms, subtropical storms are also to be included in such analyses. **Subtropical storms** are non–frontal low pressure systems comprising initially baroclinic circulations developing over subtropical waters with sustained one minute surface winds of at least 18 ms^{-1} (National Oceanic and Atmospheric Administration 1997). Such nomenclature has been utilized since 1968, though it is likely that these systems were designated and included in the database as tropical storms previously. Thus, failure to include the subtropical storms into the climate record examined would introduce an artificial bias into the database (Neumann et al. 1993).

Subsets of these tropical cyclone designations are also useful. For example, Gray (1968) showed distinctly differing environmental conditions for tropical cyclogenesis in subtropical latitudes versus those forming closer to the equator. Ideally, we would wish to stratify Atlantic tropical cyclones

Table I. Maximum sustained wind speed, minimum surface pressure, storm surge, and general damaging effects for the five Saffir–Simpson (Simpson 1974) Hurricane Scale values. The last column provides a relative value for median of normalized U.S. hurricane damages per category. A Category 1 hurricane – scaled a "1" in the table – had a median damage of $33 million (Pielke and Landsea 1998). Note how in the table the median damages go up enormously with increasing category of hurricane.

Saffir–Simpson Category	Maximum Sustained Wind Speed (m s^{-1})	Minimum Surface Pressure (mb)	Storm Surge (m)	Relative Value and Damaging Effects
1	33 to 42	≥ 980	1.0 to 1.7	1 – Minimal
2	43 to 49	979 to 965	1.8 to 2.6	10 – Moderate
3	50 to 58	964 to 945	2.7 to 3.8	50 – Extensive
4	59 to 69	944 to 920	3.9 to 5.6	250 – Extreme
5	> 69	< 920	> 5.6	500 – Catastrophic

into those forming from easterly waves and developing without impacts by middle and upper tropospheric troughs versus those with some baroclinic component, as is done in Hess et al. (1995). However, such categorization is extremely subjective and may not be reliable as even today debates occur as to the influence that upper lows play in tropical cyclone development. This uncertainty is due both to the paucity of over–ocean data and to the incompleteness of intensification theories (e.g., Elsberry et al. 1992). As an alternative that is not as physically based, but more defensible is to stratify the named/subtropical storms time series into those forming poleward (**northern named/subtropical storms**) and equatorward (**southern named/subtropical storms**) of 23.5°N.

Additionally, of significant societal concern is the number of hurricanes which affect the people of the Caribbean. Reading (1989) and Gray (1990) showed that there are substantial interdecadal variations on the number of hurricanes striking this region. Thus a time series of **Caribbean hurricanes** is constructed and analyzed based upon the presence of hurricanes within the the Caribbean Sea and hurricanes directly affecting the land masses which surround the Caribbean (e.g. Central America [including the eastern Yucatan coast of Mexico], Cuba, Jamaica, Hispaniola, Puerto Rico, the Lesser Antilles, northern Venezuela and northern Colombia).

One objective method for determining the seasonal amount of tropical cyclone activity is through the summed duration of each storm. This partially removes the subjectivity involved in categorizing the intensity of tropical cyclones. A seasonal total of **hurricane days** is the amount of days in which a hurricane existed (two hurricanes existing simultaneously for 24 hours count as two days). The computations count days in six hour increments.

In addition to these indices, two additional time series were created: the **peak intensity** – the strongest sustained winds reached by the strongest hurricane each hurricane season – and **mean intensity** – the average of the strongest winds for all of the named and subtropical storms for a season. These indices are independent of frequency variations and only measure changes in intensity.

While records are available for the entire Atlantic basin for hurricanes back to the late 1800s (Jarvinen et al. 1984) and for landfalling hurricanes along the United States coastline back to the 16th Century (Ludlum 1989), reliably knowing the intensity of such systems extends for a much briefer period of time. For the whole Atlantic basin, reliable intensity measures exist back to the commencement of routine aircraft reconnaissance in 1944 (Neumann et al. 1993), but even these data have been arbitrarily corrected to remove an overestimation bias in the winds of intense hurricanes during the 1940s through the 1960s (Landsea 1993). The winds of strong hurricanes from 1944 through 1969 have been reduced 2.5–5 m s^{-1} as a first order bias removal[1]. No estimate of the true occurrence of all–basin intense hurricanes is attempted for the era before the mid 1940s because of the lack of reliable data on the strong inner core of the hurricanes except for very infrequent measurements conducted by unfortunate ships' crews. Thus 1944 marks the beginning of all of the all–basin time series analyzed here.

For U.S. landfalling hurricanes, observations of minimum central pressure provide accurate records back to 1899 for nearly all hurricanes (Jarrell et al. 1992). Before this year, records of intensity at landfall are incomplete and can only provide rough estimates of a hurricanes' strength. Thus 1899 is the first year utilized for U.S. landfalling hurricane data.

Lastly for tropical cyclone–related datasets, Pielke and Landsea (1998) have developed a normalized mainland U.S. hurricane damage time series that takes into account inflationary, coastal county population and wealth changes. This index, independent of tropical cyclone data itself, extends from 1925 through 1996 and is an estimate of how much damage would be caused if landfalling cyclones of the past struck in 1996.

All Atlantic tropical cyclone records are presented with respect to a long–term average (1950–1990) to allow for consistent comparisons amongst indices despite differences in length of record.

A variety of environmental parameters such as Atlantic and Pacific SST, African rainfall, sea level pressures, 200 mb zonal winds and 50 mb stratospheric winds are compared with the preceding tropical cyclone records. The

---

[1] This is only a temporary solution, however. What is needed is a "reanalysis" of all available data – primarily aircraft reconnaissance – with today's analysis techniques to create an updated data set. Such efforts are currently underway (Neumann and McAdie 1997).

SST data sets are those of the 1856–1991 reconstruction of historical ship–based data of Kaplan et al. (1998). SST data from the optimally interpolated product of Reynolds and Smith (1994) for the years from 1992 to 1996 are appended to the Kaplan dataset. Box averages as well as cubic splines were used to interpolate the 1.0° Reynolds and Smith data to the Kaplan 5.0° grid. The Reynolds and Smith SST anomalies were calculated using the Kaplan as well as the Reynolds and Smith SST climatology. The smaller root mean squared difference between the two globally averaged SST anomaly time series for the common period (November 1981 to December 1991) was obtained when the Reynolds and Smith SST anomalies were calculated using the the Reynolds and Smith SST climatology. Both interpolation methods gave very similar results and we decided to utilize the cubic splines for this analysis. The El Niño–Southern Oscillation (ENSO) is represented by SSTs during August–October (ASO) in the region 5°N – 5°S and 120–170°W (the "Niño 3.4" index) as suggested by Barnston et al. (1997) as having the strongest concurrent link to Atlantic hurricanes. After Saunders and Harris (1997), we utilize ASO tropical North Atlantic SST anomalies for the region 5–25°N and 15–55°W to examine local SST effects on Atlantic hurricane activity.

To analyze the relationship of tropical cyclones with Atlantic SST anomalies in a way that is independent of the separately analyzed ENSO relationship, it is convenient to first account for and remove the teleconnected effects of ENSO on the Atlantic Ocean. Enfield and Mestas-Nuñez (1998) have done this by computing global SST anomaly modes after first extracting the global ENSO variability represented by a complex empirical orthogonal function (EOF) mode. An EOF analysis is a multivariate statistical technique that gives the most efficient representation of the variability in a data set. The main advantage of EOF analysis is that it generally allows one to capture most of the variance in the data in the first few modes. However, it is important to note that the EOFs do not necessarily represent physical modes. Each mode is comprised of a spatial pattern and an associated time series. Enfield and Mestas-Nuñez' third non–ENSO mode represents interannual to multidecadal SST variability, primarily in the Atlantic, with the multidecadal component being strongest.

The African rainfall dataset is a five station index of June–September (JJAS) 1899 through 1996 rainfall for the Western Sahel region, updated from that reported in Landsea et al. (1992). A Caribbean/Gulf of Mexico eleven station sea level pressure index was developed using the same stations and methodology as Knaff (1997) but for ASO 1950 through 1996 from the original dataset of Vose et al. (1992). For an index of Caribbean 200 mb zonal winds, data from Shea et al. (1994) for twelve stations (Table II) were compiled for the August–October 1950 through 1996 time period. ASO Caribbean 50 mb stratospheric winds were provided from the dataset collected and updated at Colorado State University (Gray 1984a, Gray et al. 1992a) for the years 1950 through 1996.

Table II. Summary of the twelve upper–air stations used for the August through October index of 200 mb zonal winds for the Caribbean. Listed along with the World Meteorological Organization (WMO) number is the station name and coordinates.

WMO number	Station	Latitude	Longitiude
76644	Merida, Mexico	21.0°N	89.5°W
78367	Guantanamo, Cuba	19.9°N	75.2°W
78384	Roberts Field, Grand Cayman	19.3°N	81.3°W
78397	Kingston, Jamaica	17.9°N	76.8°W
78501	Swan Island	17.4°N	83.9°W
78526	San Juan, Puerto Rico	18.4°N	66.0°W
78806	Balboa, Canal Zone	8.8°N	79.6°W
78866	St. Martin	18.1°N	63.1°W
78954	Seawell, Barbados	13.1°N	59.5°W
78970	Piarco, Trinidad	10.6°N	61.4°W
78988	Curacao	12.2°N	69.0°W
80001	San Andres Island, Columbia	12.6°N	81.7°W

3. Climatic Forcing of Atlantic Tropical Cyclones

Understanding tropical cyclone variability on interannual to interdecadal timescales is hampered by the relatively short period over which accurate records are available. Figure 1 presents the various observational platforms available for analyzing tropical cyclone occurrences in the Atlantic basin. Changes in the tropical cyclone databases due to observational platform improvements (and sometime degradations) can often be mistaken as true variations in tropical cyclone activity. Thus caution is urged in interpretations of quantitative tropical cyclone indices. For studies of climatic changes it is strongly suggested that one not utilize incomplete data collected prior to the dates indicated in the previous section. With these caveats in mind, some studies have had successes in the fields of interannual and longer time scales for Atlantic hurricanes, some of which are detailed below.

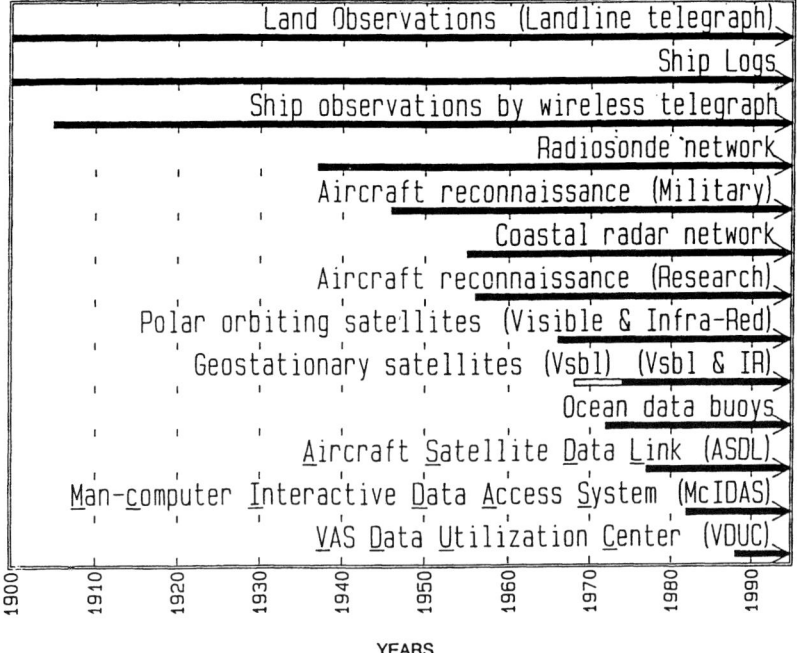

Figure 1. Technical advances in systems for observing tropical cyclones in the Atlantic basin, 1900 through the mid 1990s – from Neumann et al. (1993).

3.1. INTERANNUAL CHANGES

Globally, tropical cyclones are affected dramatically by the El Niño–Southern Oscillation (ENSO). ENSO is a fluctuation on the scale of a few years in the ocean–atmospheric system involving large changes in the Walker and Hadley Cells throughout the tropical Pacific Ocean region (Philander 1989). The state of ENSO can be characterized, among other features, by the SST anomalies in the eastern/central equatorial Pacific: warmings in this region are referred to as El Niño events and coolings as La Niña events. During El Niño events when convective activity in the equatorial Pacific is shifted to the east, the tropospheric vertical shear increases due to a pickup of the climatological westerly winds in the upper troposphere, primarily from 10 to 20°N between Africa and the Americas (hereby known as the "main development region" after Goldenberg and Shapiro 1996). In contrast, reduced 200mb westerlies and shear have been observed during La Niña events (Gray 1984a, Shapiro 1987, Goldenberg and Shapiro 1996). Strong vertical wind shear disrupts the incipient tropical cyclone and can prevent genesis, or, if a tropical cyclone has already formed, strong vertical shear can weaken or destroy the tropical cyclone by interfering with the organization of deep convection around the cyclone center (DeMaria 1996). Thus the larger (smaller) vertical shear accompanying El Niño (La Niña) events lead directly toward decreased (increased) numbers of Atlantic hurricanes. Gray (1984a) demonstrated that the United States is more at risk during non–El Niño years: a 3–to–1 difference in the frequency of landfalling intense hurricanes.

In addition to ENSO, the Atlantic basin shows systematic alterations of tropical cyclone frequency by the stratospheric Quasi-Biennial Oscillation (QBO – Gray 1984a, Shapiro 1989), an east–west oscillation of stratospheric winds that encircle the globe near the equator (Wallace 1973). Atlantic hurricane activity is enhanced during the west phase of the QBO and diminished in the east QBO years. This relationship may be due to alterations in the static stability and dynamics near the tropopause (Gray et al. 1992b, Knaff 1993).

Interannual variations in the Atlantic basin tropical cyclones have also been linked to more localized, basin–specific features such as sea level pressures, local SSTs and West African rainfall. Additional environmental factors may also be important for interannual Atlantic tropical cyclone variability – such as low level moisture availability (Landsea et al. 1998), tropospheric moist static stability (Pasch et al. 1998), and lower tropospheric jet characteristics (Molinari et al. 1997). An examination of these environmental forcings should be considered in future analyses.

Sea level pressures (SLPs) act to directly impact the strength of the vertical wind shear. In the Atlantic basin because of a relatively invariant sea level pressure field near the equator, above (below) average SLP in the main development region tightens (loosens) the local pressure gradient and strengthens (weakens) the easterly tradewinds by 1 to 3 m s^{-1}, thereby contributing to increased (decreased) vertical shear (Gray et al. 1993, 1994). Additionally, Gray et al. (1993) have suggested that below average SLP indicates a poleward shift and/or a strengthening of the ITCZ. Both situations contribute to less subsidence and drying in the main development region through which easterly waves move. Knaff (1997) indicates that low SLP is accompanied by a deeper boundary layer, a weakened tradewind inversion, a more moist middle troposphere, and weaker 200mb westerly winds (and vertical shear). Moreover, an enhanced ITCZ provides more large–scale, low level cyclonic vorticity to incipient tropical cyclones, thereby creating an environment that is more conducive for tropical cyclogenesis (Gray 1968). In contrast, above average SLP tends to be associated with opposite conditions which are unfavorable for tropical cyclogenesis.

Sea surface temperatures in the genesis regions of tropical cyclone basins have a direct thermodynamic effect on tropical cyclones through their influence on moist static stability (Malkus and Riehl 1960). SSTs also indirectly alter the vertical shear through a strong inverse relationship with surface pressures (Shapiro 1982, Gray 1984b). In particular for the Atlantic basin, warmer than average waters are usually accompanied by lower than average surface pressures, and thus, weaker tradewinds and reduced shear.

Cooler than average waters are usually accompanied by higher pressure, stronger tradewinds and increased shear (Knaff 1997). Somewhat surprisingly, interannual SST variations have relatively small contributions toward altering the total named storm frequency in the Atlantic basin (Raper 1992, Shapiro and Goldenberg 1998). However, Saunders and Harris (1997) provide substantial evidence that both preceding and during the hurricane season that Atlantic SSTs in the main development region contribute a large percentage of the variance explained (over 30% during the height of the season) with the number of hurricanes generated in that area. Indeed they argue through a partial correlation analysis that these Atlantic SSTs are the dominant physical modulator of tropical Atlantic hurricanes. The discrepancies between these studies need to be understood.

One aspect of interannual variability that has recently been uncovered is the association of the Atlantic tropical cyclone basin with the monsoon of West Africa. June through September monsoonal rainfall in Africa's Western Sahel has shown a very close association with intense hurricane activity (Reed 1988, Gray 1990, Landsea and Gray 1992, Landsea et al. 1992). Wet years in the Western Sahel (e.g. 1988 and 1989) are accompanied by dramatic increases in the incidence of intense hurricanes, while drought years (e.g. 1990 through 1993) are accompanied by a decrease in intense hurricane activity. Variations in tropospheric vertical shear and African easterly wave intensity have been hypothesized as the physical mechanisms that link the two phenomena (Gray 1990, Landsea and Gray 1992), although Goldenberg and Shapiro (1996) have demonstrated that changes in the vertical shear probably dominate. They note that wet (dry) years are associated with reduced (increased) wind shear, due to both weaker (stronger) than average lower tropospheric tradewinds and upper tropospheric westerlies throughout the main development region.

3.2. INTERDECADAL CHANGES

Gray (1990) and Landsea et al. (1992) described large multidecadal variations in the Atlantic intense hurricanes while the total number of named storms remained more constant from decade to decade. Overall, they found that the late 1920s to the 1960s were very active and 1900s to the mid 1920s as well as the 1970s to the mid 1980s were quiescent. Landsea et al. (1996) showed evidence that the quiet period of the 1970s and 1980s continued through at least the early 1990s.

These multidecadal intense Atlantic hurricane variations are attributed by Gray (1990) to changes in the Atlantic SST structure. Warmer (cooler) than average conditions in the Atlantic north of the equator coupled with cooler (warmer) than average SSTs in the South Atlantic favor increased (decreased) intense hurricane activity. Such a dipole structure of the Atlantic

SSTs also forces drought and wet periods in the North Africa's Western Sahel (e.g. Folland et al. 1986), which at least partially explains why there is a strong concurrent link between the year–to–year Sahel rainfall variations and intense Atlantic hurricanes described in the previous section. The SST dipole pattern appears to alter the overlaying tropospheric circulation such that warm North/cold South Atlantic conditions correspond to reduced vertical wind shear in the main development region favoring the formation and intensification of tropical cyclones (Gray et al. 1997). In contrast, a cool North/warm South Atlantic acts in concert with enhanced tradewind easterlies and upper tropospheric westerlies and thus increased tropospheric vertical wind shear. Additionally, these SST variations likely play a direct role in providing changes of the heat input available to the incipient tropical cyclone by changing the boundary layer moist enthalpy values (Saunders and Harris 1997, Landsea et al. 1998).

It has been hypothesized (Gray et al. 1997) that these multidecadal oceanic temperature, intense hurricane and Sahel rainfall changes are regulated by the strength of the thermohaline circulation and North Atlantic deep water formation – portions of the global "Great Ocean Conveyor" (Broecker 1991). Kushnir's (1994) analysis agrees with Gray et al. in suggesting that the thermohaline circulation is the cause of these multidecadal SST variations. Deser and Blackmon (1993), however, suggest that fresh water flux in the Labrador region alters oceanic deep water production on shorter (about 10 year) periods. Presently, the exact cause of these variations has yet to be confirmed.

4. Indices of Atlantic Basin Hurricanes

4.1. ALL–BASIN ACTIVITY

Examination of the record for the Atlantic numbers of named and subtropical storms shows substantial yearly variability, but no significant trend (Fig. 2). In contrast, the numbers of intense hurricanes (Fig. 3) have gone through pronounced multidecadal changes: active during the late 1940s through the mid 1960s, quiet from the 1970s through the early 1990s, and then a shift again to busy conditions again during the extraordinarily active years 1995 and 1996. The Atlantic hurricanes (Fig. 4) share characteristics of both of the previous time series with only a moderate interdecadal change, but substantial interannual variability. In comparison with hurricane frequency, hurricane days (Fig. 5) show a much more substantial interdecadal variability, indicating that duration of such systems is more sensitive to forcing on these timescales than is the total number of events.

Concurrent with these frequency and duration changes, Figure 6 shows that there have been periods of strong mean intensity of the Atlantic trop-

Atlantic Named & Subtropical Storms
1944-1996

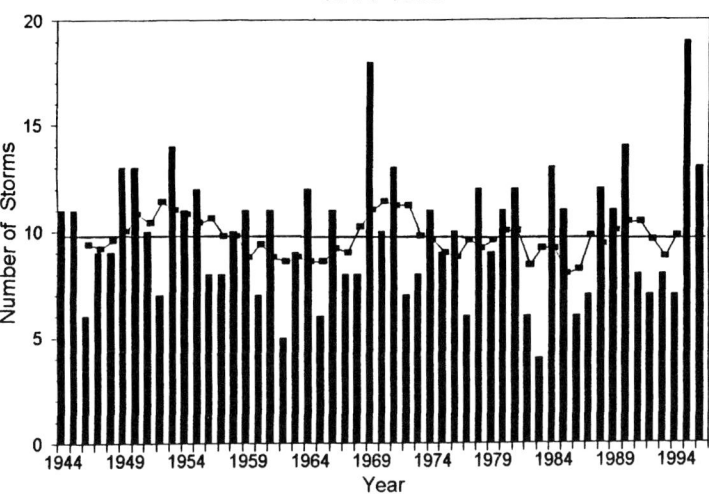

Figure 2. Annual number of named and subtropical storms over the Atlantic basin for the period of reliable record from 1944 to 1996. The long term (1950–90) average of 9.8 per year (solid line) as well as a five year running mean (light curve) are superimposed on the time series.

ical cyclones (mid 1940s–1960s and 1995–1996) and weak mean intensity (1970s–early 1990s). These long–term variations in mean intensity are primarily driven by decreases in the numbers of intense hurricanes (Fig. 3), while the total frequency of named and subtropical storms remains relatively constant (Fig. 2). Perhaps somewhat surprisingly, the time series of peak intensity (Fig. 7) reached by the strongest hurricane each year shows no significant trend or interdecadal fluctuations. However, there is evidence of more interannual variability during the 1970s through the 1990s compared with the more constant period of the 1940s to the 1960s. However, the bias–correction employed for the strong hurricane intensities (e.g. Landsea 1993) may cause an artificial smoothing of the peak intensities in the earlier decades. Thus this change in interannual variability characteristics is suspect. It is hoped that the aforementioned hurricane re–analysis effort will allow this issue to be addressed in more detail in the future.

While the named and subtropical storms have shown no trend or even any substantial multidecadal fluctuations over the last five decades, the same does not hold true when the data are categorized by latitude of formation. Figure 8 shows the annual counts of storms forming north and south of the 23.5°N latitude line. Both generally display multidecadal swings with mirroring variations: high (low) numbers of south (north)–forming storms in

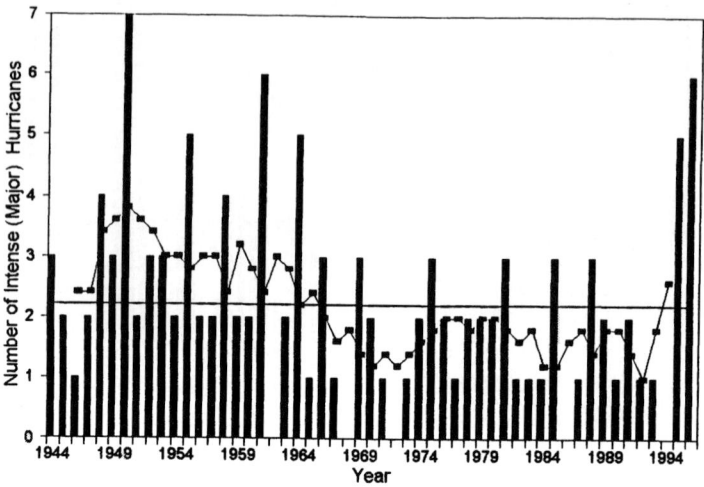

Figure 3. Annual number of intense hurricanes over the Atlantic basin for the period of reliable record from 1944 to 1996. The data for the years 1944–1969 have been revised downward to take into account the overestimation of intensities during these years. See Landsea (1993) for details. The long term (1950–90) average of 2.2 per year (solid line) as well as a five year running mean (light curve) are superimposed on the time series.

the 1940s to the 1960s and a switch to reduced (increased) values during the 1970s through the early 1990s. The southern tropical/subtropical storms match the fluctuations of the intense hurricanes, while the northern forming tropical/subtropical storms are the only class of Atlantic tropical cyclones that appears to have been at higher than usual levels during the 1970s through the early 1990s.

Little can be said of quantitative value for the all–basin tropical cyclone activity before the mid 1940s because of lack of observational networks over the open ocean. However, Fernández–Partagás and Diaz (1996) estimate that the overall Atlantic tropical storm and hurricane activity for the years 1851–1890 was 12% lower than the corresponding forty year period of 1951–1990, though nothing can be said regarding the intense hurricanes. They base this assessment upon a constant ratio of U.S. landfalling tropical cyclones to all–basin activity, which is likely a valid assumption for the multidecadal timescale even though it does not hold for year to year variations. This also assumes that Fernández–Partagás and Diaz were able to uncover all U.S. landfalling tropical cyclones back to 1851, which may be a somewhat less valid contingency.

Atlantic Hurricanes
1944-1996

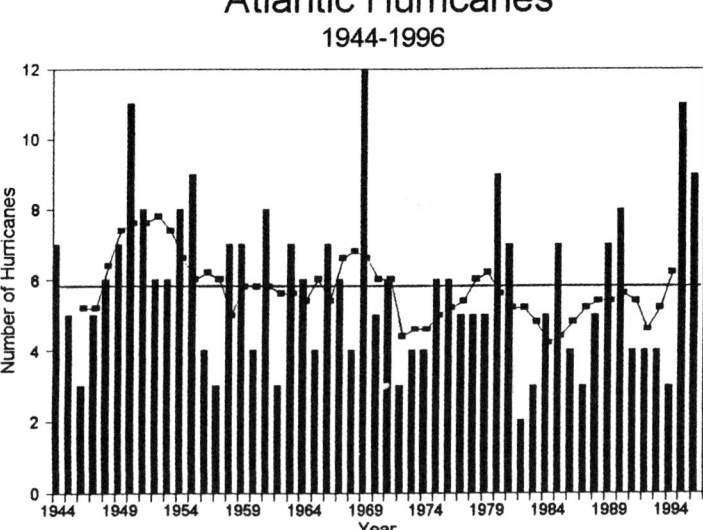

Figure 4. Annual number of hurricanes over Atlantic basin for the period of reliable record from 1944 to 1996. The long term (1950–90) average of 5.8 per year (solid line) as well as a five year running mean (light curve) are superimposed on the time series.

Atlantic Hurricane Days
1944-1996

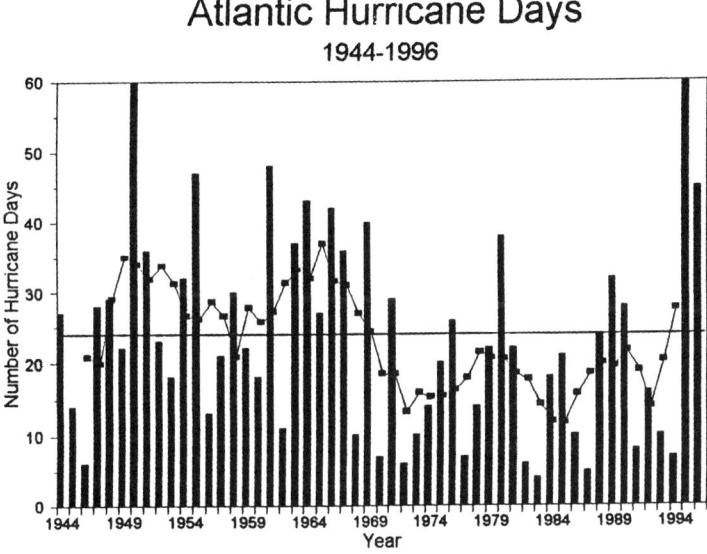

Figure 5. Annual number of hurricanes days over Atlantic basin for the period of reliable record from 1944 to 1996. The long term (1950–90) average of 23.7 days per year (solid line) as well as a five year running mean (light curve) are superimposed on the time series.

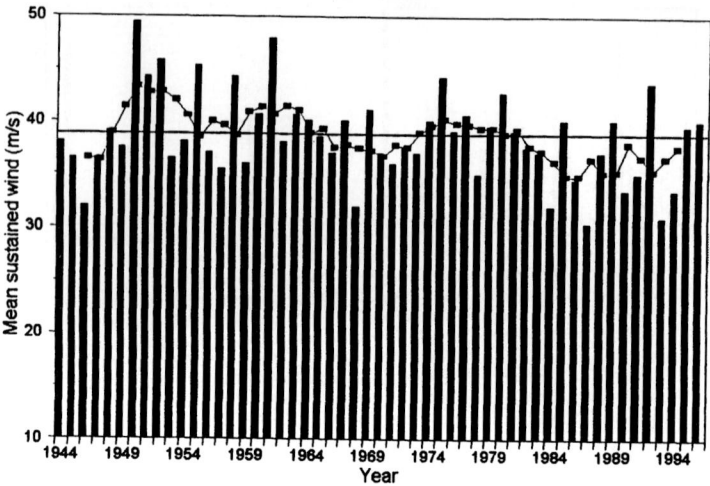

Figure 6. Time series of the Atlantic basin mean intensity – average of the highest sustained surface winds achieved by all of the storms for each year – for the period of reliable record from 1944 to 1996. The long term (1950–90) average of 39.0 ms^{-1} (solid line) as well as a five year running mean (light curve) are superimposed on the time series.

Most of these figures appear to emphasize the dominance of interdecadal fluctuations rather than linear trends, especially for the stronger tropical cyclone records. However, knowing what linear trends – if any – are contained in the data can also be a useful quantity. Table III presents the temporal correlations and the resulting regression coefficients. Note that only five tropical cyclone indices show statistically significant trends: intense hurricanes, hurricane days, mean intensity, and north–forming named and subtropical storms, and Caribbean hurricanes. All of these parameters, except for the north–forming named and subtropical storms, show significant trends toward **less** activity in recent decades.

4.2. LANDFALLING HURRICANES

The dominance of large multidecadal variations for the entire Atlantic basin is shared by many of the indices for hurricanes striking land. In particular, the region of the Caribbean Sea has shown dramatic changes in hurricane activity – averaging around 1.5 per year during the 1940s through the 1960s dropping to near 0.5 per year in the 1970s to the early 1990s (Figure 9). This is followed by the unprecedented (in this five decade time series) six hurricanes afflicting the region in 1996.

Strongest Atlantic Hurricane Windspeed
1944-1996

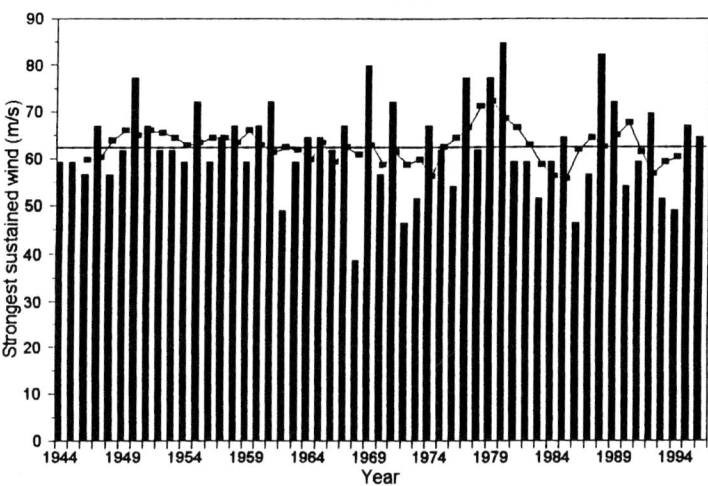

Figure 7. Time series of the Atlantic basin peak intensity – the highest sustained surface winds for the strongest hurricane each year – for the period of reliable record from 1944 to 1996. The long term (1950–90) average of 63.1 ms^{-1} (solid line) as well as a five year running mean (light curve) are superimposed on the time series.

In contrast to the Caribbean, U.S. landfalling hurricanes show a relatively small multidecadal variability in the time series with only the 1940s (active) and the 1970s (quiet) having distinctly different values than the long term average (Figure 10). This is similar to the behavior seen in the all–basin hurricane record. Few multidecadal variations are found in of the named and subtropical storms which struck the U.S. (not shown).

Where one does observe a strong multidecadal fluctuation in the United States tropical cyclones is with intense hurricanes, especially those that hit the East Coast from the Florida peninsula up to New England (Fig. 11). In this case, the quiet period in recent decades was so inactive that not a single East Coast intense hurricane made landfall from 1966 through 1984. This is in extreme contrast to the nineteen years previous (1947–1965) when fourteen intense hurricanes struck. Only in the last few years does it appear that this extremely quiescent period may be ending. This recent quiet period – well matched by the all–basin intense hurricanes described earlier – is similar to, but more extreme than, an inactive period during the first two decades of this century. In contrast, a subset of the Atlantic basin consisting of the U.S. Gulf Coast from Texas to the Florida panhandle (Fig. 12) has observed much weaker multidecadal variability in intense hurricane strikes, with sub-

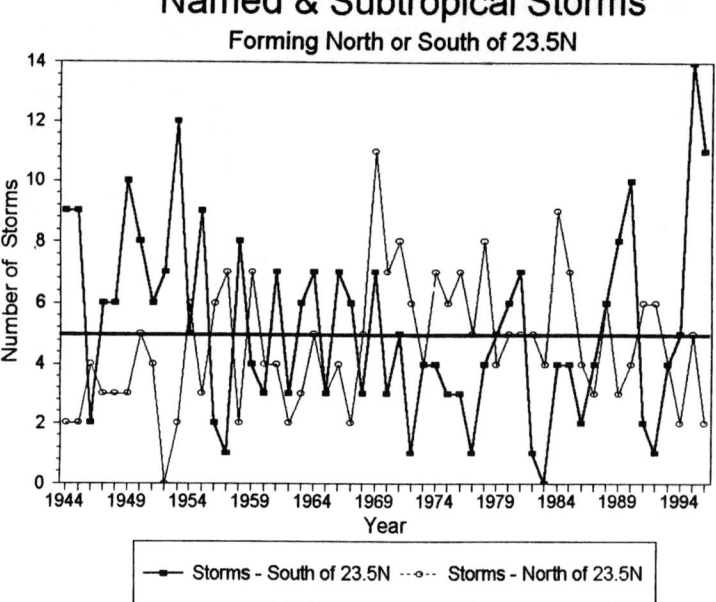

Figure 8. Annual number of Atlantic basin named and subtropical storms forming south (dark line) and north (light line) of 23.5°N for the period of reliable record from 1944 to 1996. The long term (1950–90) average of 4.9 per year (thick line) – valid for both time series – is also shown.

stantially more active conditions than average occurring only in the 1910s and quiet conditions only in the late 1940s and early 1950s.

Finally, hurricane–caused damage in the United States – when properly normalized – can also provide an independent indication of multiyear changes in tropical cyclone activity. Figure 13 shows the time series of damage normalized in terms of changes in inflation, wealth and coastal county population changes are taken into account (Pielke and Landsea 1998). Note the extreme destruction in 1926 (due to the near worst case scenario of a large Category 4 hurricane striking first the populous Miami–Ft. Lauderdale region in Florida, then striking Pensacola, Florida and Mobile, Alabama as a Category 3 hurricane), lowered values of damage in the early and mid 1930s followed by $3–7 billion damage per year for nearly every five year period from the late 1930s until the late 1960s. During the 1970s and 1980s, the normalized damage in the United States was substantially smaller ($1–3 billion per year) than in earlier decades. During the first part of the 1990s, damage again returned to higher levels due to the destructiveness of Hurricane Andrew in 1992.

Caribbean Hurricanes
1944-1996

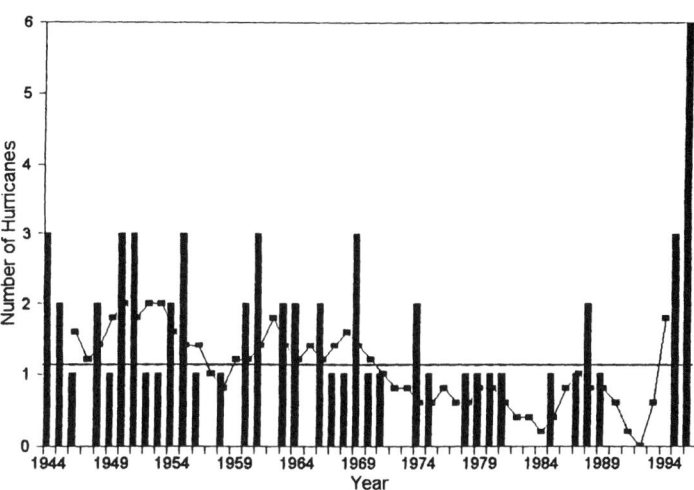

Figure 9. Time series of Caribbean Sea hurricanes for the period of reliable record from 1944 to 1996. The long term (1950–90) average of 1.1 per year (solid line) as well as a five year running mean (light curve) are superimposed on the time series.

U.S. Hurricane Strikes
1899-1996

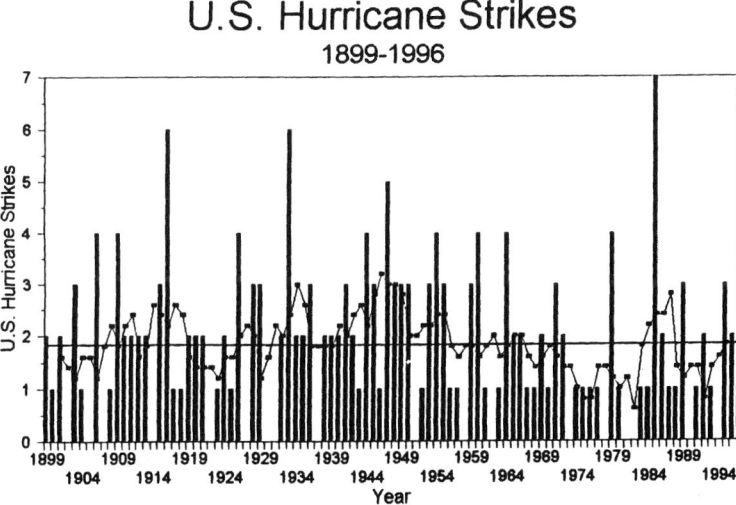

Figure 10. Annual number of hurricanes making landfall along the continental United States coastline for the period of reliable records of 1899–1996. The long term (1950–90) average of 1.66 per year (solid line) as well as a five year running mean (light curve) are superimposed on the time series.

Table III. Trends in Atlantic all–basin tropical cyclone indices. Trends calculated for the years 1944–1996 for the all–basin activity and the Caribbean hurricanes, 1899–1996 for the U.S. land-falling hurricanes, and 1925–1996 for the normalized U.S. tropical cyclone damages. Trends significant at the 0.01 level are indicated by "***", at the 0.05 level by "**", and at the 0.10 level by "*".

Tropical Cyclone Index	Linear Correlation Coefficient (r)	Regression Coefficient
Named/Subtropical Storms	+0.02	+0.04 per decade
Hurricanes	–0.11	–0.17 per decade
Intense Hurricanes	–0.23 **	–0.23 per decade
Hurricane Days	–0.18 *	–1.63 days per decade
Mean Intensity	–0.26 **	-1.34 ms^{-1} per decade
Peak Intensity	–0.04	-0.45 ms^{-1} per decade
Northern Named/Subtropical Storms	+0.25 **	+0.34 per decade
Southern Named/Subtropical Storms	–0.15	–0.30 per decade
Caribbean Hurricanes	–0.19 **	–0.15 per decade
U.S. Hurricanes	–0.07	–0.04 per decade
U.S. East Coast Intense Hurricanes	–0.02	0.00 per decade
U.S. Gulf Coast Intense Hurricanes	–0.06	–0.01 per decade
U.S. Normalized Tropical Cyclone Damage	–0.14	–$728 Million per decade

4.3. RELATIONSHIP TO CLIMATIC FORCINGS

Tables IV and V provide detailed comparisons of how the various Atlantic basin tropical cyclone indices are modulated by concurrent variations of individual environmental factors. These two tables – quartile differences and linear correlation coefficients, respectively – document the controls on inter-annual variability, though not much can be inferred here about causes on the interdecadal timescale.

The El Niño–Southern Oscillation, as represented by the Niño 3.4 SST anomalies, induces moderate–sized changes in the frequency and intensity of Atlantic tropical cyclones. La Niña events cause 36% more named/subtropical storms than El Niño events with mean intensities that are 6% stronger. ENSO's effect extends throughout the entire basin, though the effect is larger in the more southerly forming systems. The Caribbean region and the United States also see a large modulation of hurricane strikes, though the changes are only weakly significant for the U.S. Gulf Coast intense hurricanes.

Table IV. Quartile mean (median for U.S. normalized damages) differences (top 25% versus the bottom 25%) of a variety of environmental factors versus Atlantic tropical cyclone activity indices. The first value listed (or only value if just one is provided) shows the quartile differences for the period of 1950–1996 to allow a comparison of all factors based upon the same time period. Differences significant at the 0.01 level are indicated by "***", at the 0.05 level by "**", and at the 0.10 level by "*". Significance levels are adjusted to account for any serial correlation (Reid et al. 1989).

Tropical Cyclone Index (1950–1990 Mean)	ASO Niño3.4 SSTs (La Niña/El Niño)	ASO Caribbean 200 hPa Zonal Winds (Easterlies/Westerlies)
Named/Subtropical Storms (9.8)	11.6/8.5 ***	11.2/7.9 ***
Hurricanes (5.8)	6.7/4.9 ***	7.3/4.2 ***
Intense Hurricanes (2.2)	3.2/1.3 ***	3.6/1.2 ***
Hurricane Days (23.7)	29.9/19.1 ***	35.0/13.1 ***
Mean Intensity (39.0)	39.7/37.3 ***	40.5/37.1 ***
Peak Intensity (63.1)	65.4/58.8 ***	62.8/59.8 ***
Northern Named/Subtropical Storms (4.9)	5.8/5.1 *	4.2/4.8
Southern Named/Subtropical Storms (4.9)	5.8/3.4 ***	7.0/3.1 ***
Caribbean Hurricanes (1.1)	1.7/0.8 ***	2.0/0.5 ***
U.S. Hurricanes (1.7 [1899–1996: 1.8])	2.0/1.1 *** (1899–1996: 2.4/1.0 ***)	2.2/1.3 ***
U.S. East Coast Intense Hurricanes (0.32 [1899–1996: 0.36])	0.58/0.08 *** (1899–1996: 0.33/0.04 ***)	0.58/0 ***
U.S. Gulf Coast Intense Hurricanes (0.32 [1899–1996: 0.36])	0.33/0.25 (1899–1996: 0.45/0.28 **)	0.17/0.42 ***
U.S. Normalized Damage ($361 [1925–96: $1,065])	$2,358/$691 (1925–96: $3,526/$1,049 **)	$1,680/$1,418

Tropical Cyclone Index (1950–1990 Mean)	ASO Stratospheric QBO (West Phase/East Phase)	ASO Caribbean Sea Level Pressures (Low/High)
Named/Subtropical Storms (9.8)	11.8/8.8 ***	12.4/7.9 ***
Hurricanes (5.8)	7.3/5.1 ***	8.2/4.3 ***
Intense Hurricanes (2.2)	3.6/1.7 ***	3.7/1.1 ***
Hurricane Days (23.7)	35.2/17.2 ***	39.1/13.3 ***
Mean Intensity (39.0)	41.0/38.1 ***	41.5/37.0 ***
Peak Intensity (63.1)	68.4/59.8 ***	68.9/58.1 ***
Northern Named/Subtropical Storms (4.9)	5.8/4.8 **	4.0/5.0
Southern Named/Subtropical Storms (4.9)	6.0/4.0 ***	8.4/2.9 ***
Caribbean Hurricanes (1.1)	1.6/1.1 *	2.4/0.4 ***
U.S. Hurricanes (1.7 [1899–1996: 1.8])	2.5/1.7 **	2.2/1.1 ***
U.S. East Coast Intense Hurricanes (0.32 [1899–1996: 0.36])	0.33/0.33	0.50/0.17
U.S. Gulf Coast Intense Hurricanes (0.32 [1899–1996: 0.36])	0.58/0.25 ***	0.25/0.42 *
U.S. Normalized Damage ($361 [1925–96: $1,065])	$2,712/$2,256	$3,250/$902 *

Tropical Cyclone Index (1950–1990 Mean)	ASO Tropical North Atlantic SSTs (Warm/Cool)	JJAS West Sahel Rainfall (Wet/Dry)
Named/Subtropical Storms (9.8)	11.3/8.2 **	10.7/8.1 ***
Hurricanes (5.8)	7.2/5.1 ***	7.5/4.2 ***
Intense Hurricanes (2.2)	3.0/1.7 **	3.2/0.9 ***
Hurricane Days (23.7)	31.9/21.0 *	34.8/10.8 ***
Mean Intensity (39.0)	39.9/38.9	42.8/36.5 ***
Peak Intensity (63.1)	66.1/62.4	68.0/55.8 ***
Northern Named/Subtropical Storms (4.9)	3.3/5.3 ***	4.2/5.4
Southern Named/Subtropical Storms (4.9)	8.0/3.7 ***	6.4/2.7 ***
Caribbean Hurricanes (1.1)	1.8/0.8 *	1.9/0.2 ***
U.S. Hurricanes (1.7 [1899–1996: 1.8])	1.5/1.8 (1899–1996: 1.8/1.7)	2.2/1.0 ** (1899–1996: 2.0/1.3 ***)
U.S. East Coast Intense Hurricanes (0.32 [1899–1996: 0.36])	0.33/0.42 (1899–1996: 0.46/0.29)	0.83/0.08 ** (1899–1996: 0.62/0.12 ***)
U.S. Gulf Coast Intense Hurricanes (0.32 [1899–1996: 0.36])	0.25/0.42 (1899–1996: 0.42/0.17)	0.33/0.25 (1899–1996: 0.46/0.29 **)
U.S. Normalized Damage ($361 [1925–96: $1,065])	$541/$1685 (1925–96: $1453/$1534)	$5,092/$232 * (1925–96: $1,996/$682)

Table V. Linear correlation coefficients (r) of a variety of environmental factors versus Atlantic tropical cyclone activity indices. The first value listed (or only value if just one is provided) shows the correlation for the period of 1950–1996 to allow a comparison of all factors based upon the same time period. Correlations significant at the 0.01 level are indicated by "***", at the 0.05 level by "**", and at the 0.10 level by "*". Significance levels are adjusted to account for any serial correlation (Reid et al. 1989).

Tropical Cyclone Index (1950–1990 Mean)	ASO Niño3.4 SSTs	ASO Caribbean 200 hPa Zonal Winds
Named/Subtropical Storms (9.8)	–0.33 **	–0.45 ***
Hurricanes (5.8)	–0.30 **	–0.56 ***
Intense Hurricanes (2.2)	–0.40 ***	–0.62 ***
Hurricane Days (23.7)	–0.28 **	–0.60 ***
Mean Intensity (39.0)	–0.30 **	–0.38 ***
Peak Intensity (63.1)	–0.25 *	–0.25 *
Northern Named/Subtropical Storms (4.9)	–0.14	0.15
Southern Named/Subtropical Storms (4.9)	–0.24 *	–0.57 ***
Caribbean Hurricanes (1.1)	–0.29 **	–0.47 ***
U.S. Hurricanes (1.7 [1899–1996: 1.8])	–0.15 (1899–1996: –0.30 ***)	–0.24 *
U.S. East Coast Intense Hurricanes (0.32 [1899–1996: 0.36])	–0.22 (1899–1996: –0.20 **)	–0.25 *
U.S. Gulf Coast Intense Hurricanes (0.32 [1899–1996: 0.36])	–0.16 (1899–1996: –0.16)	0.11
U.S. Normalized Damage ($361 [1925–96: $1,065])	0.00 (1925–96: –0.04)	0.00

Tropical Cyclone Index (1950–1990 Mean)	ASO Stratospheric QBO	ASO Caribbean Sea Level Pressures
Named/Subtropical Storms (9.8)	0.42 ***	–0.56 ***
Hurricanes (5.8)	0.41 ***	–0.62 ***
Intense Hurricanes (2.2)	0.39 ***	–0.59 ***
Hurricane Days (23.7)	0.46 ***	–0.64 ***
Mean Intensity (39.0)	0.17	–0.41 ***
Peak Intensity (63.1)	0.23	–0.44 ***
Northern Named/Subtropical Storms (4.9)	0.15	0.06
Southern Named/Subtropical Storms (4.9)	0.33 **	–0.62 ***
Caribbean Hurricanes (1.1)	0.19	–0.54 ***
U.S. Hurricanes (1.7 [1899–1996: 1.8])	0.14	–0.27 *
U.S. East Coast Intense Hurricanes (0.32 [1899–1996: 0.36])	–0.10	–0.10
U.S. Gulf Coast Intense Hurricanes (0.32 [1899–1996: 0.36])	0.24 *	0.03
U.S. Normalized Damage ($361 [1925–96: $1,065])	–0.23 *	–0.03

Tropical Cyclone Index (1950–1990 Mean)	ASO Tropical North Atlantic SSTs	JJAS West Sahel Rainfall
Named/Subtropical Storms (9.8)	0.42 ***	0.30 **
Hurricanes (5.8)	0.38 ***	0.51 ***
Intense Hurricanes (2.2)	0.29 **	0.52 ***
Hurricane Days (23.7)	0.33 **	0.59 ***
Mean Intensity (39.0)	0.05	0.52 ***
Peak Intensity (63.1)	0.16	0.35 **
Northern Named/Subtropical Storms (4.9)	–0.32 **	–0.21
Southern Named/Subtropical Storms (4.9)	0.56 ***	0.45 ***
Caribbean Hurricanes (1.1)	0.24	0.45 ***
U.S. Hurricanes (1.7 [1899–1996: 1.8])	0.02 (1899–1996: 0.07)	0.28 * (1899–1996: 0.21 **)
U.S. East Coast Intense Hurricanes (0.32 [1899–1996: 0.36])	0.00 (1899–1996: 0.11)	0.40 *** (1899–1996: 0.23 **)
U.S. Gulf Coast Intense Hurricanes (0.32 [1899–1996: 0.36])	–0.10 (1899–1996: 0.07)	–0.03 (1899–1996: 0.09)
U.S. Normalized Damage ($361 [1925–96: $1,065])	–0.04 (1925–96: 0.07)	0.07 (1925–96: 0.07)

U.S. East Coast Intense Hurricanes
Florida Peninsula and Upper Atlantic

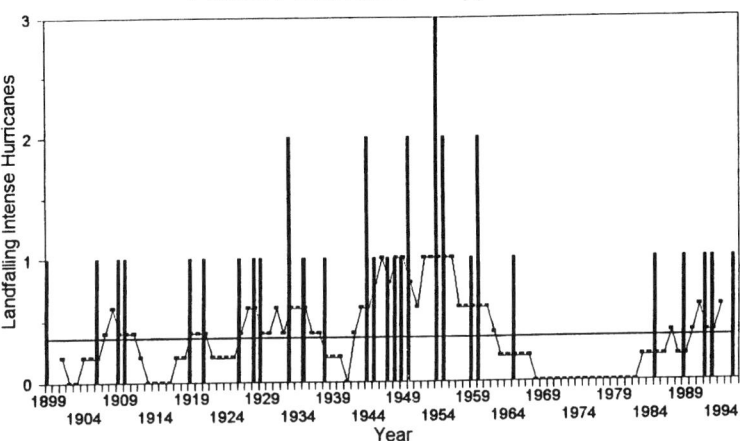

Figure 11. Annual number of intense hurricanes that have made landfall along the U. S. East Coast from the Florida peninsula to the Upper Atlantic (from Georgia to New England) for the period of reliable records of 1899–1996. The long term (1950–90) average of 0.32 per year (solid line) as well as a five year running mean (light curve) are superimposed on the time series.

The Caribbean 200 hPa zonal winds and the sea level pressures both exert a similar, very strong influence on the frequency and intensity of tropical cyclones. Easterly wind anomalies and low pressures are associated with an increase of 42% and 57% more named/subtropical storms, respectively, than in the years of westerly wind and high pressure anomalies. The mean intensity of such systems is also dramatically altered with correspondingly stronger systems by 9% and 12% for the years of easterly wind and low pressure anomalies. However, in the northerly latitudes these effects are reversed with favored formation occurring in the years of westerly wind anomalies and higher pressure, though these reversed alterations are weaker than what occurs to the southern tropical cyclone formations. The wind and pressure changes correspond to large variations in the landfalling Caribbean and U.S. hurricanes, though the U.S. Gulf Coast intense hurricanes respond to the reversed effect with more strikes in Caribbean high pressure and west wind years.

The stratospheric QBO appears to primarily have a moderate effect on the frequency of Atlantic tropical cyclones forming anywhere in the basin, but only a weak signal in the intensity of systems. The west phase of the QBO is linked to a 34% increase in total storms over the number seen in the QBO

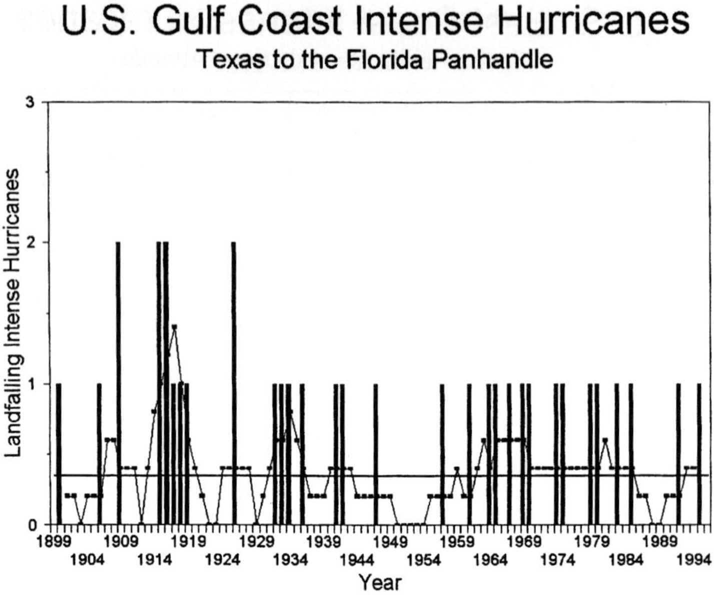

U.S. Gulf Coast Intense Hurricanes
Texas to the Florida Panhandle

Figure 12. Annual number of intense hurricanes that have made landfall along the U. S. Gulf Coast from Texas to the Florida panhandle for the period of reliable records of 1899–1996. The long term (1950–90) average of 0.32 per year (solid line) as well as a five year running mean (light curve) are superimposed on the time series.

east years. The mean intensity, however, is not significantly correlated to the QBO phase. The landfalling hurricanes in the Caribbean and U.S. are all significantly modulated by the QBO, though the effects are relatively weak.

The tropical North Atlantic SSTs appears to have the largest influence on the spatial distribution of the total number of named and subtropical storms. There is a over a two–to–one difference in southern forming storms during warm versus cold Atlantic SST years; this is nearly reversed for the northerly latitude forming storms. However, as more intense categories of tropical cyclones are considered, the tropical North Atlantic SSTs effects are diminished on this interannual timescale. Thus Atlantic SST impact upon the frequency of U.S. landfalling hurricanes is negligible and only marginally observed for Caribbean hurricanes.

Finally, the West Sahel rainfall difference appear to play a moderate role in the total frequency and have a very large impact on intensity. There is a 32% increase in named and subtropical storms in the wet Sahel years compared with the dry years, though partially this is somewhat reduced because of a compensating decrease in the number of formations in northerly latitudes. The intensity of the mean storm is strengthened by 17% in the wet versus the dry years. These alterations cause big changes in the Caribbean and U.S. hurricanes, again primarily along the U.S. East Coast intense hurricanes.

US Normalized Tropical Cyclone Damage
1925-1996

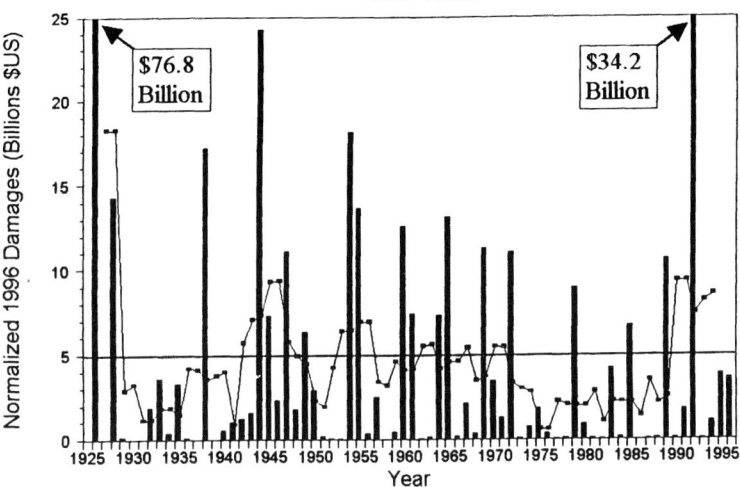

Figure 13. Time series of annual normalized tropical cyclone damage for the U.S. East and Gulf Coasts. The long–term (1950–90) average of $3.5 billion per year (solid line) as well as a five year running mean (light curve) are superimposed on the time series. Damage is normalized to 1996 dollars by inflation, coastal county population changes, and an index of personal property amounts.

In a comparative analysis, the frequency of named/subtropical storms is most strongly altered by sea level pressure variations followed by 200mb zonal winds, the stratospheric QBO and Atlantic SSTs. However, the total frequency of events distorts large regional differences based upon latitudinal variations. For southerly latitude formations the primary interannual forcing is due to sea level pressure and Atlantic SST variations. These effects are followed by changes associated with Western Sahel rainfall and 200mb zonal winds. The same factors – with the exception of Atlantic SSTs – also appear to strongly modulate Caribbean hurricane strikes. Only two factors were found that had a homogeneous, basinwide effect. ENSO and the stratospheric QBO both show reduced activity throughout the Atlantic basin during warm events and east phases, respectively. However, effects were most pronounced in the southern portion of the basin. Opposite effects between tropical cyclones forming in the southerly and northerly portion of the basin are seen most strongly for the Atlantic SSTs and the West Sahel rainfall, with a weaker signal present in the sea level pressure anomalies. Along the U.S. coastline, the factors of sea level pressures, ENSO and West Sahel rainfall have the largest effects on number of hurricane strikes. Again, this is dependent upon location, however. For the East Coast, interannual variations of intense hurricane strikes are best related to ENSO, West Sahel rainfall and 200mb zonal winds. For intense hurricane strikes along the Gulf Coast, the single largest factor is the phase of the QBO followed by the 200mb zonal winds and sea level pressure anomalies both in an opposite sense where westerly winds and high pressures in the Caribbean favor more strikes. Interannual variations in normalized U.S. damages are only well related to ENSO, with significantly less damage during El Niño events and more during La Niña events, and just marginally associated with Caribbean sea level pressures and West Sahel rainfall. These relationships show up more clearly in the composite analyses since the linear correlation analyses have difficulty in identifying a coherent signal in the rather noisy distribution of normalized U.S. damages.

Independent of frequency, mean and peak intensities of Atlantic tropical cyclones are most strongly linked to West Sahel rainfall followed by changes in Caribbean sea level pressures and 200mb zonal winds. Likewise, duration of hurricanes is nearly equally strongly associated with variations of these three previously mentioned parameters with longer lived hurricanes during years of wet West Sahel seasons, easterly 200mb zonal wind anomalies and low sea level pressures.

After investigating the interannual aspects of Atlantic tropical cyclones and their environmental controls, an analysis of multidecadal relationships of Atlantic tropical cyclones to Atlantic SSTs was attempted. Figure 14 presents the "Atlantic Multidecadal Mode" EOF time series from Enfield and Mestas–Nuñez (1998) in the bottom panel. The top panel of Fig. 14 is the correlation between the time series of the EOF and the actual SST data at every grid point. This has very similar spatial structure to the actual EOF spatial pattern and gives a measure of the local fractional variance (squared temporal correlation) accounted at every grid point. The dominant spatial relationships – in the positive phase – have warm SSTs in the tropical North Atlantic between 5 and 25°N, the far North Atlantic between 40 and 70°N, and cool SSTs in the South Atlantic. There are also secondary signals seen with warm conditions in the North Pacific and cool conditions in the Indian Ocean, the South Pacific and along the U.S. New England coastal waters. This mode, likely the same as has been previously identified by Folland et al. (1986) and Kushnir (1994), operates primarily with the most power in the multidecadal mode with distinct multiple decades of warm North/cold South Atlantic and vice versa. Mestas–Nuñez and Enfield (1998) further show that under a varimax rotation, variability outside the North Atlantic separates into distinct rotated modes; only the North Atlantic rotated mode retains the multidecadal temporal signature seen in Fig. 14, while the North and South Atlantic rotated modes are independent of each other. It is concluded that the Multidecadal mode seen in Fig. 14 is primarily a representation of North Atlantic SST variability after accounting for the ENSO influence. The duration of these various modes can approximately be broken down into the following years since 1857 when the analysis starts:

- 1857 to 1868 - Indeterminate

- 1869 to 1893 - Warm North Atlantic (25 years)

- 1894 to 1925 - Cold North Atlantic (32 years)

- 1926 to 1970 - Warm North Atlantic (45 years)

- 1971 to 1994 - Cold North Atlantic (24 years)

- 1995 to 1996 - Indeterminate

The recent change of the mode in 1995 and 1996 may signify a return to a multidecadal period of the warm North Atlantic phase, but there is not enough evidence yet to be conclusive. For the following analysis, 1971 to 1994 will be considered as part of the cold North Atlantic phase and 1995 to 1996 will be considered indeterminate.

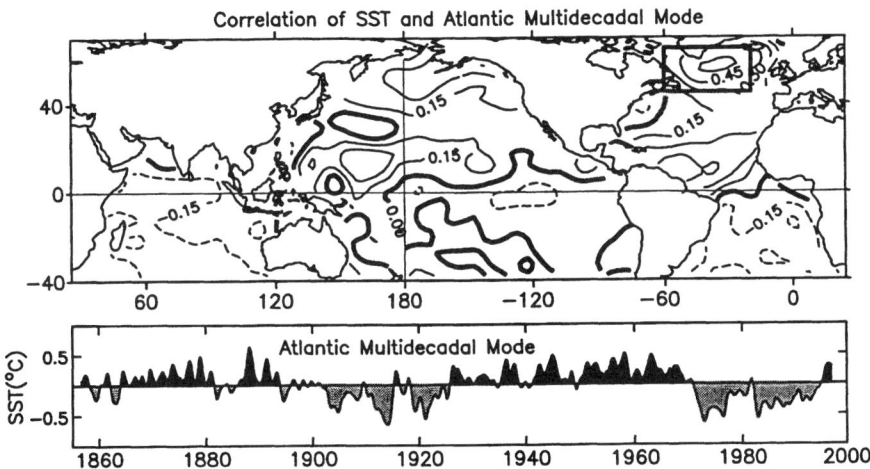

Figure 14. Top) Distribution of correlations for the years 1857 to 1996 between local monthly SST anomalies versus the third EOF "Atlantic Multidecadal Mode"; Bottom) Temporal realization of the Atlantic Multidecadal Mode computed from temporal amplitude time series and the area-average spatial loadings over the rectangular area in the North Atlantic.

Table VI presents differences in Atlantic tropical cyclone activity based upon the phase of the Atlantic Multidecadal Mode. Variations based upon these SST differences are similar to, but generally stronger than, the interannual changes shown in Table V. Decades of the warm North Atlantic phase show little change in total frequency of storms, primarily because there is a nearly compensating decrease in northerly latitude formations compared with those forming in the southerly portion of the basin. Intensity of tropical cyclones are modulated strongly with 80% more intense hurricane occurring in the warm North Atlantic years over the cold North Atlantic decades. These impacts are strongly felt in the Caribbean basin with 200% more intense hurricanes occurring in the warm North Atlantic decades and for intense hurricanes striking the U.S. East Coast with an increase of 165%. Differences for the U.S. Gulf Coast intense hurricanes appear insignificant. Consequently, normalized damage for the U.S. is also dramatically altered with a median value of $1,747 million per year in the warm North Atlantic regime versus only $317 million per year in the cold North Atlantic decades. Figure 15 provides a spatial comparison of intense hurricane activity throughout the whole basin for the two multidecadal regimes. Note how frequencies of intense hurricane are decreased throughout the entire basin with the exception of the Gulf of Mexico during the cold North Atlantic decades in comparison with the warm North Atlantic decades.

Table VI. Mean (median for U.S. normalized damage) values for Atlantic tropical cyclone activity based upon Atlantic Multidecadal Mode variations. Differences between warm North Atlantic (all years) and cold North Atlantic (all years) are tested for significance in the last column. Differences significant at the 0.01 level are indicated by "***", at the 0.05 level by "**", and at the 0.10 level by "*".

Tropical Cyclone Index (Years available)	Cold North 1899–1925	Warm North 1926–70	Cold North 1971–94	Warm North All Years	Cold North All Years	Significance
Named/Subtropical Storms (1944–94)	–	9.9	9.2	9.9	9.2	
Hurricanes (1944–94)	–	6.2	5.0	6.2	5.0	**
Intense Hurricanes (1944–94)	–	2.7	1.5	2.7	1.5	***
Hurricane Days (1944–94)	–	27.7	16.5	27.7	16.5	***
Mean Intensity (1944–94)	–	39.5	37.4	39.5	37.4	*
Peak Intensity (1944–94)	–	62.5	62.0	62.5	62.0	
Northern Named/Subtropical Storms (1944–94)	–	4.0	5.3	4.0	5.3	**
Southern Named/Subtropical Storms (1944–94)	–	5.9	3.9	5.9	3.9	***
Caribbean Hurricanes (1944–94)	–	1.5	0.5	1.5	0.5	***
U.S. Hurricanes (1899–1994)	1.7	2.1	1.4	2.1	1.6	*
U.S. East Coast Intense Hurricanes (1899–1994)	0.22	0.53	0.17	0.53	0.20	**
U.S. Gulf Coast Intense Hurricanes (1899–1994)	0.41	0.36	0.29	0.36	0.35	
U.S. Normalized Damage (1925-94)	$0	$1,747	$528	$1,747	$317	

5. Policy Relevance

Indices of Atlantic basin hurricane activity underlie important policy decisions in the public and private sectors. For example, insurance and reinsur-

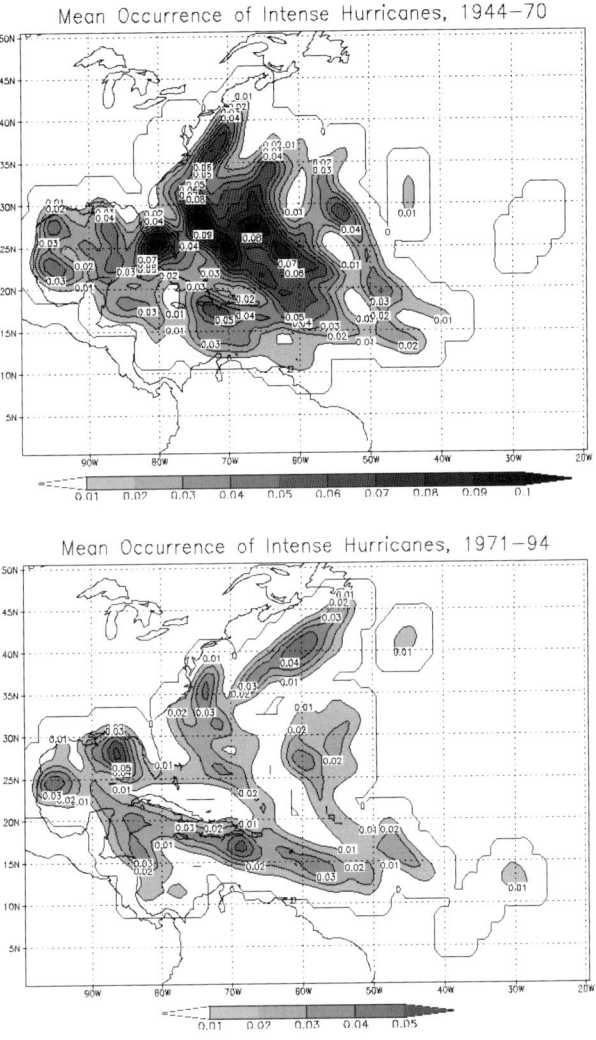

Figure 15. Observed annual frequency of intense hurricanes in a 1.0 by 1.0 latitude/longitude grid field smoothed with a nine point binomial filter twice for (top) 1944-70 (warm North Atlantic) and (bottom) 1971-94 (cold North Atlantic)

ance companies evaluate risk, in part, based on their interpretation of the historical record. More broadly, public attention and resources are directed to the hurricane problem based upon assessments of past tropical cyclone incidences and impacts (Pielke and Pielke 1997a). In recent years, some in the media and in policy positions have identified hurricane activity as a signal of human-induced climate change (Pielke 1997). Consequently, an examination of past trends in hurricane incidence is directly relevant to both climate change policies and natural hazards policies.

5.1. CLIMATE CHANGE POLICY

In 1996, Working Group III of the IPCC estimated increased worldwide damages and loss of life related to hurricane impacts in a doubled CO_2 world at \$630 million and 8,000 additional lives lost (Watson et al. 1996). Working Group III concluded that these economic losses and lives lost would be prevented with the adoption of emissions reductions policies. There is an obvious inconsistency between the projections by IPCC Working Group III of increased impacts and the conclusions of Working Group I, which stated that "the state of the science does not allow assessment of future changes" in tropical cyclone indices.

Setting aside for the moment this inconsistency, the logic of the IPCC Working Group III is fundamentally flawed. Even if there were valid theoretical reasons to expect more tropical cyclones in the future related to human–caused climate change, the climatological record gives no indication that society can modulate hurricane impacts through energy policies. That is, as atmospheric CO_2 levels have increased, "there is currently no evidence that there has been systematic changes in the observed tropical cyclones around the globe" (Landsea 1998). The suggestion by the IPCC that a reduction in greenhouse gas emissions will lead to less or less intense tropical cyclones and therefore less impacts to society begs several further questions of relevance to the policy community which have thus far gone unasked and unanswered:

- Can the scientific community reliably differentiate future hurricane frequencies and magnitudes based on the various scenarios of greenhouse gas emissions and concentrations (i.e., IS92a–f from Houghton et al. 1992)?

The analysis of climatological information presented in this paper suggests that for many decades to come, detection of a human–forced signal in the tropical cyclone record will be extremely difficult to detect because of both the relatively modest size of the predicted changes in MPI and the rather large apparently natural multidecadal variability (cf. Henderson–Sellers et al. 1998). Therefore, it is unrealistic for policy makers to expect in the near term (i.e., in the next few years) that the scientific community will be able to reliably predict future hurricane incidences differentiated by various emissions scenarios. Henderson–Sellers et al. (1998) noted that "global and mesoscale model-based predictions for tropical cyclones in greenhouse conditions have not yet demonstrated prediction skill."

- Is there reason to believe that policy makers should expect the policy actions now being contemplated (e.g., the Kyoto Accord to the Framework Convention on Climate Change) will reduce the number of and intensities of future hurricanes that will impact society?

There is no evidence to suggest that society can intentionally modulate tropical cyclone frequencies and magnitudes through energy policies[2]. Therefore, policy responses to hurricanes ought to focus on the reduction of society's vulnerability to hurricanes, rather than on prevention of the storms themselves (Pielke and Pielke 1997b). For instance, in the context of insurance, Henderson–Sellers et al. (1998) recommend a focus on "appropriate reserves and restrictive underwriting" rather than on accurate predictions, or by extension, on controlling future hurricane incidences.

Answers to these questions do not exclude the possibility that an anthropogenic forcing might lead to changes (Henderson-Sellers et al. 1998). They do strongly suggest that reliable prediction of future hurricane indices (much less societal impacts) differentiated by various emissions scenarios is beyond the capabilities of the scientific community. Further, if a policy objective is to reduce society's vulnerability to hurricane impacts, then decision makers would be wiser to consider better adapting to documented variability, rather than preventing storms from occurring (Pielke 1998)[3].

5.2. NATURAL DISASTER POLICY

One of the most striking features of the information presented in section four of this paper are the 19 years which passed between intense hurricane landfalls on the U.S. East Coast from 1966 through 1984. These decades saw much of the population growth and development of coastal communities. Overall, the 19 years prior to 1966 saw 14 intense hurricanes strike the U.S. East Coast. Most of the historical economic losses are the result of storms striking the U.S. East Coast rather than the Gulf Coast (Table VII). Consider also that over the seven year period 1944 to 1950, the state of Florida saw $44.2 billion (normalized to 1995 values, see Pielke and Landsea 1998) in losses, or more than $6 billion per year, while the 46 year period 1951 to 1997 saw a similar total amount of normalized damages, $49.3 billion or about $1.1 billion per year. Most of the damages of the latter period were the result of Hurricane Andrew in 1992.

The review of indices for hurricane climatic changes reveals that from the perspective of societal impacts, recent decades are indeed anomalous. But

---

[2] This is not to say that energy policies CANNOT affect hurricanes, only that there is no proof that society can purposely modulate hurricane activity via energy policies. See Pielke (1998) for a broader discussion of the Kyoto Protocol.

[3] A more comprehensive discussion of mitigation and adaptation in the context of global climate policy is found in Pielke (1998).

Table VII. Correlations of normalized U.S. hurricane damages versus indices of Atlantic basin hurricane activity. Correlations are calculated for the years 1944–1996 for the all–basin activity and the Caribbean hurricanes and 1925–1996 for the U.S. landfalling hurricanes. Correlations significant at the 0.01 level are indicated by "***", at the 0.05 level by "**", and at the 0.10 level by "*". Significance levels are adjusted to account for any serial correlation (Reid et al. 1989).

Tropical Cyclone Index	Linear Correlation Coefficient (r)
Named/Subtropical Storms	0.00
Hurricanes	+0.12
Intense Hurricanes	+0.08
Hurricane Days	+0.15
Mean Intensity	+0.26 **
Peak Intensity	+0.21 *
Northern Named/Subtropical Storms	0.00
Southern Named/Subtropical Storms	+0.02
Caribbean Hurricanes	+0.16
U.S. Hurricanes	+0.51 ***
U.S. East Coast Intense Hurricanes	+0.61 ***
U.S. Gulf Coast Intense Hurricanes	+0.29 ***

contrary to conventional wisdom of some in the media, public, and policy communities, recent decades are unique because of the relative infrequency of U.S. landfalls of strong hurricanes, and not because of any upsurge in strong storms (cf. Landsea et al. 1996). Hurricanes arguably are the natural hazard with the greatest potential for economic disruption in the United States, and further, the potential for a large loss of life related to a hurricane's landfall is increasing with coastal development (Pielke and Pielke 1997a). Because the nation's hurricane policies have been typically developed in the immediate aftermath of a disaster (Birkland 1997, Simpson 1998), it would be prudent for the policy community to assess whether or not the lack of hurricane impacts in recent decades has led to an atrophying of the nation's hurricane policies. Some questions to consider include:

- Are national, state, and local hurricane policies supported by public and private decision makers in a manner commensurate with the documented vulnerability of society?

- How prepared is the U.S. east coast for 14 intense hurricanes in 19 years as occurred in the 1940s–1960s?

- How prepared is the nation, and Florida specifically, for a recurrence of the hurricanes of the late 1940s?

- Is the time ripe for the United States to develop a national hurricane policy?

Asking and answering questions like these are important steps in reducing the nation's vulnerabilities to hurricane impacts. One benefit of past hurricane impacts is that society has learned many lessons. These lessons provide a basis of experience on which to reduce the nation's vulnerability to hurricane impacts (Pielke and Pielke 1997a). What seems to be lacking is awareness of whether the nation's risk is matched by its response.

6. Discussion and Summary

Atlantic hurricane variability can be characterized as having a lack of strong linear trends, but comprised of robust multidecadal variations. Such decade to decade changes are not evident when examining the entire named/subtropical storm database over the last five decades, but such variability becomes evident upon stratification by latitudinal regime and by intensity of the tropical cyclones.

In particular, we found that the hurricanes, especially those reaching sustained winds of 50 ms^{-1} – the intense hurricanes – were very common in the 1940s through the 1960s and much reduced in occurrence from the 1970s through the early 1990s. The years of 1995 and 1996 showed an intriguing return to high levels of activity more reminiscent of the earlier, active decades. The duration of hurricanes also showed similar variations with longer–lived (around 25–40 days per year) systems in the 1940s through the 1960s and rather short–lived hurricanes (around 10–25 days per year) in the decades since.

Such variability was not uniform throughout the basin. In fact, for the named/subtropical storms forming north of 23.5°N, the 1970s to the early 1990s actually showed an increase of activity, while there was a nearly equal decrease in formations south of that latitude. It is quite possible that some of this increase in named/subtropical storms in the northern latitudes was due to the availability of geostationary satellite imagery starting in the mid 1960s, as suggested by Elsner et al. (1996). However, it does appear that the decrease in activity in the southern latitudes is a real change. Inhabitants the Caribbean and the U.S. East Coast, in particular, were quite fortunate during the last few decades as these regions experienced many fewer damaging hurricanes than in earlier decades. Consequently, normalized hurricane damages in the U.S. were substantially lower in the 1970s and 1980s than in previous decades.

The one region with accurate enough records to extend the analysis back to the turn of the century – the U.S. East and Gulf Coasts – shows that the quiet period of recent decades is similar to the first two and a half decades of the century, though this is more true for the East Coast than the Gulf Coast which shows less multidecadal variability.

Analyses were also performed to contrast the impact that various environmental factors have upon Atlantic tropical cyclone variability. Figure 16 provides a schematic summarizing these interannual and interdecadal forcings of Atlantic tropical cyclones. The largest interannual variations appear to be associated with Caribbean sea level pressures and 200mb zonal winds: years of low pressures and easterly 200mb wind anomalies corresponded with more frequent and more intense tropical cyclones. Of course, the existence of more tropical cyclones will naturally directly contribute toward lower surface pressures and easterly wind anomalies. However, Knaff (1997) showed that the direct impact on the sea level pressures was small and that these SLP changes did not significantly alter the correlations derived. Additionally, Landsea et al. (1998) showed that, for at least the hurricane season of 1995, the lowered sea level pressures and 200mb easterly anomalies preceded the hurricane activity by several months and thus in this case could not have been caused by the hurricanes themselves.

The other environmental factors considered here – El Niño–Southern Oscillation, the stratospheric Quasi–Biennial Oscillation, West Sahel rainfall and Atlantic sea surface temperatures – also showed moderate to strong influences on Atlantic tropical cyclone activity, confirming previous studies. A new finding is that some of the environmental factors including the Caribbean sea level pressures (consistent with Knaff 1997), Atlantic SSTs and West Sahel rainfall did induce a weakly, opposite forcing of increased activity in the northerly portion of the basin during years of high pressure, cool SSTs and dry West Sahel seasons. In contrast, two of the environmental factors appeared to cause consistent basinwide alterations – ENSO and the QBO – in particular, though their effects in the southerly latitudes were strongest. Thus the hypothesis from Goldenberg and Shapiro (1996) that there is an "out-of-phase" relationship between the southerly and northerly latitude storm formations because of vertical shear proves true in the case of West Sahel rainfall, but not for ENSO. In addition, the key dependency of the QBO on the U.S. Gulf Coast intense hurricanes mirrors the findings of Lehmiller et al. (1997) in their scheme for forecasting Gulf of Mexico intense hurricane activity by 1 August.

Results in this study may be able to somewhat resolve the aforementioned discrepancies regarding the influence of Atlantic SSTs on interannual basinwide tropical cyclones. It was found here that the SSTs appear to ex-

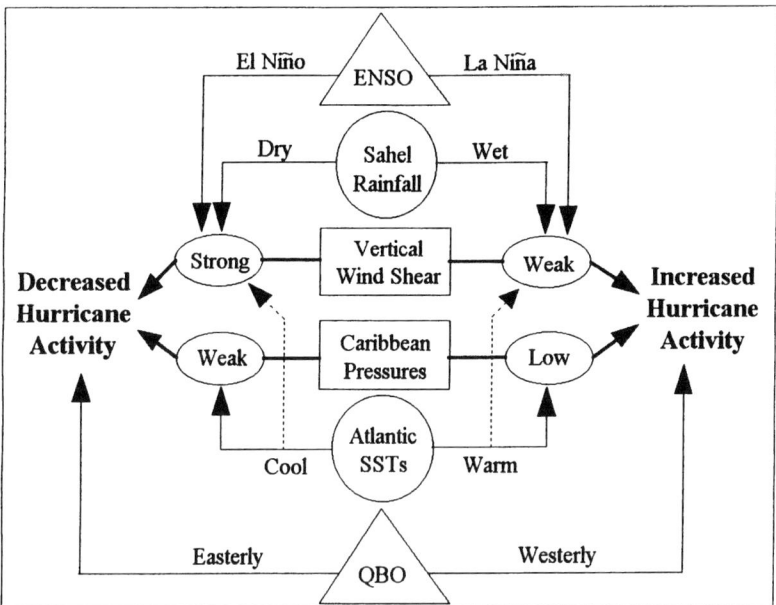

Figure 16. Schematic of environmental forcings of Atlantic tropical cyclones on primarily interannual (triangles) and interdecadal (circles) timescales. The rectangles indicate the physical factors directly responsible for tropical cyclone variations.

ert the most dominant "dipole" type signal in Atlantic tropical cyclones with more (fewer) storms forming in the southern (northern) latitudes during warmer than usual years. However, this signal is diluted when the entire tropical cyclone dataset is considered as a whole. Additionally, the relationship weakens as more intense tropical cyclones are considered. Thus Saunders and Harris (1997) are correct that Atlantic SSTs are the dominant mechanism when analyzing formations just in the eastern tropical North Atlantic. However, when looking at the entire basin and for more intense hurricanes, Raper (1992) and Shapiro and Goldenberg (1998), respectively, are also correct that the Atlantic SSTs play a weaker interannual role. The key is that the dataset of tropical cyclones must be broken down into spatial location and by intensity when doing the analysis.

The analysis here is not meant to imply that these various environmental factors independently affect Atlantic tropical cyclones, without interacting with one another. Indeed, Knaff (1997) demonstrated that the Caribbean SLP variations were strongly linked and were possibly due to a feedback with the tropical upper tropospheric trough and the strength of the associated 200 mb westerlies. However, of the six environmental factors considered here, only five pairs (out of 15 possible) have a covariance exceeding 25% of the variability: West Sahel rainfall and Caribbean SLPs (27%), Caribbean SLPs and Caribbean 200 mb zonal wind anomalies (29%), West Sahel rainfall and Caribbean 200 mb zonal wind anomalies (32%), tropical North Atlantic SSTs and Caribbean 200 mb zonal wind anomalies (34%), and West Sahel rainfall and tropical North Atlantic SSTs (38%). The remaining ten pairs of combinations had relationships explaining less than one–fifth of the variance between the two environmental factors. While most of these relatively small values of co–variability suggest that the environment factors may be independently related to Atlantic tropical cyclones, further study is certainly warranted in investigating the interdependence of these conditions.

While the Atlantic SSTs were a quite weak interannual environmental factor for intense hurricanes, the multidecadal mode of Atlantic SSTs in contrast corresponds strongly to the observed decade to decade changes in Atlantic intense hurricanes. In particular, the quiet regimes of 1899–1925 and 1971–94 are well related to an SST regime of cold North Atlantic conditions. The years of 1926–70, which were distinctly warm in the North Atlantic, correspond to active conditions for Atlantic hurricanes. There is an 80% increase in intense hurricanes during the warm North Atlantic decades compared with the cold North Atlantic years. These changes have direct impacts on the hurricanes which strike the Caribbean islands as well as the United States, especially intense hurricanes making landfall from the Florida peninsula to New England. The U.S. normalized damages between the two SST regimes are striking, with a factor of four increase in median damages during the warm North Atlantic decades.

The discrepancy between the interannual and interdecadal response of Atlantic tropical cyclones to Atlantic SSTs may be explained by the SSTs relationship to the other environmental controls. On the interdecadal timescale, Gray (1990) showed that there are long–term variations in the vertical wind shear amounts, which are in phase with the Atlantic SST mode: reduced vertical shear during the warm North Atlantic decades favoring more intense hurricane activity and increased vertical shear during the cold North Atlantic years inhibiting strong hurricanes. In contrast to this, the influence of Atlantic SSTs is occasionally not in phase with the overlying tropospheric circulation on the interannual timescale. A good example of this occurred in 1997 with very warm tropical North Atlantic SSTs, yet the intense hurricanes were much reduced from preceding years because of the strong vertical shear induced by the 1997 El Niño event (Bell and Halpert 1998).

The lack of a distinct multidecadal variation of intense hurricanes in the Gulf of Mexico is likely due to local conditions that dominate over these basinwide SST changes (Landsea et al. 1992). Since 1967 when satellite monitoring made it possible, all of the U.S. East Coast intense hurricanes were spawned from easterly waves. In contrast, baroclinically–initiated tropical cyclones (e.g. stationary frontal boundaries or upper–tropospheric cutoff lows) occasionally have developed into intense hurricanes that make landfall along the U.S. Gulf Coast. Hurricane Alicia, which struck the Texas coast in 1983, is a notable example of this latter phenomena. Additionally, vertical shear changes in the Gulf are not correlated highly with variations of ENSO or West Sahel rainfall, unlike the main development region (Goldenberg and Shapiro 1996).

The years of 1995 and 1996 showed at least a temporary return of warm North Atlantic SST conditions, and with it, a return to numerous intense hurricanes. An analysis is currently underway to evaluate whether this change is likely one of a multidecadal nature. If so, this holds the potential of providing a reliable multidecadal forecast of Atlantic hurricane activity.

Knowledge of future incidences of hurricanes holds the promise of economic benefits and enhanced response. Nevertheless, decision makers do not need to know the future with certainty to begin to stimulate improved responses to hurricanes, they need only to understand and appreciate the past. An understanding of trends in hurricane indices is an important step in that direction.

Acknowledgements

The authors wish to thank Joe Cione, David Enfield and Todd Kimberlain for their constructive comments on an earlier draft of this paper. We also greatly appreciate the discussions and interaction with Dr. Bill Gray, who has inspired much of this work. The fourth author thanks Tom Vonder Haar and Jim Purdom for time to work on this project. Partial financial support for the lead and third authors comes from NOAA's Climate and Global Change Program in Pan American Climate Studies (proposal number GC97-306). NCAR (second author) is sponsored by the National Science Foundation.

References

Barnston, A. G., Chelliah, M., and Goldenberg, S. B.: 1997, 'Documentation of a highly ENSO–related SST region in the equatorial Pacific', *Atmos.–Ocean* **35**, 367–383.

Bell, G. D., and Halpert, M. S.: 1998, 'Climate assessment for 1997', *Bull. Am. Meteorol. Soc.* **79**, S1–S50.

Birkland, T. A.: 1997, 'Factors inhibiting a national hurricane policy', *Coastal Management* **25**, 387–401.

Broecker, W. S.: 1991, 'The great ocean conveyor', *Oceanography* **4**, 79–89.

Chan, J. C. L.: 1985, 'Tropical cyclone activity in the Northwest Pacific in relation to the El Nino / Southern Oscillation phenomenon', *Mon. Wea. Rev.* **113**, 599–606.

DeMaria, M.: 1996, 'The effect of vertical shear on tropical cyclone intensity change', *J. Atmos. Sci.* **53**, 2076–2087.

Deser, C., and Blackmon, M. L.: 1993, 'Surface climatic variations over the North Atlantic during winter: 1900–1989', *J. Clim.* **6**, 1743–1753.

Elsberry, R. L., Holland, G. J., Gerrish, H., DeMaria, M., Guard, C. P., and Emanuel, K.: 1992, 'Is there any hope for tropical cyclone intensity prediction? – A panel discussion', *Bull. Am. Meteorol. Soc.* **73**, 264–275.

Elsner, J. B., Lehmiller, G. S., and Kimberlain, T. B.: 1996, 'Objective classification of Atlantic hurricanes', *J. Clim.* **9**, 2880–2889.

Emanuel, K. A.: 1986, 'An air–sea interaction theory for tropical cyclones. Part I: Steady–state maintenance', *J. Atmos. Sci.* **43**, 585–604.

Emanuel, K. A.: 1995, 'Comments on "Global climate change and tropical cyclones": Part I', *Bull. Am. Meteorol. Soc.* **76**, 2241–2243.

Enfield, D. B. and Mestas–Nuñez, A. M.: 1998, 'Multiscale variabilities in global sea surface temperatures and their relationships with tropospheric climate patterns', *J. Clim.* in press.

Fernández–Partagás, J., and Diaz, H. F.: 1996, 'Atlantic hurricanes in the second half of the 19th Century', *Bull. Am. Meteorol. Soc.* **77**, 2899–2906.

Folland, C. K., Palmer, T. N., and Parker, D. E.: 1986, 'Sahel rainfall and worldwide sea temperatures, 1901–1985', *Nature* **320**, 602–607

Goldenberg, S. B. and Shapiro, L. J.: 1996, 'Physical mechanisms for the association of El Niño and West African rainfall with Atlantic major hurricane activity', *J. Clim.* **9**, 1169–1187.

Gray, W. M.: 1968, 'Global view of the origins of tropical disturbances and storms', *Mon. Wea. Rev.* **96**, 669–700.

Gray, W. M.: 1984a, 'Atlantic seasonal hurricane frequency: Part I: El Niño and 30 mb quasi-biennial oscillation influences', *Mon. Wea. Rev.* **112**, 1649–1668.

Gray, W. M.: 1984b, 'Atlantic seasonal hurricane frequency: Part II: Forecasting its variability', *Mon. Wea. Rev.* **112**, 1669–1683.

Gray, W. M.: 1990, 'Strong association between West African rainfall and US landfall of intense hurricanes', *Science* **249**, 1251-1256.

Gray, W. M., Landsea, C. W., Mielke, Jr., P. W., and Berry, K. J.: 1992a, 'Predicting Atlantic seasonal hurricane activity 6–11 months in advance', *Wea. Forecasting* **7**, 440–455

Gray, W. M., Landsea, C. W., Mielke, Jr., P. W., and Berry, K. J.: 1993, 'Predicting Atlantic basin seasonal tropical cyclone activity by 1 August', *Wea. Forecasting* **8**, 73–86.

Gray, W. M., Landsea, C. W., Mielke, Jr., P. W., and Berry, K. J.: 1994, 'Predicting Atlantic basin seasonal tropical cyclone activity by 1 June', *Wea. Forecasting* **9**, 103–115.

Gray, W. M., Sheaffer, J. D., and Knaff, J. A.: 1992b, 'Influence of the stratospheric QBO on ENSO variability', *J. Meteor. Soc. Japan* **70**, 975–995.

Gray, W. M., Sheaffer, J. D., and Landsea, C. W.: 1997, 'Climate trends associated with multidecadal variability of Atlantic hurricane activity', in Diaz, H. F., and Pulwarty, R. S. (eds.), *Hurricanes, Climate and Socioeconomic Impacts*, Springer, Berlin, pp. 15–53.

Henderson–Sellers, A., Zhang, H., Berz, G., Emanuel, K., Gray, W., Landsea, C., Holland, G., Lighthill, J., Shieh, S–L., Webster, P., and McGuffie, K.: 1998, 'Tropical cyclones and global climate change: A post–IPCC assessment', *Bull. Am. Meteorol. Soc.* **79**, 19–38.

Hess, J. C., Elsner, J. B., and LaSeur, N. E.: 1995, 'Improving seasonal predictions for the Atlantic basin', *Wea. Forecasting* **10**, 425–432.

Holland, G. J.: 1993, 'Ready Reckoner – Chapter 9', *Global Guide to Tropical Cyclone Forecasting*, WMO/TC–No. 560, Report No. TCP–31, World Meteorological Organization, Geneva.

Holland, G. J.: 1997, 'The maximum potential intensity of tropical cyclones', *J. Atmos. Sci.* **54**, 2519–2541.

Houghton, J. T., Jenkins, G. J., and Ephramus, J. J. (eds.): 1990, *Climate Change: The IPCC Scientific Assessment*, Cambridge University Press, New York, 364 pp.

Houghton, J. T., Callander, B. A., and Varney, S. K. (eds.): 1992, *Climate Change 1992: The Supplementary Report to the IPCC Scientific Assessment*, Cambridge University Press, New York, 198 pp.

Houghton, J. T., Meira Filho, L. G., Callander, B. A., Harris, N., Kattenberg, A., and Maskell, K. (eds.): 1996, *Climate Change 1995: The Science of Climate Change*, Contribution of WGI to the Second Assessment Report of the Intergovernmental Panel on Climate Change. Cambridge University Press, New York, 572 pp.

Jarrell, J. D., Hebert, P. J., and Mayfield, M.: 1992, 'Hurricane experience levels of coastal county populations from Texas to Maine', *NOAA Tech. Memo.* NWS NHC 46, Coral Gables, Florida, 152 pp.

Jarvinen, B. R., Neumann, C. J., and Davis, M. A. S.: 1984, 'A tropical cyclone data tape for the North Atlantic Basin, 1886–1983: Contents, limitations, and uses', *NOAA Tech. Memo.* NWS NHC 22, Coral Gables, Florida, 21 pp.

Kaplan, A., Cane, M. A., Kushnir, Y., Clement, A. C., Blumenthal, M. B., and Rajagopalan, B.: 1998, 'Analysis of global sea surface temperatures 1856–1991', *J. Geophys. Res.* **103**, 18,567–18,589.

Knaff, J. A.: 1993, 'Evidence of a stratospheric QBO modulation of tropical convection', Atmospheric Science Paper No. 520, Colorado State University, Ft. Collins, CO, 80523, 91 pp.

Knaff, J. A.: 1997, 'Implications of summertime sea level pressure anomalies in the tropical Atlantic region', *Mon. Wea. Rev.* **125**, 789–804.

Knutson, T. R., Tuleya, R. E., and Kurihara, Y.: 1998, 'Simulated increase of hurricane intensities in a CO_2–warmed climate', *Science* **279**, 1018–1020.

Kushnir, Y.: 1994, 'Interdecadal variations in North Atlantic sea surface temperature and associated atmospheric conditions', *J. Clim.* **7**, 141–157.

Lander, M.: 1994, 'An exploratory analysis of the relationship between tropical storm formation in the Western North Pacific and ENSO', *Mon. Wea. Rev.* **122**, 636–651.

Landsea, C. W.: 1993, 'A climatology of intense (or major) Atlantic hurricanes', *Mon. Wea. Rev.* **121**, 1703–1713.

Landsea, C. W.: 1998, 'Climate Variability of Tropical Cyclones: Past, Present and Future', in Pielke, Sr., R. A., and Pielke, Jr., R. A. (eds.), *Storms*, Routledge Press, in press.

Landsea, C. W., and Gray, W. M.: 1992, 'The strong association between Western Sahel monsoon rainfall and intense Atlantic hurricanes. *J. Clim.* **5**, 435–453.

Landsea, C. W., Gray, W. M., Mielke, Jr., P. W., and Berry, K. J.: 1992, 'Long–term variations of Western Sahelian monsoon rainfall and intense U.S. landfalling hurricanes', *J. Clim.* **5**, 1528–1534.

Landsea, C. W., Nicholls, N., Gray, W. M., and Avila, L. A.: 1996, 'Downward trends in the frequency of intense Atlantic hurricanes during the past five decades', *Geo. Res. Lett.* **23**, 1697–1700.

Landsea, C. W., Bell, G. D., Gray, W. M., and Goldenberg, S. B.: 1998, 'The extremely active 1995 Atlantic hurricane season: Environmental conditions and verification of seasonal forecasts', *Mon. Wea. Rev.* **126**, 1174–1193.

Lehmiller, G. S., Kimberlain, T. B., and Elsner, J. B.: 1997, 'Seasonal prediction models for North Atlantic basin hurricane location', *Mon. Wea. Rev.* **125**, 1780–1791.

Lighthill, J., Holland, G., Gray, W., Landsea, C., Craig, G., Evans, J., Kurihara, Y., and Guard, C.: 1994, 'Global climate change and tropical cyclones', *Bull. Am. Meteorol. Soc.* **75**, 2147–2157.

Ludlum, D. M.: 1989, *Early American Hurricanes 1492–1870*. Lancaster Press, Inc., Lancaster, Pennsylvania, 198 pp.

Malkus, J. S., and Riehl, H.: 1960, 'On the dynamics and energy transformations in steady-state hurricanes', *Tellus* **12**, 1–20.

Molinari, J., Knight, D., Dickinson, M., Vollaro, D., and Skubis, S.: 1997, 'Potential vorticity, easterly waves, and Eastern Pacific tropical cyclogenesis', *Mon. Wea. Rev.* **125**, 2699–2708.

Neumann, C. J., Jarvinen, B. R., McAdie, C. J., and Elms, J. D.: 1993, *Tropical cyclones of the North Atlantic Ocean, 1871–1992*, National Climatic Data Center in cooperation with the National Hurricane Center, Coral Gables, FL, 193 pp.

Neumann, C. J., and McAdie, C. J.: 1997, 'The Atlantic tropical cyclone file: A critical need for a revision', *Preprints, 22nd Conference on Hurricanes and Tropical Meteorology*, American Meteorological Society, Boston, pp. 401-402.

National Oceanic and Atmospheric Administration (NOAA): 1997, *Tropical Cyclone Program*, WSOM Chapter–41, W/OM12. NOAA, Washington, D. C., 59 pp.

Nicholls, N.: 1979, 'A possible method for predicting seasonal tropical cyclone activity in the Australian region', *Mon. Wea. Rev.* **107**, 1221–1224.

Pasch, R. J., Avila, L. A., and Jiing, J.-G.: 1998, 'Atlantic tropical systems of 1994 and 1995: A comparison of a quiet season to a near–record–breaking one', *Mon. Wea. Rev.* **126**, 1106–1123.

Philander, S. G. H.: 1989, *El Niño, La Niña, and the Southern Oscillation*, Academic Press, New York, 293 pp.

Pielke Jr., R. A.: 1997, 'Reframing the U.S. Hurricane Problem', *Society and Natural Resources* **10**, 485–499.

Pielke, Jr., R. A.: 1998, 'Rethinking the role of adaptation in climate policy', *Global Environmental Change*, in press.

Pielke, Jr. R. A., and Landsea, C. W.: 1998, 'Normalized Atlantic hurricane damage, 1925–1995', *Wea. Forecasting* **13**, 621–631.

Pielke, Jr., R. A., and Pielke, Sr., R. A.: 1997a, *Hurricanes: Their Nature and Impacts on Society*, John Wiley and Sons Press, London, 279 pp.

Pielke Jr., R. A., and Pielke Sr., R. A.: 1997b, 'Societal Vulnerability to Hurricanes: Considerations on the Use of Long-Range Forecasts', in Diaz, H. F., and Pulwarty, R. S. (eds.), *Hurricanes, Climatic Change and Socioeconomic Impacts*, Springer, Berlin, pp. 147–184.

Raper, S.: 1992, 'Observational data on the relationships between climate change and the frequency and magnitude of severe tropical storms', in Warrick, R. A., Barrow, E. M., and T. M. L. Wigley (eds.), *Climate and sea level change: Observations, projections and implications*, Cambridge University Press, pp. 192–212.

Reading, A. J.: 1989, 'Caribbean tropical storm activity over the past few centuries', *Int. J. Climatology* **10**, 365–376.

Reed, R. J.: 1988, 'On understanding the meteorological causes of Sahelian drought', *Pontificiae Academiae Scientarvm Scripta Varia* **69**, 179–213.

Revell, C. G., and Goulter, S. W.: 1986, 'South Pacific tropical cyclones and the Southern Oscillation', *Mon. Wea. Rev.* **114**, 1138–1145.

Reynolds, R. W., and Smith, T. M.: 1994, 'Improved global sea surface temperature analyses using optimum interpolation', *J. Clim.* **7**, 929–948.

Reid, G. C., Gage, K. S., and McAfee, J. R.: 1989, 'The thermal response of the tropical atmosphere to variations in equatorial Pacific sea surface temperature', *J. Geophys. Res.* **94**, 14,705–14,716.

Saunders, M. A., and Harris, A. R.: 1997, 'Statistical evidence links exceptional 1995 Atlantic hurricane season to record sea warming', *Geo. Res. Lett.* **24**, 1255–1258.

Shapiro, L. J.: 1982, 'Hurricane climatic fluctuations. Part II: Relation to large–scale circulation', *Mon. Wea. Rev.* **110**, 1014-1023.

Shapiro, L. J.: 1987, 'Month–to–month variability of the Atlantic tropical circulation and its relationship to tropical storm formation', *Mon. Wea. Rev.* **115**, 2598-2614.

Shapiro, L. J.: 1989, 'The relationship of the quasi–biennial oscillation to Atlantic tropical storm activity', *Mon. Wea. Rev.* **117**, 2598-2614.

Shapiro, L. J., and Goldenberg, S. B.: 1998, 'Atlantic sea surface temperatures and tropical cyclone formation', *J. Clim.* **11**, 578–590.

Shea, D. J., Worley, S. J., Stern, I. R., and Hoar, T. J.: 1994, 'An introduction to atmospheric and oceanographic data', National Center for Atmospheric Research Tech. Note 404, 136 pp. [Available from Climate and Global Dynamics Division, National Center for Atmospheric Research, Boulder, CO 80307.]

Simpson, R. H.: 1974, 'The hurricane disaster potential scale', *Weatherwise* **27**, 169 and 186.

Simpson, R. H.: 1998, 'Stepping stones in the evolution of a national hurricane policy', *Wea. Forecasting* **13**, 617–620.

Vose, R. S., Schmoyer, R. L., Steuer, P. M., Peterson, T. C., Heim, R.,. Karl, T. R., and Eisheid, J. K.: 1992, 'The global historical climatology network: Long-term monthly temperature, precipitation, sea level pressure, and station pressure data', Carbon Dioxide Information Analysis Center Publ. 3912, 189 pp. [Available form Carbon Dioxide Information and Analysis Center, Oak Ridge National Laboratory, Oak Ridge, TN 37831.]

Wallace, J. M.: 1973, 'General circulation of the tropical lower stratosphere', *Rev. Geophys. Space Phys.* **11**, 191–222.

Watson, R. T., Zinyoweya, M. C., and Moss, R. H. (eds.): 1996, *Climate Change 1995: Impacts, Adaptations and Mitigation of Climate Change: Scientific–Technical Analyses*, Contribution of WGIII to the Second Assessment Report of the Intergovernmental Panel on Climate Change. Cambridge University Press, New York, 879 pp.

(Received 14 September 1998; in revised form 18 December 1998)

THE USE OF INDICES TO IDENTIFY CHANGES IN CLIMATIC EXTREMES

P.D. JONES[1], E.B. HORTON[2], C.K. FOLLAND[2], M. HULME[1], D.E. PARKER[2] and T.A. BASNETT[2]

[1]Climatic Research Unit, University of East Anglia, Norwich, NR4 7TJ, U.K.
[2]Hadley Centre, Meteorological Office, Bracknell, RG12 2SY, U.K.

Abstract. Changes in the frequencies of extremes are investigated by a variety of methods using daily temperature data from the British Isles, and monthly 5° latitude x 5° longitude grid-box temperatures over the land and marine regions of the world. The 225 year long daily Central England Temperature record shows no significant increase in very warm days in recent years but there is a marked decrease in the frequency of very cold days. Thus the rise in temperature in the last two decades is principally associated with a reduction in very cold days. Temperatures on days with particular wind circulation or pressure pattern types over the British Isles show multidecadal variations. Analyses using monthly gridded temperature data around the world since 1951 indicate that the recent rise in global surface temperatures is accompanied both by reductions in the areas affected by extremely cool temperatures and by increases in the areas with extremely warm temperatures.

1. Introduction

Conventional readily-available climatological databases (Hulme, 1994; Jones, 1994, Peterson and Vose, 1997 and New et al., 1998) generally archive time series of monthly averages or totals. Recent improvements have extended the archives from monthly mean temperatures, sea level pressures and total precipitation to include monthly mean maximum and minimum temperatures. Monthly data are useful for monitoring changes in extremes and variability on the same or longer timescales. Such timescales are also adequate for the study of droughts. However, for most other societally-sensitive extremes and related changes in their variability (e.g. record temperatures, storms/winds, heavy rain, floods, day-to-day temperature changes) daily data are essential.

Long time series of daily data of comparable length to the available monthly series have generally been put together for specific regions or research projects (e.g. Parker et al., 1992, for Central England Temperature from 1772; Razuvayev et al., 1993, 223 stations for the former Soviet Union). Such series will generally be considerably less extensive compared to the 'monthly' data, less likely to be available in real time and more likely to be subject to commercial pricing considerations. Availability of the required daily data, therefore, varies greatly from region to region. For some regions, especially Africa and parts of southeast Asia, historical daily data may only exist in manuscript form.

Climatic Change **42**: 131–149, 1999.

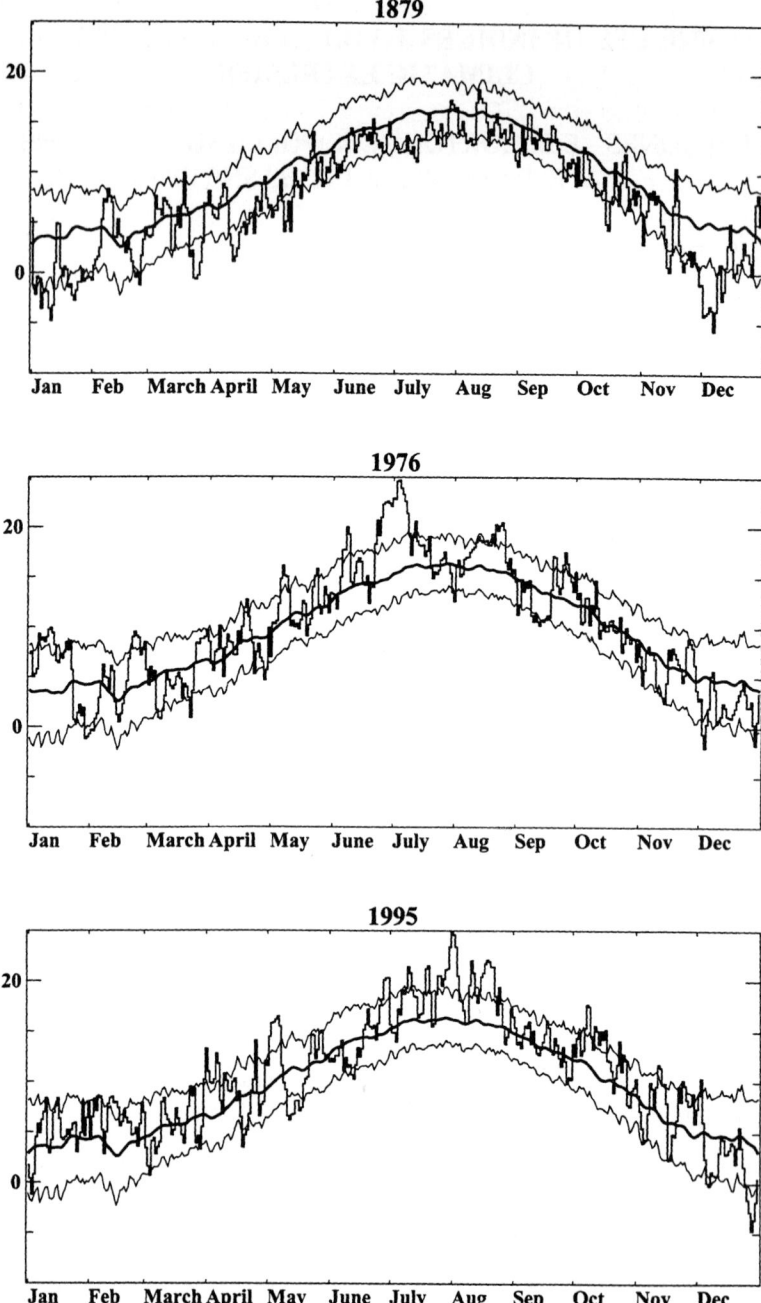

Figure 1. Unsmoothed daily Central England temperature for 1879, 1976 and 1995. Also shown is the average daily temperature for 1961 to 1990, calculated as a centred 11 term binomially filtered average of the 1961-90 daily mean with half power at 8.5 calendar days. The thin solid lines are similar calculations of the 10th and 90th percentiles estimated using gamma distributions as described later in section 2.1.

The lack of a consistent and globally extensive daily database is a serious limitation to the study of the changing frequency of extremes. Synoptic messages from 1977-1991 (NCDC, 1994) provide an all too brief climate database of daily data, and these data are difficult to quality control. Homogeneity of the daily climatic database is another major issue. Whilst a number of subjective and relatively objective criteria (e.g. Jones et al., 1986, Easterling and Peterson, 1995 and Alexandersson and Moberg, 1997) exist for testing monthly data, all would agree that daily data entail another magnitude of complexity and that monthly-based corrections applied to daily data would only be a first guess. Note that even the "objective" criteria generally require considerable subjective evaluation.

In this paper we consider two issues related to the changing frequency of extremes. First we assess what can be gleaned about extremes from monthly and daily temperature and air pressure data, using examples from the British Isles where long and homogeneous records of monthly and daily climatic records exist. Secondly, we analyse extremes at the monthly scale using a global surface temperature data set developed on a 5°x 5° grid-box basis from land-based station and sea surface temperature anomalies (Jones, 1994 and Parker et al., 1995).

2. British Isles

2.1. TEMPERATURE

This region has both the longest monthly (1659→) and daily (1772→) temperature record of any region in the world. The record is known as the 'Central England' temperature (CET) series and details of its construction are given in Manley (1974) and Parker et al. (1992). Careful quality control has been performed on these data with a correction for urbanization. Since 1878, the daily series is the average of three stations, but for previous years it is based on a single site that varies in position with time. So the day-to-day variance of the single-site part of the series was appropriately reduced to enable absolute comparisons of variance to be made across the complete time series (Parker et al., 1992). Figure 1 shows three of the years with extreme temperature spells (1879, 1976 and 1995). Notably cold and warm days or spells need not be characteristic of the season as a whole (see Figure 1) and may not be reflected in the average seasonal or monthly temperature. Extreme 30-day periods that straddle months will not register as exceptional in a calendar monthly series and yet may have just as much impact.

Both CET series have been quite extensively studied (e.g. Parker et al., 1992; Jones and Hulme, 1997 and Hulme, 1997), in particular with regard to relationships between extreme features of winter or summer and average winter/summer conditions (see Table 1). Figures 2a,b show the various series listed in Table 1 from 1772. The choices of temperature thresholds of 0°C in winter and

20°C in summer are somewhat arbitrary, but 0°C is about 4.5°C below the average winter temperature and 20°C is about 4.5°C above the average summer temperature. Correlations between counts of cold/warm days and average seasonal temperatures are higher than between degree day counts below/above the same thresholds and average seasonal temperatures. Correlations between both cold/warm day counts and degree day counts and seasonal averages are higher in winter than in summer. This implies that mean winter temperatures are a better index of extremes than are mean summer temperatures. Whilst mean temperature/extreme day count correlations are higher in winter, the temperature of the winter is not always well related to its 'snowiness' (see Table 1). The 'snowiness' indices are based on the subjective assessments of L.C.W. Bonacina (e.g. Jackson, 1976) and are discussed in Jones et al. (1997a).

Table 1

Correlations between various measures of winter severity and summer heat and average winter (DJF) and summer (JJA) temperatures.

Winters (dated by January)

Measure	Period	Correlation with temperature
Days below 0°C	1772-1997	-0.90
Degree days below 0°C	1772-1997	-0.84
Winter snow index[1]	1876-1997	-0.65
Summers		
Days greater than 20°C	1772-1997	0.73
Degree days above 20°C	1772-1997	0.63

[1]For definition see Jones et al. 1997a.

Hot summer months need not always be associated with extremely high temperatures (e.g. July 1983) and a few extremes (e.g. the warmest day in the CET record on 29 July 1948) need not lead to a warm month. Two summers, 1976 and 1995, clearly stand out in Figure 2b. They were also the driest summers in the England and Wales precipitation series (Jones and Conway, 1997) which extends back to 1766 and the second and third sunniest summers since 1909 (Hulme, 1997). The combination of dry air and clear skies in a brief northerly airflow led to both ground and air frosts in eastern England in early August 1976, a surprising occurrence in the eleventh warmest ever August, but evident in Figure 1. Hot summers are not necessarily dry. The most recent summer analysed here in 1997

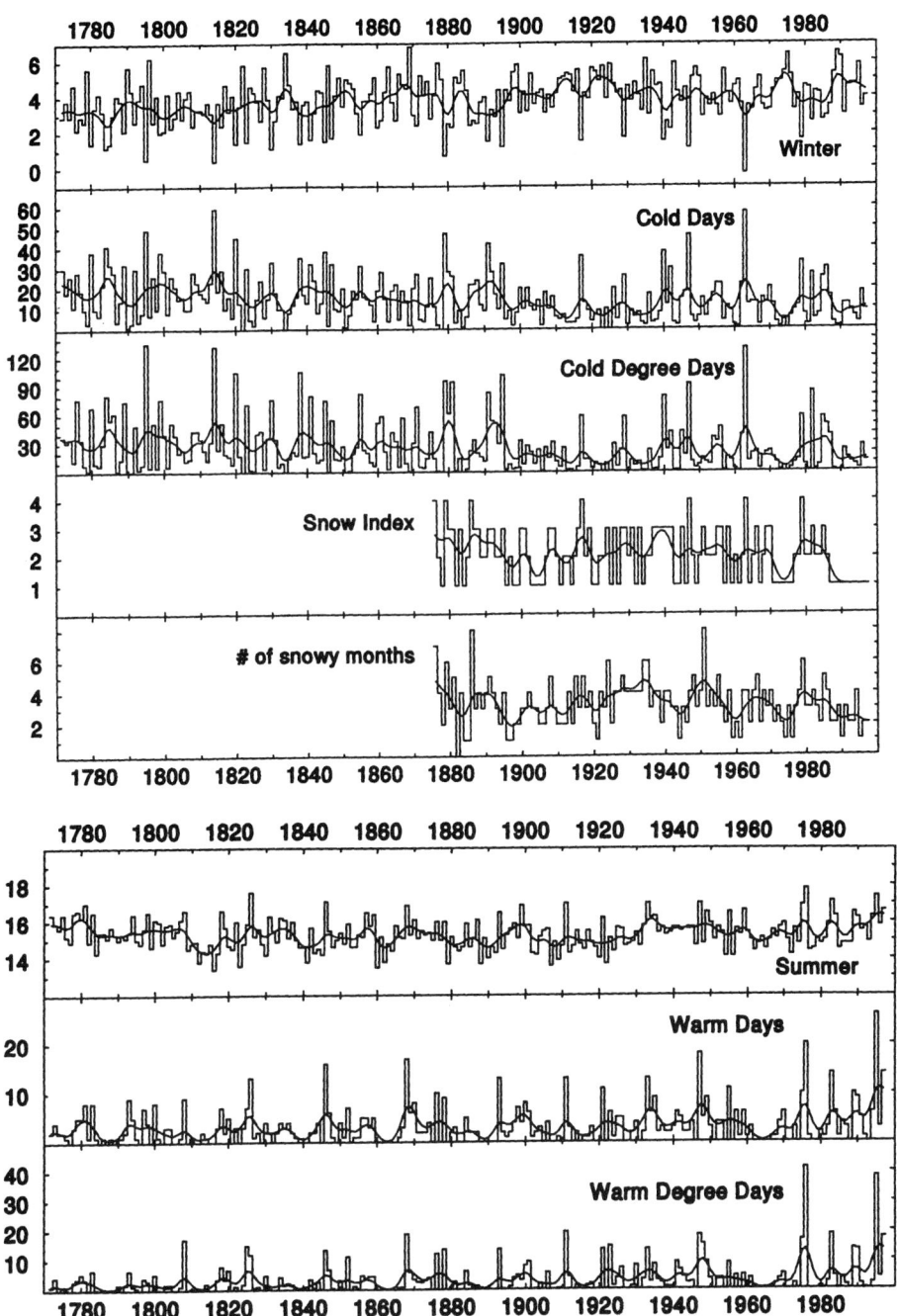

Figure 2. a) Central England temperatures for winters (DJF), numbers of days with daily mean below 0°C, degree days below 0°C, and winter snowiness indices. For record lengths see Table 1. b) Central England temperatures for summers (JJA), numbers of days and degree days above 20°C. The smooth line highlights variations on the decadal timescale using a 10 term Gaussian filter.

was among the warmest, with 14 days above 20°C, but was also wetter than average.

Another method of assessing extremes in the CET series is to use seasonally varying thresholds rather than the fixed temperature thresholds used in Figure 2 and Table 1. In this method, both the annual cycle of temperature and changes in variability during the year are removed from the basic data by transforming each day's temperature to percentiles using the 3-parameter gamma function (Horton et al., 1998).

As a first step, an average daily temperature value for each day of the year is derived from the period 1961-90. Our first guess at this average for 1st January is the mean value of the thirty January 1st values available. Values for February 29th in non leap years are estimated by calculating the mean of February 28th and March 1st. The 366 values thus calculated do not produce a smooth annual cycle of daily means, because 30 years will not be long enough to remove random interannual variations from the result. So a more realistic series can be produced by smoothing these calendar daily means. However, the smoothing is not carried out over too long a calendar period so as not to bias the new estimate. We use a centred 11 term binomial filter with a half power at about 8.5 days, to include significant information for approximately the week centred on the nominal date. This smoothed daily mean is shown in all plots in Figure 1 as the thicker line.

In the second step, a 3 parameter gamma distribution is fitted, using maximum likelihood methods to the daily anomaly values relative to the smoothed daily mean. The gamma distribution for a given calendar day is calculated from 150 anomalies for five equally spaced days centred on the specific calendar day. The days are spaced at five day intervals to provide a set of 150 almost independent anomalies. This gives a more robust fit with little change in the shape of the distribution between successive calendar days. The distribution of anomaly values is significantly negatively skewed during most of the year. This is the main reason for using a 3 parameter gamma distribution (Horton et al., 1998). The gamma distribution procedure transforms each day's temperature anomaly to a percentile or return period, thus expressing how unusual the temperature is relative to that expected over the period 1961-90. Examples of the 90th and 10th percentile fits are shown on Figure 1 as the thin lines.

Figure 3 shows time series from 1772 to 1997 of the annual counts of days above the 90th percentile and days below the 10th percentile. The >90th (<10th) percentile counts are strongly positively (negatively) correlated with mean annual temperature (not shown). 1779 has the highest number of >90th percentile days (87) while 1814 has the highest number of days below the 10th percentile value (117). The increase in mean temperature in recent decades has not been accompanied by a significant increase in >90th percentile days (solid curve). In contrast, a marked decline in the number of days <10th percentile value has occurred since the late 19th century. However some caution is needed when

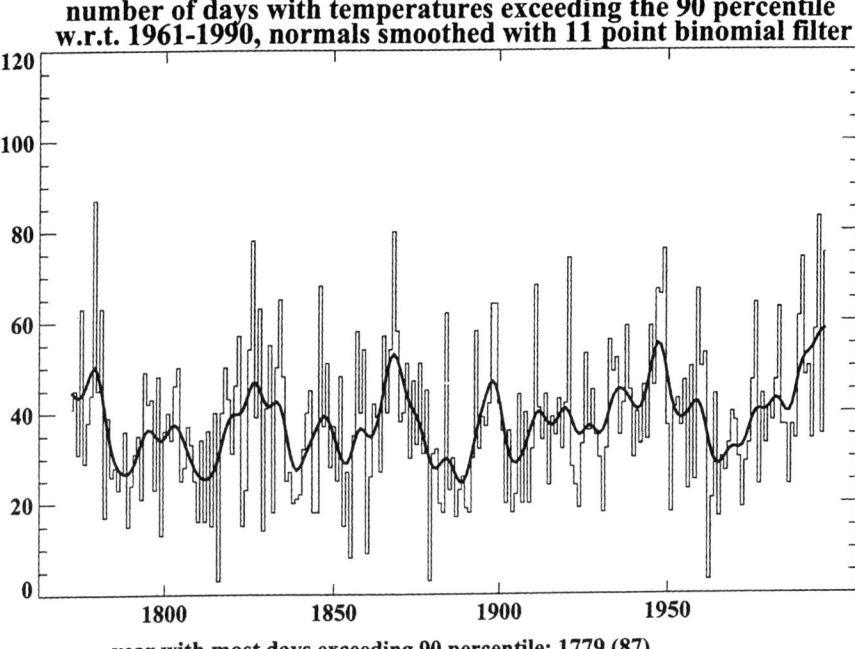

year with most days exceeding 90 percentile: 1779 (87)

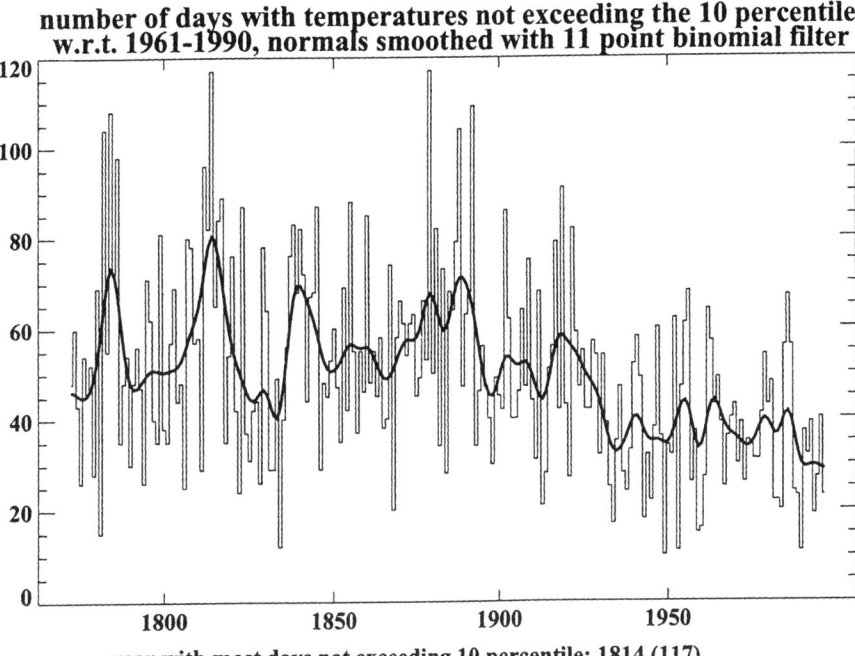

year with most days not exceeding 10 percentile: 1814 (117)

Figure 3. Counts of days each year, 1772-1997, in CET above/below percentile thresholds a) above the 90th and b) below the 10th. Also shown is a 21 point binomial filtered average (half power near 10 years).

drawing conclusions from data before 1880; the use of different types of screen before then (or no screen at all before the mid 19th century) may influence these results, probably by increasing the likelihood of large extremes (Parker, 1994).

Bearing this problem in mind, addition of the two series in Figure 3 provides an extremes index (Figure 4). This can be used to assess whether extremes as a whole have changed in frequency relative to the 1961-90 period. Overall, the frequency of extremes reduces through time. Years with high values in Figure 4 may be due to warmth (coldness) throughout the year as in 1779 (1814) or years when extremes of both sign occurred but cancelled each other in the annual mean temperature value. For example, in 1947 there were 127 days with extreme warm or cold temperatures, but the annual mean temperature anomaly was only 0.14°C. The higher numbers of extremes, relative to 1961-90, before the 1930s is principally due to cooler temperatures in these years and hence greater numbers of very cool days (see Figure 3).

The results in Figures 3 and 4 are moderately insensitive to the choice of base period. Fairly similar results are achieved using the 1931-60 base period. This is not surprising as the mean annual temperatures for these two periods are very similar, 9.6 (1931-60) and 9.5 (1961-90). When the extreme index was recalculated relative to the much colder period of 1881-1910 there was also little change in the average number of days with extreme temperatures with time.

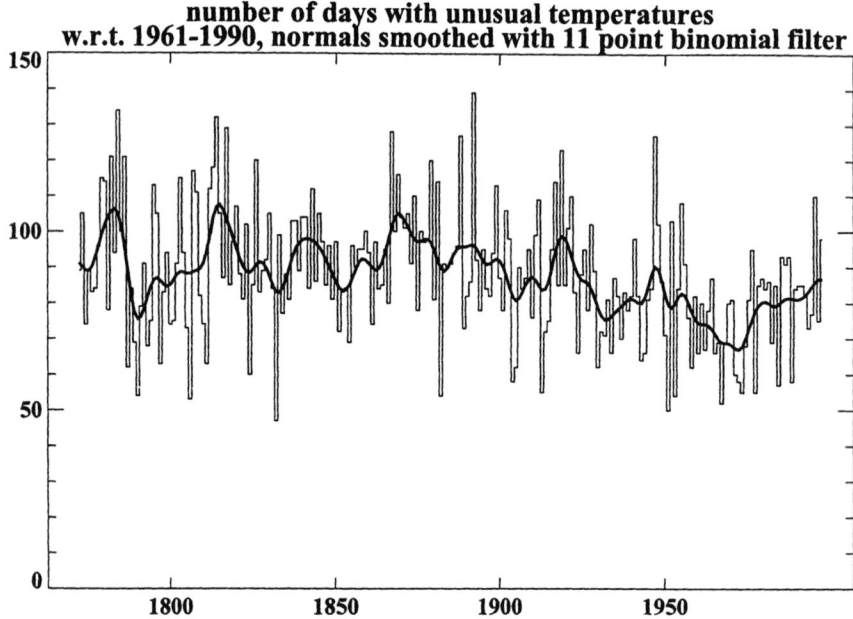

Figure 4. CET extreme index defined as the sum of the number of days above the 90th and number of days below the 10th percentiles, 1772-1997. The smooth line is a 21 point binomial filtered average (half power near 10 years).

2.2. CIRCULATION TYPING AND GALE FREQUENCY

Other important extreme phenomena include very strong winds, which in the extratropical North Atlantic are generally related to deep baroclinic depressions. Estimates of changes in the occurrence of such events from anemometer records suffer from numerous site and instrumental-change problems. This makes the development of long homogeneous wind time series difficult. An alternative method of calculating time series of gale statistics is to use grid-point mean-sea-level pressure (MSLP) data. The sources of the daily 5° latitude by 10° longitude MSLP grid-point data used here are described in Jones (1987). The data have been adjusted so that the monthly average MSLP values agree with a recent version of the Global MSLP data set, updated from that in Allan et al. (1996). We use the adjusted daily MSLP to calculate the strength of the geostrophic wind flow (F) and the vorticity (Z) (see Jenkinson and Collison, 1977, and Hulme and Jones, 1991), yielding a gale index of

$$G = (F^2 + (0.5Z)^2)^{\frac{1}{2}} \tag{1}$$

While the relatively coarse grid of 5° latitude by 10° longitude and a daily sampling interval might not resolve all small intense cyclonic systems, such features are not very typical in these latitudes, the storm of 16 October 1987 being a notable exception. Smith (1982) used a gale index based solely on F, developing regression equations to reconstruct wind speeds before modern anemometer instruments became available around 1960. The incorporation of the vorticity term in the gale index takes some account of subgrid-scale variability and ageostrophic flow, and so improves the usefulness of the index on the scales of the British Isles.

Figure 5 shows the annual number of severe gales for each year over the United Kingdom for the period 1881 to 1997. A severe gale day is defined for most years as having G=>40. However, owing to the varying methods of chart construction, the threshold value of G is reduced to 37 for 1899-1939 and 1949-59 and to 32 for 1960-65 to maintain long-term homogeneity (see Hulme and Jones, 1991, for details). Around 1990 the frequency of severe gales reached a higher level, averaged over a decade, than that recorded previously, though individual years at earlier times give the maximum values of G in the time series (e.g. 1887, 1916). The high recent level of gale activity over the United Kingdom, and in adjacent regions of similar latitude, is related to the very strong westerly phase of the North Atlantic Oscillation during the late 1980s and in the early 1990s (Jones et al., 1997b).

The use of a grid-based method of estimating changes in gale frequency enables equivalent indices to be derived from the control runs of GCMs. These are rarely validated at this level of regional detail, Hulme et al. (1993) being an exception.

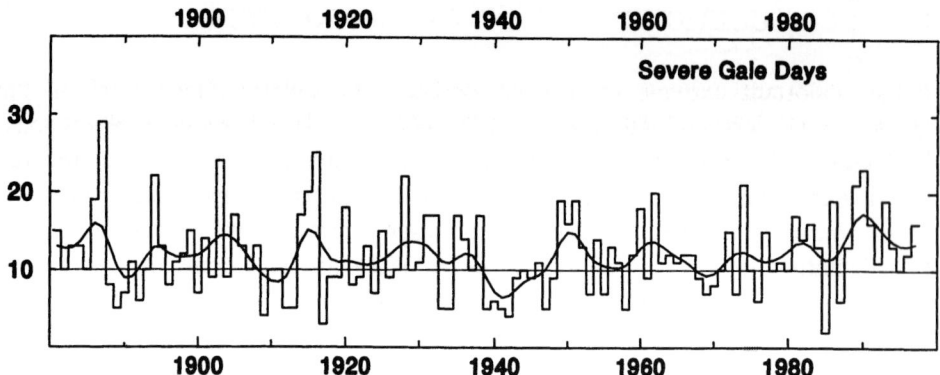

Figure 5. The number of severe gales over the United Kingdom for each year 1881 to 1997. The smooth line highlights decadal timescale variations.

That analysis showed that both the GCMs analysed underestimated gale frequency by up to 50%, particularly during the winter season.

Jenkinson and Collison (1977) also used daily mean MSLP data to classify daily atmospheric circulation objectively (see also Jones et al., 1993) to match the subjectively-derived Lamb weather types (Lamb, 1972). In the subjective Lamb typing scheme, a strong decline in the number of westerly days (the dominant wind direction) was highlighted by Lamb (e.g. Lamb, 1972). With the objective classification scheme of Jenkinson and Collison (1977), however, there is no evidence of a long-term decline (see Figure 6a) though multi-decadal fluctuations are readily apparent. The objective classification scheme has higher numbers of southerly days, particularly in autumn and winter (see Jones et al., 1993). Keeping these uncertainties in mind, circulation classification schemes can be used together with temperature and precipitation series to assess whether overall trends, for example to warmer/drier conditions, are due to changes in the frequency of certain circulation types or to changes of temperature/precipitation for specified weather types. For example, a cooling in April CET between 1961-1987 appears to be due to an increase in blocking and thus more northeasterly winds. Conversely, on days throughout the year with generally westerly flow types over UK, temperatures in the Scilly Isles (near 50°N, 6°W) are correlated well with SST anomalies in the

North Atlantic (Fig 6b) over five year averages (adapted from Parker and Folland, 1988). This diagram is striking in showing the existence of appreciable multidecadal variations in the mean temperature of days with specific circulation types over the UK. It is the kind of "within circulation type" diagnostic that should be extended to studies of extremes.

The two air flow characteristics, F and Z also have considerable potential for use in GCM downscaling (Conway et al., 1996). However, calculations over diverse regions of Europe (Conway and Jones, 1996) for a run of the Hadley Centre (HADCM2) model with projected increases in greenhouse gases and changes in anthropogenic sulphate aerosols indicate no change in either F or Z, and hence no increase in gale frequency over Europe. Model projections of changes in storminess on larger scales such as the North Atlantic are beginning to be assessed (e.g. Carnell et al., 1996).

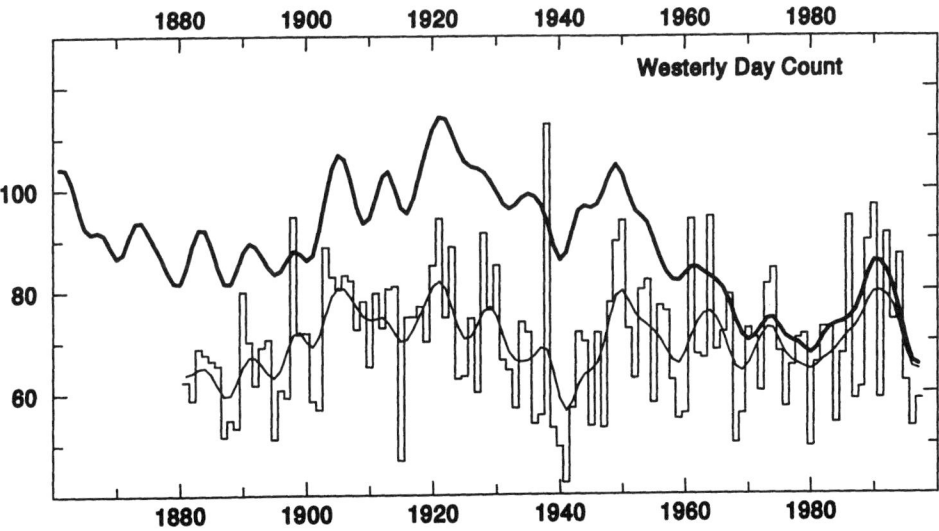

Figure 6a. Annual time series of the number of westerly days over the United Kingdom calculated from MSLP data using the objective (Jenkinson and Collison (1987) classification for the period 1881 to 1997. The thin smooth line highlights decadal timescale variations while the thicker smooth line is from the Lamb (1972) subjective classification for 1861 to 1996.

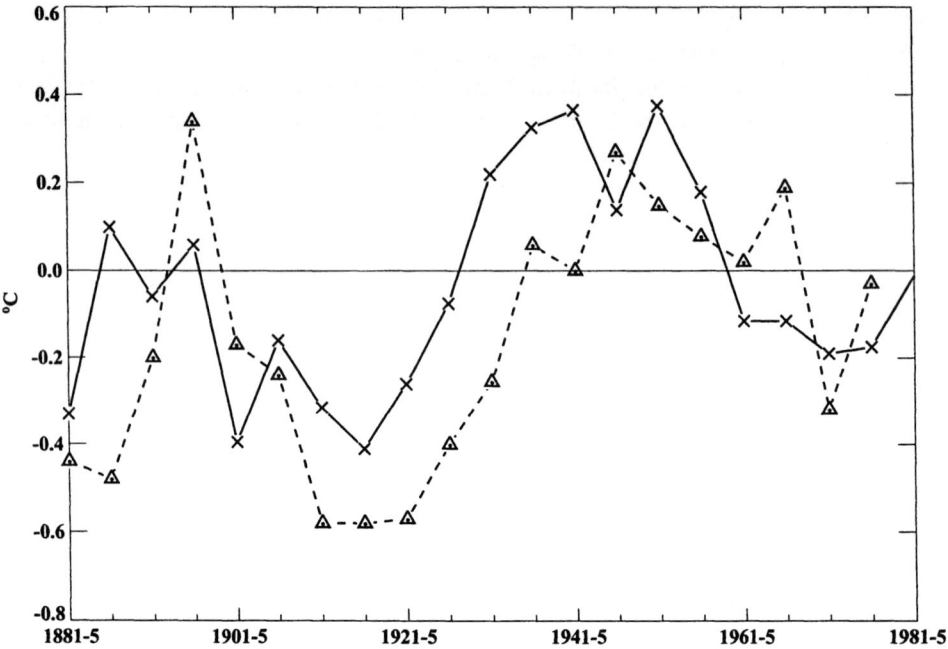

Figure 6b. Five year averaged Scilly Isles (approximately 50°N, 6°W) temperature anomalies (from a 1951-80 average) on days with Lamb's North Westerly, South Westerly, Cyclonic South Westerly, Westerly and Cyclonic Westerly atmospheric circulation types over the UK, excluding isolated days and the first day of a spell (solid line). Five year averaged SST anomalies for the North Atlantic north of 35°N (dashed line), calculated by weighting the constituent seasonal values by the total number of accepted days with the above circulation types. Both series extend from 1881-1885 to 1981-1985. The correlation between them over this period is 0.59.

3. Global Temperature

In this analysis, we use monthly 5°x 5° grid-box temperature anomalies derived from land surface air (Jones, 1994) and sea surface (Parker et al., 1995) temperatures. The gamma function distribution method used for the CET analysis enables each monthly, seasonal or annual grid-box temperature to be transformed to a percentile. This allows a direct comparison of the rarity of each grid-box value with respect to distributions based on 1961-90 averages. Interpretation of the results of such an analysis should be made with care. Each grid-box time series is likely to be composed, over time, from either a variable number of station series for land areas or from a varying number of individual sea surface temperature observations over the oceans. Correction for changes in the variance of the grid-box time series is possible using the methods discussed in Jones et al. (1997c) but has not been attempted here. For this reason, we restrict the analyses to the period 1951-97, when changes in station and observation densities are likely to be minimal.

Figure 7 shows the annual percentile map for 1997 together with the more usual anomaly form for comparison. Large parts of the eastern Pacific and the northeast Atlantic were warmer than the 98th percentile. Land regions with extreme warmth included southwestern Europe and parts of central Asia and China (Parker et al., 1998). 17% of the monitored area of the globe exceeded this percentile in 1997, showing that recent "global warming" is widespread.

Figure 8 shows a time series for 1951-97 of the global areas warmer than the 90% value and colder than the 10% value. These series can be calculated in several different ways of which we have used two: a) fitting gamma distributions to annual average temperatures (as done in Figure 7) or b) fitting gamma distributions to the monthly temperatures and averaging the resulting areas warmer than the 90% value or colder than the 10% value for each of the 12 months. The two different methods of calculation give different results. Method a) will stress lower frequency anomalous temperatures influencing a region on the annual timescale while b) highlights higher frequency anomalies (and, in addition, any appreciable random errors or very unrepresentative, sparse data). Thus an extremely variable year globally, but whose anomalies cancel when averaged to a whole year, will be reflected in relatively large areas covered by extreme percentiles in method (b) but a considerably smaller area covered in method a). Conversely, a year where the individual months are not extreme but most tend to be warm or cold will give a larger area of extreme percentiles covered by method a.

Figure 8 indicates that both methods show increases in the monitored areas of the world having warm extremes and decreases in areas with cool extremes. However, calculation of the area index using annual data (method a) leads to greater long-term variation in the areas covered by extreme percentiles than using the average of monthly calculated areas (method b). This implies that warming on the truly annual timescale is very widespread.

In a similar manner to calculations for CET (Figure 4), the sum of both warm and cold extremes provides an extremes index (Figure 9). This shows slightly higher levels of extremes before 1960 and after the mid-1980s. The muted variations in the extremes index relative to its components in Fig 8 reflect the fact that cold extremes are decreasing at the same time as warm extremes have increased. Thus temperatures worldwide do not show any appreciable tendency to become more variable in very recent decades but there is a marked tendency for an increased frequency of warm extremes which mainly reflects a general averaged warming. A slightly different conclusion might have resulted if the extremes were

Surface Temperature Anomaly Percentiles (w.r.t. 1961-90)
1997
(Anomalies fitted to Gamma Distributions)

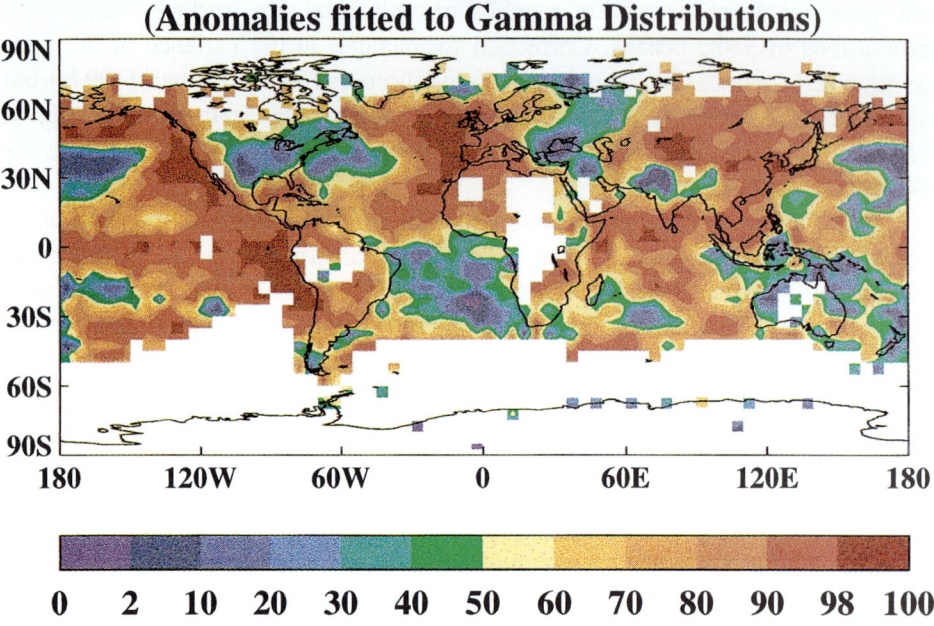

0   2   10   20   30   40   50   60   70   80   90   98   100

Surface Temperature Anomalies (^{O}C, w.r.t. 1961-90)
1997

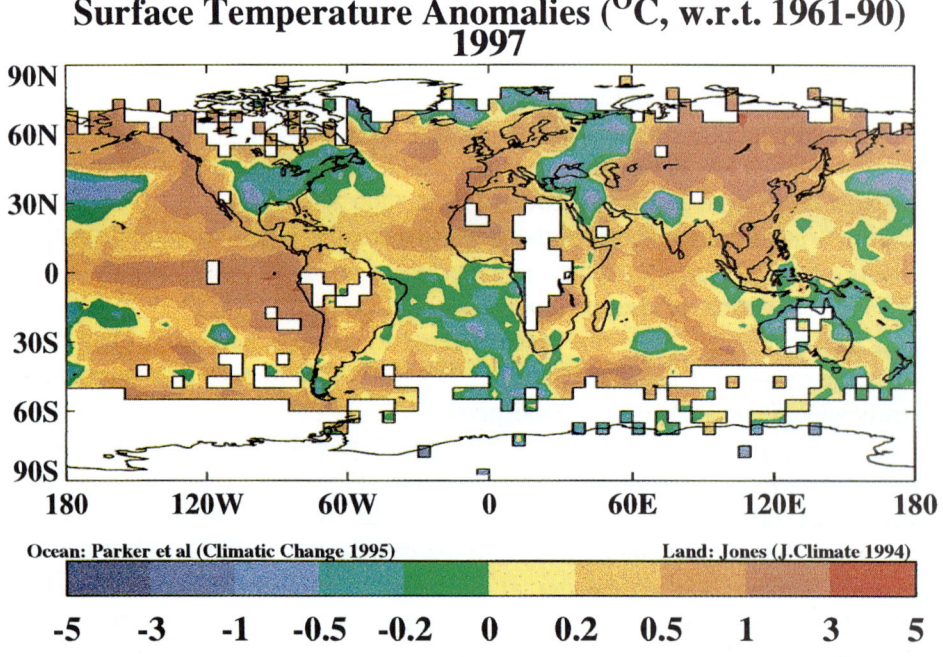

Ocean: Parker et al (Climatic Change 1995)   Land: Jones (J.Climate 1994)

-5   -3   -1   -0.5   -0.2   0   0.2   0.5   1   3   5

Figure 7. Surface temperatures for 1997, relative to the 1961-90 average, expressed a) as percentiles and b) as anomalies. The percentiles were defined by fitting gamma distributions to the 1961-90 annual deviations relative to the 1961-90 average, for all 5° x 5° boxes with at least 21 years of annual data in this period.

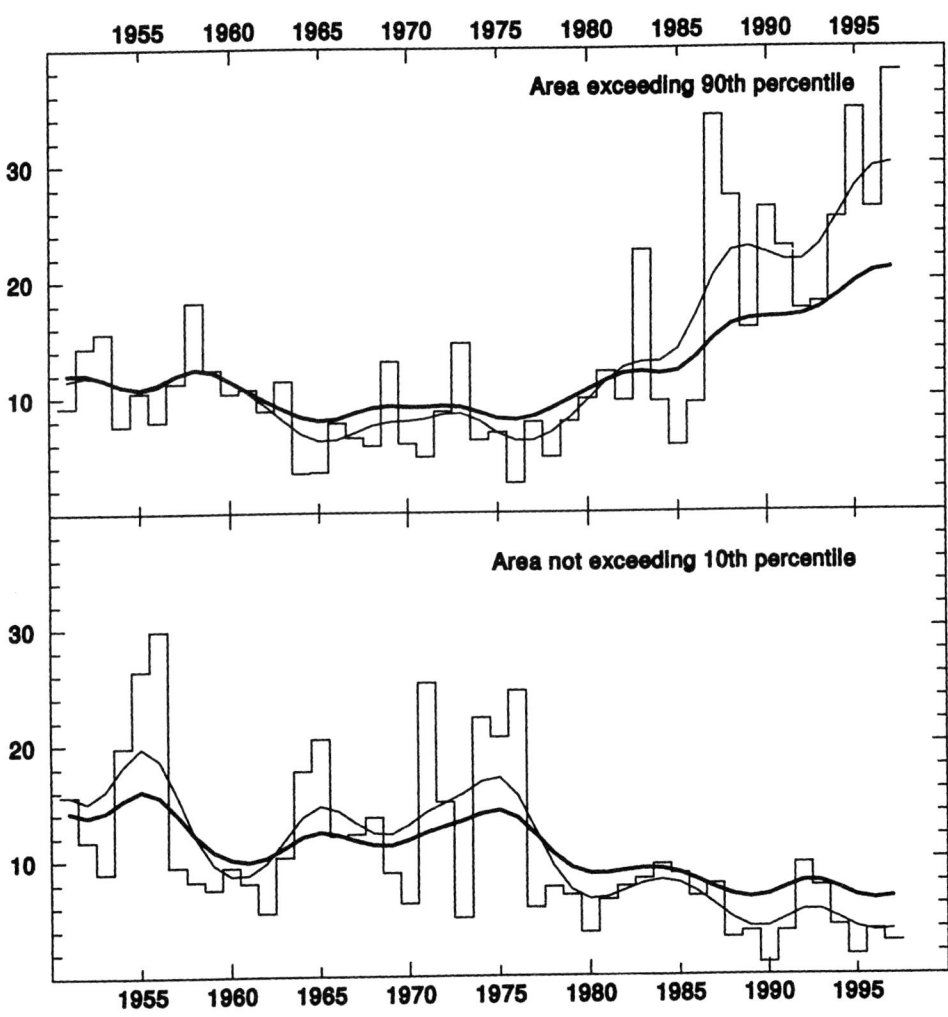

Figure 8. The percentage of the monitored area of the globe, for each year 1951-97, with annual surface temperatures (method a in the text) above the 90th percentile and below the 10th percentile. Percentiles were defined by fitting gamma distributions, based on the 1961-90 period, using all 5° x 5° boxes with at least 21 years of data in this period. The thin smooth line highlights decadal timescale variations while the thicker smooth line is calculated using monthly data (method b in text).

calculated relative to a changing base, i.e., one that reflected the gradual increase in temperatures during the last 45 years. Estimated in this way, any changes in the frequency of extremes would not be dependent upon the choice of base period, but would depend instead upon the degree of smoothing used for the changing base.

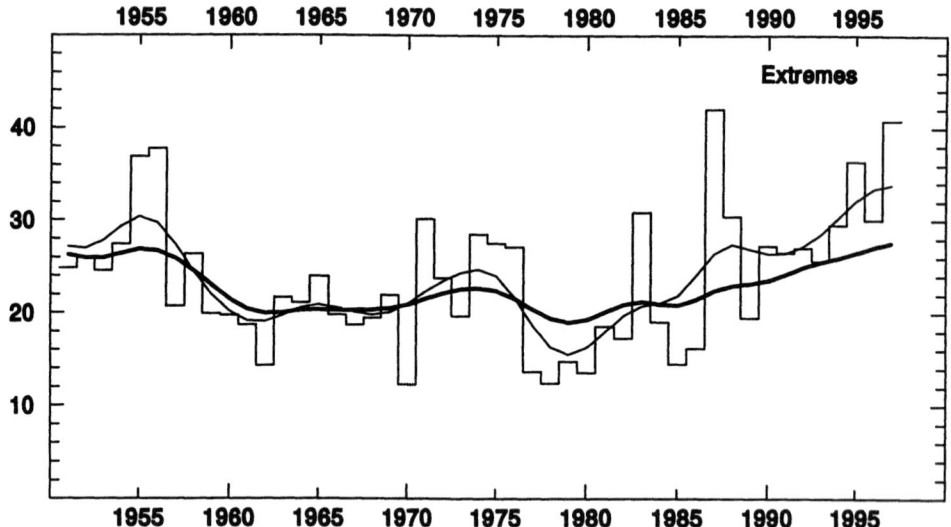

Figure 9. Annual areas of the monitored globe with 5° x 5° box values above the 90th and below the 10th percentiles (method a in text). The thin smooth line highlights near decadal timescale variations while the thicker smooth line, also a near decadal average, is calculated using monthly data (method b in text).

Conclusions

The 3-parameter gamma distribution provides a useful method of analysing not only long daily temperature series but also monthly gridded 5° x 5° grid box temperatures. The method enables extremes in both types of data to be studied without resort to arbitrary thresholds and takes account of differences in variability. The distribution enables the rarity of a particular day or month or year's temperature to be specified as a percentile value or alternatively as a return period. Use of the maximum likelihood version of the method is to be preferred because, in principle, biases due to the use of a rather short (30 year) fitting period are eliminated especially in the presence of appreciable skewness (e.g. as on the monthly timescale over continents in winter).

Although the main aim of this paper has been exploratory, we find that the statistical structure of CET and the global scale monthly gridded $5° \times 5°$ temperatures has changed over the last 10 to 15 years as temperatures have risen. Slight increases in warm extremes are evident in the CET series, but the larger impact on extremes has been a reduction in the number of extremely cool days. Globally, the areal coverage of warm (cold) extremes has increased (decreased). Traditional analyses of daily and monthly temperatures using thresholds are likely to miss these subtle changes. Our work also hints that linking the analysis of temperature (or other parameters) to that of regional atmospheric circulation variations may produce potentially insightful analyses of changes in the mean and extremes of climatic variables within weather types. This may be an especially good way of studying the emerging local signals of human-induced climate change.

Acknowledgements

This work has been supported by the U.K. Dept. of Environment, Transport and the Regions (Contract Nos. EPG 1/1/48 and PECD/7/12/37), the U.K. Public Meteorological Service (Contract MSG-2/97), and the United States Dept. of Energy, Atmospheric and Climate Research Division under grant No. DE-FG02-86ER60397. We would also like to acknowledge the help of Matthew O'Donnell.

References

Alexandersson, H. and Moberg, A., 1997: 'Homogenization of Swedish temperature data, Part 1: Homogeneity test for linear trends', *Int. J. Climatol.* **17**, 25-34.

Allan, R.J., Lindesay, J.A. and Parker, D.E., 1996: *El Niño Southern Oscillation and Climatic Variability*, CSIRO Publishing, Melbourne, 405pp.

Carnell, R.E., Senior, C.A. and Mitchell, J.F.B., 1996: 'An assessment of measures of storminess: simulated changes in northern hemisphere winter due to increasing CO_2', *Climate Dynamics* **12**, 467-476.

Conway, D. and Jones, P.D., 1996: POPSICLE - Production of Precipitation Scenarios for Impact Assessment of Climate Change in Europe. *Final Report to the European Community Environment Research Programme (EV5V-CT94-0510)*.

Conway, D., Wilby, R.L. and Jones, P.D., 1996: 'Precipitation and airflow indices over the British Isles',*Climate Research* **7**, 169-183.

Easterling, D.R. and Peterson, T.C., 1995: 'A new method for detecting and adjusting for undocumented discontinuities in climatological time series', *Int. J. Climatol.* **15**, 369-377.

Horton, E.B., Folland, C.K. and Parker, D.E., 1998: 'The incidence of extremes in worldwide and Central England temperatures', *Climatic Research Technical Note* **83**, Hadley Centre, Meteorological Office, Bracknell, U.K.

Hulme, M., 1994: 'Validation of large-scale precipitation fields in General Circulation Models', in Desbois, M. and Désalmond, F. (eds.), *Global Precipitations and Climate Change*, Springer-Verlag, Berlin, 387-405.

Hulme, M., 1997: 'The climate in the UK from November 1994 to October 1995',*Weather* **52**, 242-257.

Hulme, M. and Jones, P.D., 1991: 'Temperatures and windiness over the UK during the winters 1988/89 and 1989/90 compared to previous years', *Weather* **46**, 126-135.

Hulme, M., Briffa, K.R., Jones, P.D. and Senior, C.A., 1993: 'Validation of GCM control simulations using indices of daily airflow types over the British Isles', *Climate Dynamics* **9**, 95-105.

Jackson, M.C., 1976: 'A classification of the snowiness of 100 winters - a tribute to the late L.C.W. Bonacina', *Weather* **32**, 91-97.

Jenkinson, A.F. and Collison, B.P., 1977: 'An initial climatology of gales over the North Sea', *Synoptic Climatology Branch Memorandum No.* **62**, UK Met. Office, Bracknell, 18pp.

Jones, P.D., 1987: 'The early twentieth century Arctic high - fact or fiction?',*Climate Dynamics* **1**, 63-75.

Jones, P.D., 1994: 'Hemispheric surface air temperature variability - a reanalysis and an update to 1993', *J. Climate* **7**, 1794-1802.

Jones, P.D. and Conway, D., 1997: 'Precipitation in the British Isles: an analysis of area-average data updated to 1995', *Int. J. Climatol.* **17**, 427-438.

Jones, P.D. and Hulme, M., 1997: 'The changing temperature of 'Central England'', in Hulme, M. and Barrow, E., *Climates of the British Isles: present, past and future*, Routledge, London, 173-196.

Jones, P.D., Raper, S.C.B., Bradley, R.S., Diaz, H.F., Kelly, P.M. and Wigley, T.M.L., 1986: 'Northern Hemisphere surface air temperature variations, 1851-1984', *J. Clim. Appl. Met.* **25**, 161-179.

Jones, P.D., Hulme, M. and Briffa, K.R., 1993: 'A comparison of Lamb circulation types with an objective classification scheme', *Int. J. Climatol.* **13**, 656-663.

Jones, P.D., Conway, D. and Briffa, K.R., 1997a: 'Precipitation variability and drought', in Hulme, M and Barrow, E., *Climates of the British Isles: present, past and future*, Routledge, London, 197-219.

Jones, P.D., Jónsson, T. and Wheeler, D., 1997b: 'Extension to the North Atlantic Oscillation using early instrumental pressure observations from Gibraltar and SW Iceland', *Int. J. Climatol.* **17**, 1433-1450.

Jones, P.D., Osborn, T.J. and Briffa, K.R., 1997c: 'Estimating sampling errors in large-scale temperature averages', *J. Climate* **10**, 2548-2568.

Lamb, H.H., 1972: 'British Isles weather types and a register of the daily sequence of circulation patterns, 1861-1971', *Geophysical Memoirs* **116**, HMSO London, 85pp.

Manley, G., 1974: 'Central England temperatures: monthly means 1659-1973', *Quart. J. Roy. Met. Soc.* **100**, 389-405.

NCDC, 1994: *Global Daily Summary (TD9950)*, CDROM produced by National Climatic Data Center, Asheville, NC.

New, M., Hulme, M. and Jones, P.D., 1998: 'Representing twentieth century climate, Part 1: Construction of a 1961-90 mean monthly terrestrial climatology', *J. Climate* (in press).

Parker, D.E., 1994: 'Effects of changing exposure of thermometers at land stations', *Int J.Climatol.* **14**, 1-31.

Parker, D.E., and Folland, C.K., 1988: 'The climate of the world, Part 9: Climatic change in the instrumental period', *Long-Range Forecasting and Climate Research* **LRFC27**, Meteorological Office, U.K., Unpublished document available from the National Meteorological Library, Met. Office, Bracknell, UK, RG12 2SZ.

Parker, D.E., Legg, T.P. and Folland, C.K., 1992: 'A new daily Central England Temperature Series, 1772-1991', *Int. J. Climatol.* **12**, 317-342.

Parker, D.E., Folland, C.K. and Jackson, M., 1995: 'Marine surface temperature: observed variations and data requirements', *Climatic Change* **31**, 559-600.

Parker, D.E., Horton, E.B. and Gordon, M., 1998: 'Global and regional climate in 1997', *Weather* **53**, 166-175.

Peterson, T.C. and Vose, R.S., 1997: 'An overview of the Global Historical Climatology Network temperature data base', *Bulletin of the American Meteorological Society* **78**, 2837-2849.

Razuveyev, V.N., Apasova, E.G., Martuganov, R.A., Vose, R.S. and Steurer, P.M., 1993: 'Daily temperature and precipitation data for 223 USSR stations', *Numerical Data Package* **NDP-040**, Carbon Dioxide Information Analysis Center, Oak Ridge, TN, 47pp plus two appendices.

Smith, S.G., 1982: 'An index of windiness for the United Kingdom', *Meteorological Magazine* **111**, 232-247.

(Received 5 November 1997; in revised form 7 August 1998)

PROGRESS IN THE STUDY OF CLIMATIC EXTREMES IN NORTHERN AND CENTRAL EUROPE

HEINO, R.[1], BRÁZDIL, R.[2], FØRLAND, E.[3], TUOMENVIRTA, H.[1], ALEXANDERSSON, H.[4], BENISTON, M.[5], PFISTER, C.[6], REBETEZ, M.[7], ROSENHAGEN, G.[8], RÖSNER, S.[8], AND WIBIG, J.[9]

1 Finnish Meteorological Institute
P.O.Box 503, FIN-00101 Helsinki, Finland
2 Masaryk University of Brno, Czech Republic
3 Norwegian Meteorological Institute, Norway
4 Swedish Meteorological and Hydrological Institute, Sweden
5 University of Fribourg, Switzerland
6 University of Bern, Switzerland
7 Swiss Federal Institute for Forest, Snow and Landscape Research, Switzerland
8 Deutscher Wetterdienst, Germany
9 University of Lodz, Poland

Abstract. A study of the long-term changes of various climatic extremes was made jointly by a number of European countries. It was found that the changes in maximum and minimum temperatures follow, in broad terms, the corresponding well-documented mean temperature changes. Minimum temperatures, however, have increased slightly more than maximum temperatures, although both have increased. As a result, the study confirms that the diurnal temperature range has mostly decreased during the present century in Northern and Central Europe. Frost has become less frequent. Two extreme-related precipitation characteristics, the annual maximum daily precipitation and the number of days with precipitation $\geq$ 10 mm, show no major trends or changes in their interannual variability. An analysis of return periods indicated that in the Nordic countries there were high frequencies of 'extraordinary' 1-day rainfalls both in the 1930s and since the 1980s. There have been no long-term changes in the number of high wind speeds in the German Bight. Occurrences of thunderstorms and hails show a decreasing tendency in the Czech Republic during the last 50 years. Finally, using proxy data sources, a 500-year temperature and precipitation event graph for the Swiss Mittelland is presented. It shows large interdecadal variations as well as the exceptionality of the latest decade 1986-1995.

1. Introduction

Climate fluctuations in Europe during the 20th century have been the topic of many investigations. There are well-documented long-term changes in the averages of climatic elements, such as a precipitation increase in north-western parts of Europe between the last two normal periods (Førland et al., 1996a), a springtime warming in Fennoscandia that started at the end of the last century (Tuomenvirta and Heino, 1996), and a warming of winters in Central Europe since the 1940s (e.g. Böhm, 1992). Some of the most comprehensive reviews

Climatic Change **42**: 151–181, 1999.
© 1999 *Kluwer Academic Publishers. Printed in the Netherlands.*

dealing with climate variations over large areas of Europe are given by Brázdil and Kolář (1994), Heino (1994a), ECSN (1995) and Heikinheimo (1995).

In connection with the observed global warming, hypotheses about a possible increase in weather and climatic extremes have been formulated. However, as stated by Nicholls et al. (1996) 'overall, there is no evidence that extreme weather events, or climate variability, has increased, in a global sense, through the 20th century, although data and analyses are poor and not comprehensive. On regional scales there is clear evidence of changes in some extremes and climate variability indicators. Some of these changes have been toward greater variability, some have been toward lower variability'. Because the majority of available climatic change investigations has been oriented towards the average climate, we need more studies dealing with weather and climate extremes in different parts of the globe.

One of the basic objectives of the study in the area of climatic extremes is to analyse long time series of climatic elements by focusing on the tails of the frequency distributions. These often correspond to the events causing the greatest hazards to society and ecosystems, e.g. the greatest flood catastrophe of the 20th century in July 1997 in the Czech Republic, Poland and eastern Germany due to heavy precipitation. From this follows that the results of extreme-related studies have many practical applications. The probabilities of various extreme events and estimates of their return periods are of great value for the planning process of societies in order to avoid human or material losses. With good quality data, it is possible to study temporal variations of climatic extremes, and to improve the reliability of estimates related to the extremes.

Houghton et al. (1996) made some tentative assessments concerning future larger-scale extreme events based on climate modeling results. However, current climate models lack the accuracy at smaller scales and the integrations have not been long enough to permit analysis of local weather extremes. Also, feedbacks including the surface are important. For example, Beniston et al. (1995) found that as a result of the changes in seasonal mean values of temperature and precipitation between a $1 \times CO_2$ and a $2 \times CO_2$ simulation, there was decreased soil moisture content which was reflected in the statistical distribution of daily precipitation; July simulations for the $2 \times CO_2$ indicated a sharp increase in the number of dry days with respect to current climate. These simulations made use of the Max-Planck-Institute ECHAM4 General Circulation Model (at T-106 resolution, i.e., $1.1 \times 1.1°$ latitude/longitude resolution) and, nested within the GCM, the NCAR RegCM2 model at 20 km resolution.

It will be important to discover, whether the expected trends in the latest IPCC assessment (e.g. changes in the frequencies of extremely hot or cold days and an increase in precipitation amounts and intensity) can already be seen in the observational data records. Earlier studies have not provided many answers to these questions. Even small changes in the mean climate and its variability can be accompanied by relatively large changes in the frequency of extreme events. This has been seen both in theory (Katz and Brown, 1992) and in climate modelling (Kattenberg et al., 1996).

The topic of this paper is fluctuation and trend analyses for some characteristics of selected climatic and weather extremes. The time series of some selected stations from Northern (Finland, Norway, Sweden) and Central (Czech Republic, Germany, Poland, Switzerland) Europe are used for the investigations.

2. Data and Homogeneity

Climatic extremes can be defined as the occurrence of values of meteorological parameters, which lie above or below specific threshold values and occur with relatively low probability or which have significant impact on society or ecosystems. The corresponding threshold values can be determined more-or-less objectively by using a theoretical distribution of the given parameter (e.g. a normal distribution for maximum air temperatures, using values of the standard deviation) or subjectively, based on experience or knowledge of any interrelations (e.g. threshold values for the thermal stress of crops).

Traditionally, long-term climatic data sets in Europe have consisted mostly of monthly (mean) values. However, the work of the North Atlantic Climatological Dataset (NACD) and of the European Climate Support Network (ECSN) have produced new extensive European datasets which will support analyses of climatic extremes (e.g. Frich et al., 1996). Long-term records of climatic extreme elements for the Nordic countries are being compiled by the REWARD-project (Førland et al., 1998) and there are plans to compile a similar dataset for a larger area of Europe. In addition, vast climate extreme-related statistics are available for most European countries for the normal period 1961-1990.

We have compiled annual series of the following parameters (not all data were available for each of the seven investigated countries):

- mean daily maximum (Tmax) and minimum (Tmin) temperatures
- diurnal temperature range (DTR = Tmax - Tmin)
- number of frost days (Tmin < 0°C)
- maximum 1-day (24 h) precipitation
- number of days with precipitation greater than or equal to 10 mm
- wind speed based on geostrophic wind
- number of occurrences of thunderstorms and hails.

The stations used in this study are shown in Figure 1. The study period has been 1901-1995, as far as possible. The data for this study were partly compiled from existing monthly series and partly from long-term daily climatic records in digital form. In future studies, the wider use of daily data would enable free selection of the length of events and of the limits of threshold values to be studied.

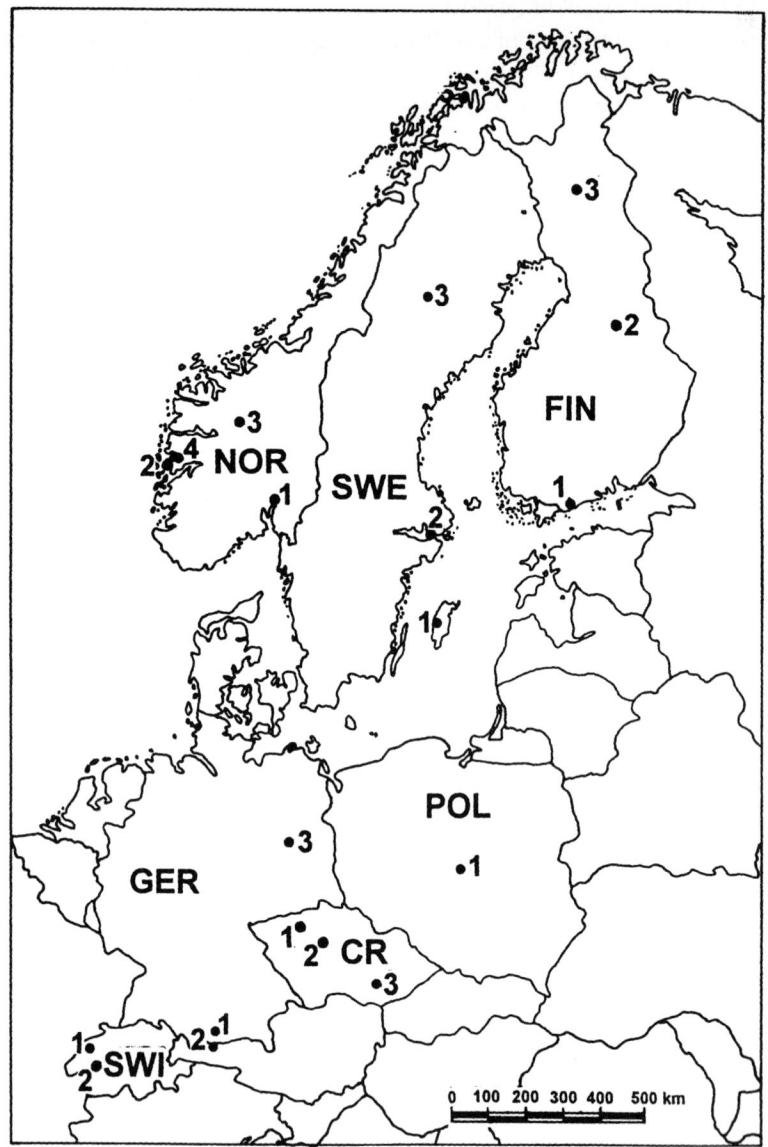

Figure 1. Geographical distribution of the stations used (cf. also Table I).

Czech Republic (CR): 1 Milešovka, 2 Prague-Klementinum, (Prague-Karlov, very close to Prague-Klementinum), 3 Brno-airport; *Finland* (FIN): 1 Helsinki, 2 Kajaani, 3 Sodankylä; *Germany* (GER): 1 Hohenpeissenberg, 2 Zugspitze, 3 Potsdam; *Norway* (NOR): 1 Oslo, 2 Bergen, 3 Skjaak, 4 Samnanger, 5 Tromsø; *Poland* (POL): 1 Lodz; *Sweden* (SWE): 1 Visby, 2 Stockholm, 3 Stensele; *Switzerland* (SWI): 1 Chateau-d'Oex, 2 Neuchatel.

In climatology, the homogeneity of the data is important, especially in the study of climatic changes. Long-term changes due to real causative factors have been (and probably will be) small and slow and are hidden under large interannual variability. Climatic records, at least those which are readily available, are normally mixtures of both apparent and real variations. It is obvious that the apparent variations should be eliminated in order to get meaningful research results.

In practice, it is difficult to obtain long homogeneous data records. Various factors, such as changes in instruments and their exposure, observers, observation times, averaging methods, observation sites and their environments introduce inhomogeneities into the data. Many of the inhomogeneities are step-like changes (e.g. Figure 2) and typically alter the average value only. Application of relative homogeneity tests (e.g. Alexandersson, 1986; Easterling and Peterson, 1995) helps to find statistically significant inhomogeneities and adjust corresponding time series. But in the case of climatic extremes, application of the tests mentioned can be problematic because of the great temporal and spatial variability of many extremes.

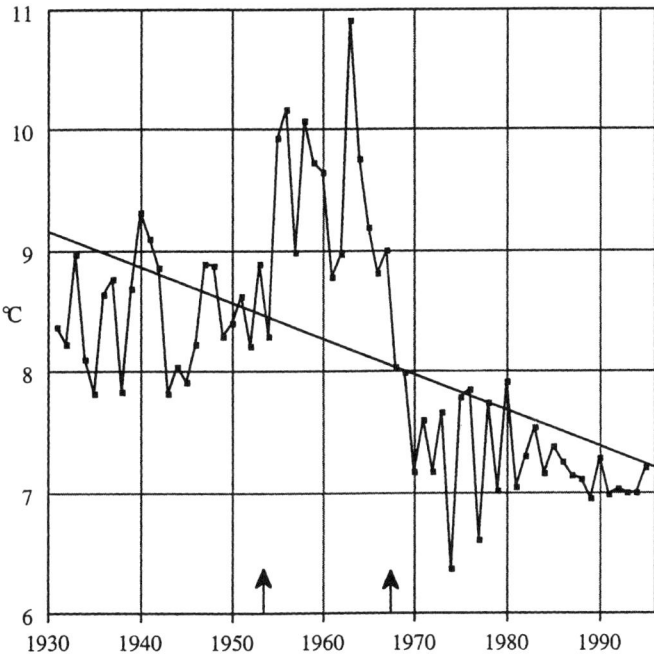

Figure 2. Difference between annual mean maximum and minimum temperatures (°C) at Mikkeli, Finland (61°44'N, 27°18'E, altitude 138 m) in 1931-1995. This difference is strongly related to the station relocations indicated by the arrows. Any statistical analysis, like the linear trend line drawn, would give misleading results without the necessary adjustments, which can be done relatively easily with the help of the available metadata.

An inhomogeneity may, moreover, also imply changes in variability or in other distribution parameters. In the automation of station networks, changes in maximum and minimum temperatures may become apparent, while the mean values may remain unchanged. In practice, the inhomogeneity of a secular time series is usually a combination of many factors. Information on the history of the measurements and the stations (metadata) is thus essential for a successful study of the data homogeneity and in making decisions about adjusting a series (cf. Heino, 1996). In this study the temperature data were in part adjusted for the most serious inhomogeneities; the other parameters were used as such, without any adjustments (cf. Table I).

Furthermore, several data series are from cities and may be progressively influenced by urban effects and their intensification. Upward temperature trends in urban stations are stronger than in the rural ones (cf. Heino, 1979, Brázdil, 1993, Brázdil et al., 1994). Because the increasing influence of the city has more or less gradual character and the estimation of its value is uncertain, it is problematic to remove it from temperature series.

Table I

Stations of this study with considerations of their homogeneity (- no major homogeneity problems, + some inhomogeneities; further considerations on the data homogeneity are presented in connection with the analysis and results).

	Coordinates and altitude			adjusted records	urban environment	homogeneity estimation
Milešovka (CR)	50°33'N	13°56'E	835 m	no	no	-
Prague-Klementinum (CR)	50 05	14 25	191	no	yes	-
Helsinki (Fin)	60 10	24 57	4	no	yes	+
Kajaani (Fin)	64 17	27 41	134	yes	no	+
Sodankylä (Fin)	67 22	26 39	179	yes	no	-
Hohenpeissenberg (Ger)	47 25	11 01	977	yes	no	-
Zugspitze (Ger)	47 25	19 59	2960	yes	no	-
Potsdam (Ger)	52 23	13 04	81	yes	yes	-
Oslo (Nor)	59 57	10 43	94	no	yes	+
Bergen (Nor)	60 25	05 20	41	no	yes	+
Skjaak (Nor)	61 54	08 10	432	no	no	-
Samnanger (Nor)	60 28	05 54	370	no	no	-
Tromsø (Nor)	69 39	18 56	100	no	no	+
Lodz (Pol)	51 44	19 24	184	no	yes	+
Visby (Swe)	57 40	18 20	42	no	no	+
Stockholm (Swe)	59 20	18 03	44	no	yes	+
Stensele (Swe)	65 04	17 09	325	no	no	-
Chateau-d'Oex (Swi)	46 29	07 08	980	no	no	-
Neuchatel (Swi)	47 00	06 57	487	no	no	-

3. Selected Climatic Extremes in Northern and Central Europe during the 20th Century

3.1. AIR TEMPERATURE

3.1.1. Maximum and minimum daily temperatures

Daily maximum and minimum temperatures (Tmax, Tmin) are basic climatic parameters, which have been observed since the beginning of the 20th century in most European countries. However, these long-term records have not yet been fully used.

The motivation for many recent European studies devoted to investigations of Tmax and Tmin (for a list of them see e.g. Brázdil et al., 1996) was the paper by Karl et al. (1993). They showed that the Tmin of 37 % of the global land mass increased three times more quickly than the corresponding Tmax during the period 1951-1990 (0.84°C and 0.28°C, respectively). For Central Europe (0.7 % of the global land mass) the corresponding values were 0.60°C for Tmin and 0.52°C for Tmax (Brázdil et al., 1996).

Beniston et al. (1994) have shown that mean annual minimum temperatures in Switzerland have undergone a strong increase during the 20th century (about 2°C/100 years) while the increase in mean annual maximum temperatures has been much smaller.

Heino (1994b) studied the changes of Tmax and Tmin in Finland since the 1950s. The features of both cold (the 1960s and 1980s) and warm (mid-1950s and recent years) episodes are well reflected in the results. In addition, minimum temperatures showed a greater increase than maximum temperatures on an annual basis.

Brázdil et al. (1996) made a wide study of ten selected countries and regions in Central and Southeast Europe during the period 1951-1990, and also some analysed secular series covering the 20th century. They concluded that there had been an increase in both annual maximum and minimum temperatures, for some regions a statistically significant one; however, minimum temperatures had increased more.

Gerstengarbe and Werner (1993) made a study of extreme climate events at several European stations up to 1990. The investigation into the frequency of cold/ hot summers showed that e.g. during the period 1901-1927 there had been more than normal cold summers, while in 1928-1953 more than normal hot summers had been experienced. For 1954-1980 the numbers of cold and hot summers were nearly the same. In addition, the 95th percentile of the daily maximum temperature showed no significant trend.

There is, quite naturally, a substantial coupling between series of mean temperature and of mean maximum and mean minimum and also of absolute maximum and absolute minimum temperatures. To illustrate this Table II gives correlation coefficients on a monthly and an annual basis for one station in northern Sweden.

Table II

Correlation coefficients (in hundredths) between monthly mean temperatures and various other temperatures (left column) at Stensele (Swe), 1860-1996.

	J F M A M J J A S O N D Y
Mean max.	98 98 92 88 93 94 93 92 90 96 97 98 93
Mean min.	98 98 98 90 83 84 82 77 86 97 98 98 95
Abs max.	58 68 54 60 64 67 57 69 58 59 40 56 19
Abs min.	78 80 79 61 51 45 48 57 35 81 76 84 70

The somewhat lower correlations in spring and summer between mean temperature and mean minimum could be explained by the fact that anticyclonic months with sunny and warm days and high mean temperatures often include periods with fairly low night temperatures. For the connections with the absolute highest and lowest temperature the most surprising figures occur for the year as a whole with a very modest value for the maximum but considerably higher one for the minimum. This can be explained by the fact that the winter months with their large year-to-year variations dominate the annual means, and it is of course also during the winter months that the absolute minima occur.

3.1.2. Diurnal temperature range

As a result of the trends in annual maximum and minimum temperatures Karl et al. (1993) detected a decreasing trend in diurnal temperature range (DTR) during 1951-1990 over a large area of the globe. No human-induced local effects could provide a satisfactory basis for the widespread decrease in the DTR, while among many possibly natural variables, changes in cloudiness gave the best explanation of the decline. Recently, Easterling et al. (1997) came to the same conclusions with a more complete data set.

Heino (1994b) also shows a slight decrease in the DTR in Finland by approximately 0.5°C since the 1950s. The corresponding seasonal changes were also studied. It appeared that the decrease in the annual DTR is mostly explained by decreases in spring and summer, while the DTR in winter has remained at about the same level despite large inter-decadal changes. Cloudiness data from the stations used in the DTR calculations were found to be strongly correlated (about -0.8) with DTR on a monthly and seasonal basis. Since cloudiness has experienced a recent increase and the duration of sunshine a corresponding decrease, this appears to be a reasonable explanation of the DTR changes in Finland as well.

Brázdil et al. (1996) similarly made a wide review of the annual and seasonal DTR for Central and South-Eastern Europe. However, no significant trends were found for the annual DTR in the period 1951-1990 in most of the analysed regions (Germany, Poland, Slovak Republic, Austria, Hungary, Slovenia, lowland-Croatia and Bulgaria) except for the Czech Republic and the Swiss Mittelland. The trends in the DTR can be explained by cloudiness, since the correlation coefficients of annual DTR and cloudiness showed significant values, fluctuating between -0.60 (Bulgaria) and -0.83 (Hungary).

The evolution of the DTR in Northern and Central Europe was studied here with data sets covering the whole century (Table III and Figure 3).

Table III

Trend of the diurnal temperature range (°C/10 years) at each station in 1910-1995 (Prague 1910-1992), together with the statistical significance. Left column is the linear trend tested with two-sided t-test. Right column is Sen's nonparametric estimate of slope (Gilbert, 1987) and the significance of the trend is evaluated with nonparametric Mann-Kendall test (Sneyers, 1990); (* 5%, ** 1%, *** 0.1%).

station	linear trend	Sen's estimate
Sodankylä (Fin)	-0.001	-0.063*
Stensele (Swe)	-0.057*	-0.056*
Kajaani (Fin)	-0.062**	-0.068**
Bergen (Nor)	-0.025	-0.021
Helsinki (Fin)	0.001	0.007
Stockholm (Swe)	-0.001	0.004
Visby (Swe)	0.048**	0.044*
Potsdam (Ger)	-0.055**	-0.059**
Milešovka (CR)	-0.038*	-0.039
Prague-Klementinum (CR)	0.042*	0.040*
Hohenpeissenberg (Ger)	0.095**	0.024
Zugspitze (Ger)	0.052***	0.050***
Neuchatel (Swi)	-0.236***	-0.241***

In addition to the linear trend a more resistant trend calculation by Sen's method (median slope between all possible pairs of values) was used with significance estimation as in Gilbert (1987). Both methods, however, gave quite similar results.

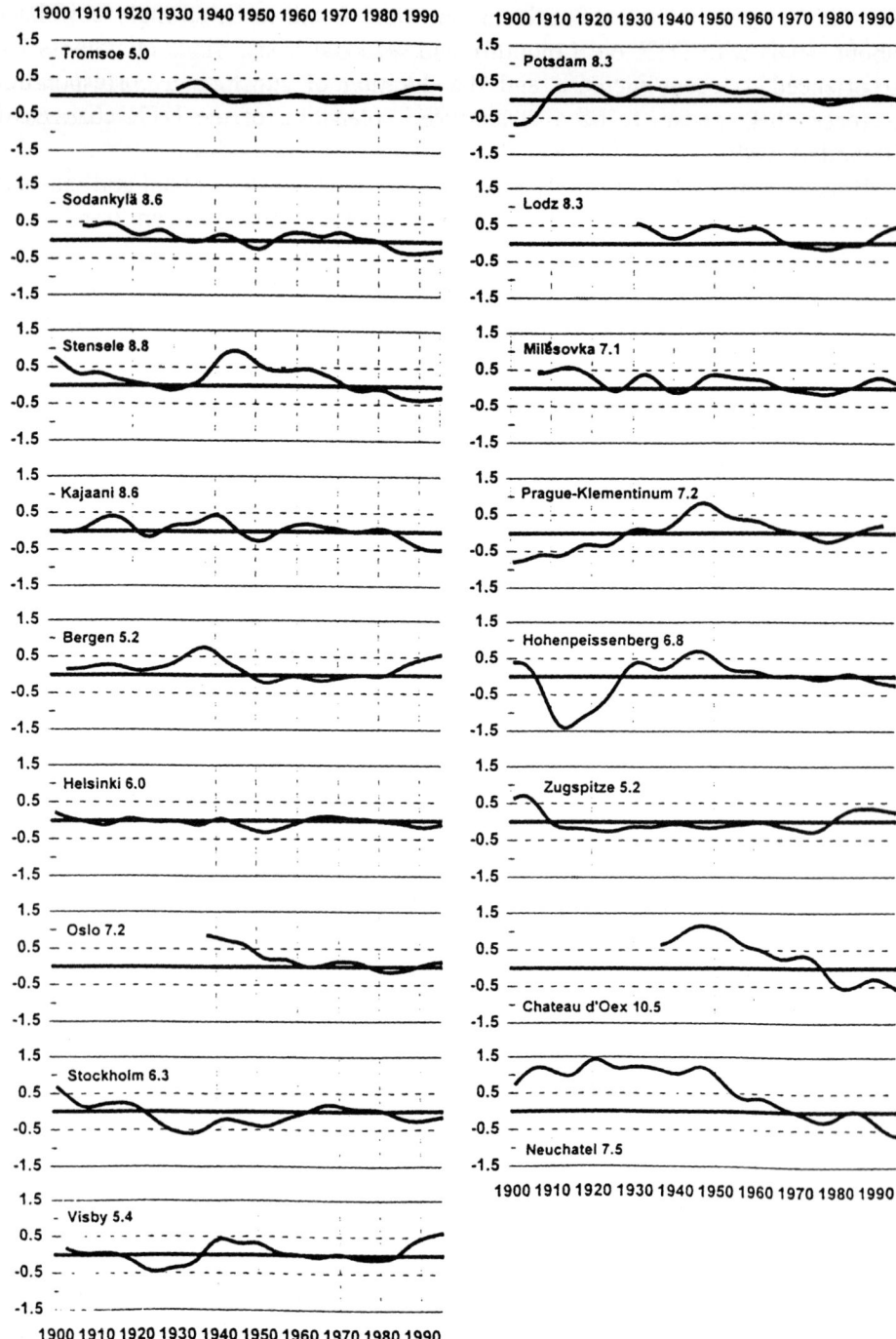

Figure 3. Annual anomalies of diurnal temperature range (DTR, °C); corresponding mean (°C) after the station name. Curves are smoothed with a low-pass filter with Gaussian weighting coefficients equivalent to 10 year moving averages. Reference period: 1961-1990

A decreasing trend since the 1940-50s is evident in most of the records, but at a few stations a recent increase is also seen. Before the 1940s there are more variable features (see also Brázdil et al., 1996).

In Hohenpeissenberg and Zugspitze the records are quite different from the others, but the increase of the DTR is statistically significant. In the Swiss data a general decreasing trend in the DTR, i.e. of approx. 2°C over the last 50 years, has been observed since about 1950 with a few higher values around 1990 at both sites. At Lodz there is also a recent statistically significant decreasing trend.

It should be emphasised that the DTR is more sensitive to non-homogeneities than is the mean temperature. For example, a relocation in a coastal area can give higher (inland relocation) or lower (relocation towards the coast) DTR-values, while the mean temperature would be less affected. In fact, this kind of relocation could explain the recent upward trend of the DTR-trends at Bergen and Visby, where the stations were moved inland in 1984 and 1986.

In addition, there has been a very strong trend towards a less open landscape that easily can give rise to false DTR-decreases for a number of stations. This tendency has many reasons (e.g. access to other fuels than wood, more buildings today around the houses than earlier, more fruit-trees, bushes etc. in gardens nowadays).

3.1.3. Number of frost days

The occurrence of frost has a considerable impact on many human activities, including the agricultural and construction industries. Since meteorological yearbooks contain standard monthly statistics on the number of days of maximum and minimum temperatures above/below certain limits, monthly counts of e.g. Tmin < 0°C are easily available. The year-to-year changes of some of these counts were also collected in this study.

For the detection of changes of climatic extremes, however, these indices are problematic as the counts depend on the local climate as well as threshold temperature chosen. Thus, while changes in mean temperatures were well reflected in the number of frost days (e.g. the warm 1930s and cold 1940s), in cold regions the counts reflected the changes in mean autumn and spring temperatures while in warm regions they reflected the change in winter temperatures, or even hardly appeared at all. For this reason the year-to-year variability is also smaller in the north than in the south.

However, it is very evident that some decrease of frost days has taken place since the 1930s (Table IV). For example, at both Swiss sites there are nowadays nearly 50 frost days less than in the early part of the record. This is linked to the very strong increase in winter minimum temperatures. The Finnish and Czech series also show decreasing tendencies.

Table IV

Average of the annual number of frost days (Tmin<0°C) and linear trends in days/10 years, at some stations in 1931-1995, together with the statistical significance. Left column is the linear trend tested with two-sided t-test. Right column is Sen's nonparametric estimate of slope (Gilbert, 1987) and the significance of the trend is evaluated with nonparametric Mann-Kendall test (Sneyers, 1990); (* 5%, ** 1%, *** 0.1%).

station	average	linear trend	Sen's estimate
Sodankylä (Fin)	225	-0.79	-0.87
Kajaani (Fin)	197	0.10	0.23
Helsinki (Fin)	136	-1.45	-1.17
Milešovka (CR)	146	-2.15*	-2.00*
Prague-Karlov (CR)	85	-3.11**	-3.04**
Neuchatel (Swi)	72	-7.46***	-7.40***
Chateau d'Oex (Swi)	154	-6.21***	-6.25***

3.1.4. Frequency distribution of extreme temperatures

For the reasonable use of any time series it is essential to know the nature of the frequency distribution, but several statistical methods and models can be applied only if the time series considered is stationary, i.e. averages, variances and also higher moments do not vary with time. If the distribution of the climatic element in question is approximately normal (i.e. Gaussian), arithmetic means and standard deviations alone describe the distribution quite well. But Heino (1994b) has demonstrated that the skewness and kurtosis of frequency distributions are not fixed, but may vary considerably in the course of time.

An example of a change in the frequency distribution of temperature accompanying a shift in the mean is given in Figure 4. It illustrates the change in the probability density function of daily minimum temperatures between two distinct periods at Davos (eastern Alps, Switzerland). The period 1904-1910 corresponds to the coldest part of the century at this location, while 1988-1994 corresponds to the warmest part of the century. The shift of the entire distribution towards higher temperatures (and therefore lower frequencies of extreme low minimum daily temperatures) is quite obvious. Moreover, the skewness of the distribution shifts from approximately zero skewness (i.e., close to a Gaussian distribution) to negative skewness. This implies that there has been a gain in the upper range of the probability density function at the expense of the lower range of minimum temperatures.

Frequency [%]

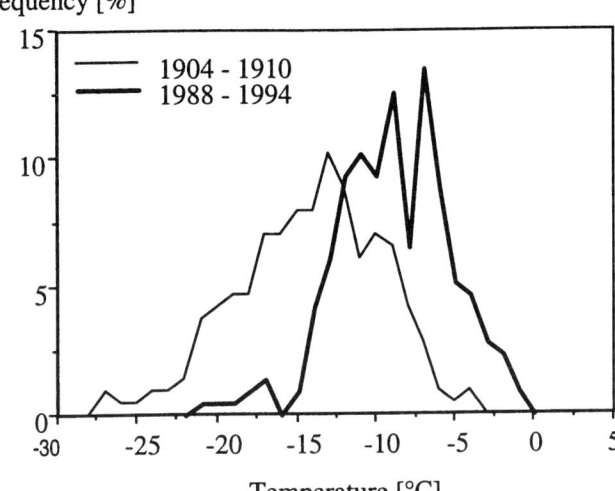

Figure 4. Frequency distribution of daily minimum temperatures (°C) for two distinct periods at Davos, Switzerland (46°50'N, 09°51'E, altitude 1590 m).

3.2. PRECIPITATION

3.2.1. Maximum daily precipitation

3.2.1.1. Introduction. Every year flooding connected with heavy rainfall causes losses of human lives and enormous damage. Heavy 1-day rainfall plays a crucial role in the flooding in small watersheds and especially in urban areas. Monitoring maximum 1-day rainfall is therefore important both to deduce return period values and to detect any long-term trends in their occurrence. Longer-term totals of heavy precipitation are also needed, and for this kind of analysis, series of daily observations are also necessary.

The consequence of extreme high daily precipitation, mainly in the form of downpours or cloudbursts, can be flooding on a small scale. If heavy precipitation falls for several days, greater rivers and areas can be influenced by floods. This was the case for the catastrophic flood in July 1997 in Central Europe. It was the flood of the century in the Czech Republic and also affected the Oder river in Poland and eastern Germany.

The reason for the flooding was the heavy precipitation during 5-8 July 1997, when many stations in northern Moravia and Silesia recorded totals which were 2-3 times higher than the normal sum for the whole of July. On four occasions daily totals higher than 200 mm were reached in that area (e.g. on 6 July 1997, Mt. Lysá hora at an altitude of 1324 m received 233.8 mm and Ostravice-Šance at an altitude of 445 m 230.2 mm). In the Tatra Mountains high values were also recorded (e.g. 223 mm at Hala Gasiennicowa on 8 July).

The flood wave had not yet reached the lower course of the Oder river when, between July 18 and 21, heavy rain (over wide areas more than 100 mm, at Mt. Praděd at an altitude of 1492 m 188 mm, and at Mt. Lysá hora, 164 mm) again poured down, seriously aggravating the situation at the already soaked dams.

3.2.1.2. Analysis. In the countries investigated the highest-ever 1-day precipitation amounts recorded during the 20th century are the following:

- Czech Republic: 240.2 mm at Nová Červená Voda (north Moravia) 9 July 1903
- Finland: 198.4 mm at Lahnus (close to Helsinki) 21 July 1944
- Germany: 260 mm at Zelthain (Saxony) 6 July 1906 and at Stein (Upper Bavaria) 8 July 1954
- Norway: 229.6 mm at Indre Matre (Western Norway) 26 November 1940
- Poland: 300 mm at Hala Gasiennicowa (Tatra Mountains) 30 June 1973
- Sweden: 237 mm at Karlaby (southern Sweden) 6 August 1960 (the next highest value, 198 mm, was observed in Northern Sweden in July 1997).

Within the Nordic REWARD-project, a dataset of monthly values of maximum 1-day precipitation is being established. The dataset mainly covers the period 1890-1996, and contains more than 80 series from the Nordic countries. Preliminary analyses of these series indicate that there are large differences in the trend patterns even for neighbouring stations (Førland et al., 1998). For example, based on more than 150 series, it was possible to divide Norway into five regional groups with very similar trends in annual precipitation (Hanssen-Bauer et al., 1997). But even within these well-defined regions, quite different trend patterns were found for annual maximum 1-day precipitation (Førland et al., 1998).

Just as in the temperature section, an example of the connection between mean values (monthly precipitation totals) and extremes (24-hour precipitation maximum) is given in Table V.

Table V

Correlation coefficients (in hundredths) between monthly precipitation totals and 24-hour maxima at Stensele (Swe), 1860-1996.

J	F	M	A	M	J	J	A	S	O	N	D	Y
68	71	76	84	75	69	72	72	71	76	69	73	24

While there is a strong coupling on a monthly basis (the wettest day of the month contributes considerably to the total sum), the correlation is small (0.24) on an annual basis. However, studies of other series indicate that 0.30 is a more typical value.

Examples of annual maximum 1-day precipitation from the analysed countries are shown in Figure 5 and Table VI. No major trends are seen and no changes in the year-to-year variability are found, either. Rising trends are evident at the German mountain stations, but only the trend at Zugspitze is statistically significant. At the Swiss and Finnish stations slight increasing trends are found, which are not, however, statistically significant.

Table VI

Linear trend of annual anomalies of 1-day maximum precipitation (mm/10 years) at each station for 1910-1995, together with the statistical significance. Left column is the linear trend tested with two-sided t-test. Right column is Sen's nonparametric estimate of slope (Gilbert, 1987) and the significance of the trend is evaluated with nonparametric Mann-Kendall test (Sneyers, 1990); (* 5%, ** 1%, *** 0.1%).

station	linear trend	Sen's estimate
Sodankylä (Fin)	0.08	0.00
Tromsø (Nor)	-0.24	-0.26
Stensele (Swe)	-0.78	-0.46
Kajaani (Fin)	0.52	0.59
Skjaak (Nor)	-0.17	-0.15
Samnanger (Nor)	0.13	0.00
Helsinki (Fin)	0.42	0.18
Visby (Swe)	-0.26	-0.09
Potsdam (Ger)	0.63	0.47
Milešovka (CR)	-0.10	0.06
Hohenpeissenberg (Ger)	0.49	0.50
Zugspitze (Ger)	2.36**	2.37**
Neuchatel (Swi)	0.37	-0.05

There are several reasons for these diverging trend patterns for maximum 1-day precipitation:

- There are usually large local precipitation gradients in heavy rainfall events. Thus, even for long secular series, some stations may have experienced several 'accidental hits' of heavy rainfall, whereas neighbouring stations have no such 'hits' of precipitation at all. These local 'accidental hits' may be caused either by slow-moving frontal systems, by local quasi-stationary convective cells, or by local orographic rainfall enhancement.

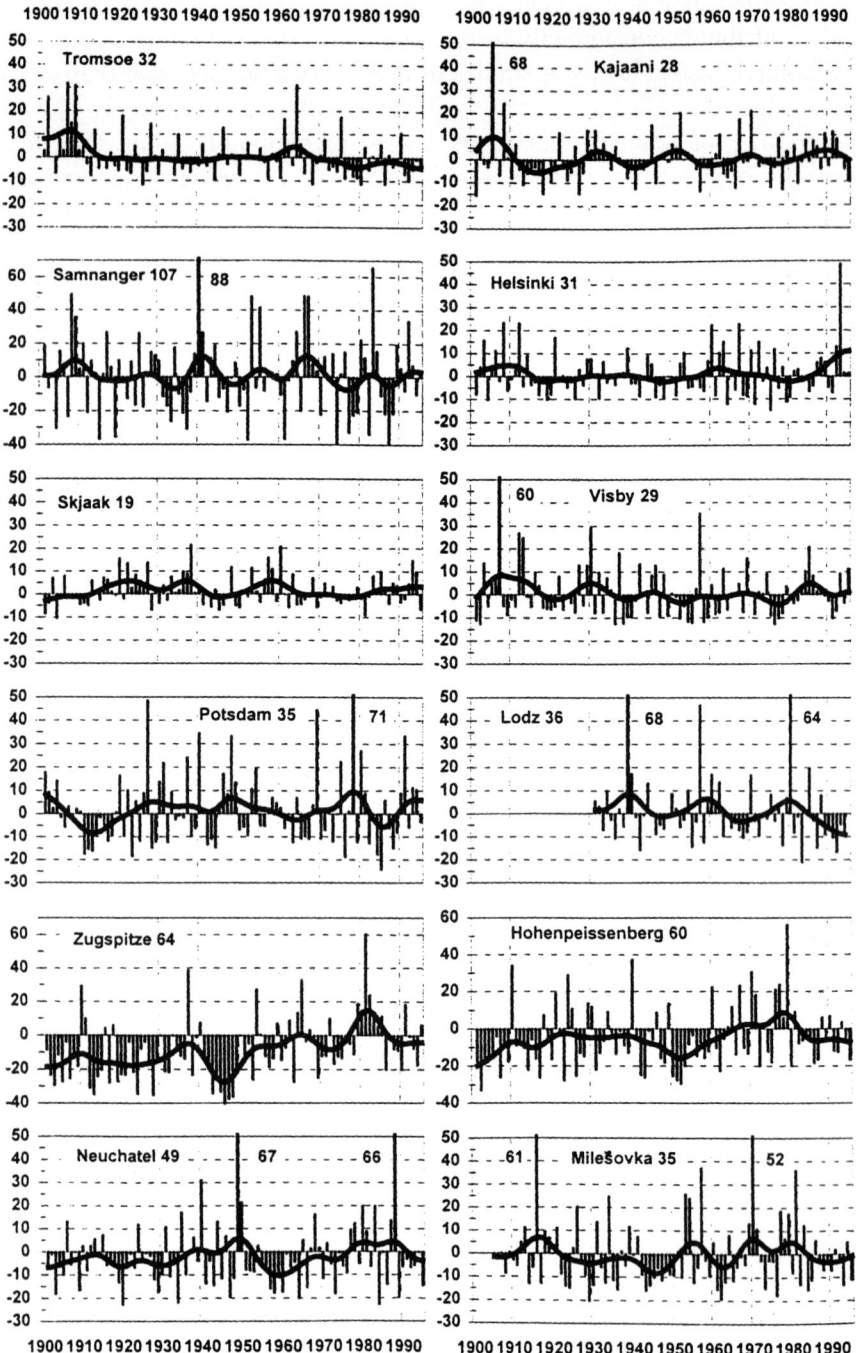

Figure 5. Annual anomalies of maximum 1-day precipitation (mm) for selected stations (corresponding mean after the station name), smoothed with a low pass filter with Gaussian weighting coefficients equivalent to 10 year moving averages. Reference period: 1961-1990. Data period 1901-1995 where possible. Single columns exceeding 50 mm are indicated by their values.

- One local 'accidental' outlier may influence filtered trend curves for several years.
- 1-day precipitation values are very vulnerable to misreading and to erroneous sampling interval, e.g. the reading represents rainfall from a longer period than 24 hours. Most of these erroneous values are corrected by quality control systems, but in some cases it is difficult to judge whether an observation is true or false.
- An extreme 24-hour rainfall event may be split into two parts by the regular observing hours. Thus the trend in 1-day precipitation may be influenced by number of 'accidental hits' of extreme 24h rainfall events being focused between 06-06 UTC.
- Inhomogeneities may seriously influence trends in series of annual precipitation (cf. Hanssen-Bauer and Førland, 1994; Heino, 1994b). However, for most of the causes of inhomogeneities (e.g. small relocations, installation of a wind shield, changes in instruments, changes in the environment), the adjustment factor will be smaller for heavy 1-day rainfall than for annual precipitation. This is because inhomogeneities mainly affect the measurement of snowfall and drizzle or small raindrops which are more influenced by airflow near the gauge.

Accordingly, a series of 1-day precipitation from a single station is not an ideal indicator for trend analyses in extreme rainfall. Conclusions concerning regional trends of maximum daily precipitation should be based on trend studies from a dense network of stations.

Rebetez et al. (1997) have shown for Switzerland that increases in extreme 3-day precipitation in August and September are highly significant, implying that even if trends cannot be established in annual sums of precipitation (Beniston et al., 1994), they may be apparent in series for particular seasons or months.

For stations exposed to humid westerly winds in north-western Europe there is a coupling between precipitation totals and the North Atlantic Oscillation (NAO) index. Some of the monthly 1-day maxima also show a fairly strong connection with this large-scale index, as shown by data from the extremely wet Samnanger station in western Norway (Table VII).

Table VII

Correlation coefficients (in hundredths) between the NAO-index (monthly pressure difference between Ponta Delgada, Azores and Stykkisholmur, Iceland) and the monthly maximum 1-day precipitation at Samnanger (Nor), 1901-1996.

J	F	M	A	M	J	J	A	S	O	N	D	Y
38	52	40	14	30	14	38	28	22	20	11	28	4

On the average there is a coupling on monthly basis. A vigorous Icelandic low (high value of the NAO-index) favours large daily precipitation maxima in western Norway. However, on an annual basis this connection is practically zero. Shorter interludes of strong cyclonic activity in the Iceland-Norway area can also give occasions of very large 1-day precipitation during a year having a low value on the NAO-index.

3.2.1.3. Applications. Precipitation amounts and intensity constitute important information for planning engineers. For example, in Germany, based on extreme event statistics maximum 15 min precipitation amounts with return periods of 1 year have been calculated for a certain location or area and have traditionally been used for the design of drainage systems. Today return periods of from 0.5 to 100 years for up to 18 duration times are used in the design of drainage systems. An increase in maximum daily precipitation may be a hint that the capacity of the drainage systems may be exceeded more often than expected. It is clear that upgrading a drainage system will be really costly.

In Germany extreme precipitation events have been investigated within the KOSTRA project. Data of 5 min precipitation amounts cover the period 1951-1980 (Bartels et al., 1997). In order to extract different return periods, Gumbel distributions (WMO, 1981) have been fitted to the data. The study shows that in general precipitation amounts increase from north-west to south-east. Looking e.g. at the 24h precipitation amounts expected once a year, values vary from less than 35 mm in most parts of Germany to more than 80 mm in the Alps. A new project has just been set up to investigate the impact of climate change on hydro-meteorology in Germany.

For the insurance industry in Norway, a precipitation event is characterised as 'extraordinary' if the return period is higher than 5-10 years, depending on the weather situation (frozen ground, combination with snow melt, preceding heavy precipitation, etc.).

For calculating return period values of 1-day precipitation, the Gumbel distribution is still that most commonly used (WMO, 1981). This distribution has also been used for analyses of maximum 1-day precipitation in the former Czechoslovakia (Šamaj et al., 1982) and for the Carpathian countries (Šamaj et al., 1985). By applying the Gumbel distribution to annual series, return period values for the series included in this study were calculated. For the Nordic countries, additional series from the REWARD-dataset were also used. Table VIII shows the frequencies of 1-day precipitation exceeding the 5-year return period value.

For the Nordic countries, Table VIII indicates higher frequencies of 'extraordinary' rainfalls in the 1930s and since the 1980s. Periods of high 1-day precipitation values seem to coincide well with hot summers. For the Central European series, the station density is too low to draw certain conclusions.

Table VIII

Number of cases (per station and year) with maximum annual 1-day precipitation larger than the 5 year return period value in each decade.

	FIN	SWE	NOR	GER	POL	CR	SWI
No of stations	10	20	24	3	1	2	2
1900-09	0.18	0.15	0.26	0.07	-	-	-
1910-19	0.11	0.19	0.13	0.20	-	-	-
1920-29	0.16	0.18	0.15	0.20	-	0.25	-
1930-39	0.25	0.25	0.25	0.20	0.11	0.25	0.15
1940-49	0.10	0.21	0.17	0.27	0.20	0.20	0.25
1950-59	0.10	0.19	0.15	0.10	0.10	0.15	0.10
1960-69	0.19	0.18	0.17	0.23	0.20	0.10	0.15
1970-79	0.17	0.16	0.16	0.33	0.10	0.35	0.25
1980-89	0.18	0.24	0.22	0.30	0.20	0.20	0.25
1990-96*	0.29	0.11	0.27	0.26	0.00	0.07	0.07

*) 1990-94 for Germany and Poland

3.2.2. Precipitation days with totals ≥ 10 mm

Meteorological yearbooks have typically contained standard statistics on the number of days with precipitation exceeding specific amounts (e.g. 0.1, 1.0, 10.0 mm). Series of their monthly values are therefore more easily available than complete series of daily precipitation.

The number of days with precipitation exceeding 10 mm (RD10) is one of the key elements for studying climatological extremes. However, it should be noted that RD10 is not a perfect indicator of extreme precipitation, either. At some stations daily precipitation of more than 10 mm occurs rather seldom, while it is no extreme at all at other stations, e.g. on the leeward side of the Norwegian mountains RD10 is less than 5 days/year at some stations, while in areas with orographically-enhanced precipitation in the western mountains it is more than 115 days/year (even every second day during the months September-December).

Examples of the variability and trends of RD10 are shown in Figure 6 and Table IX. Any long-term changes are difficult to see, except the increasing trends for the German mountain stations as well as that in Western Norway due to orography and circulation changes.

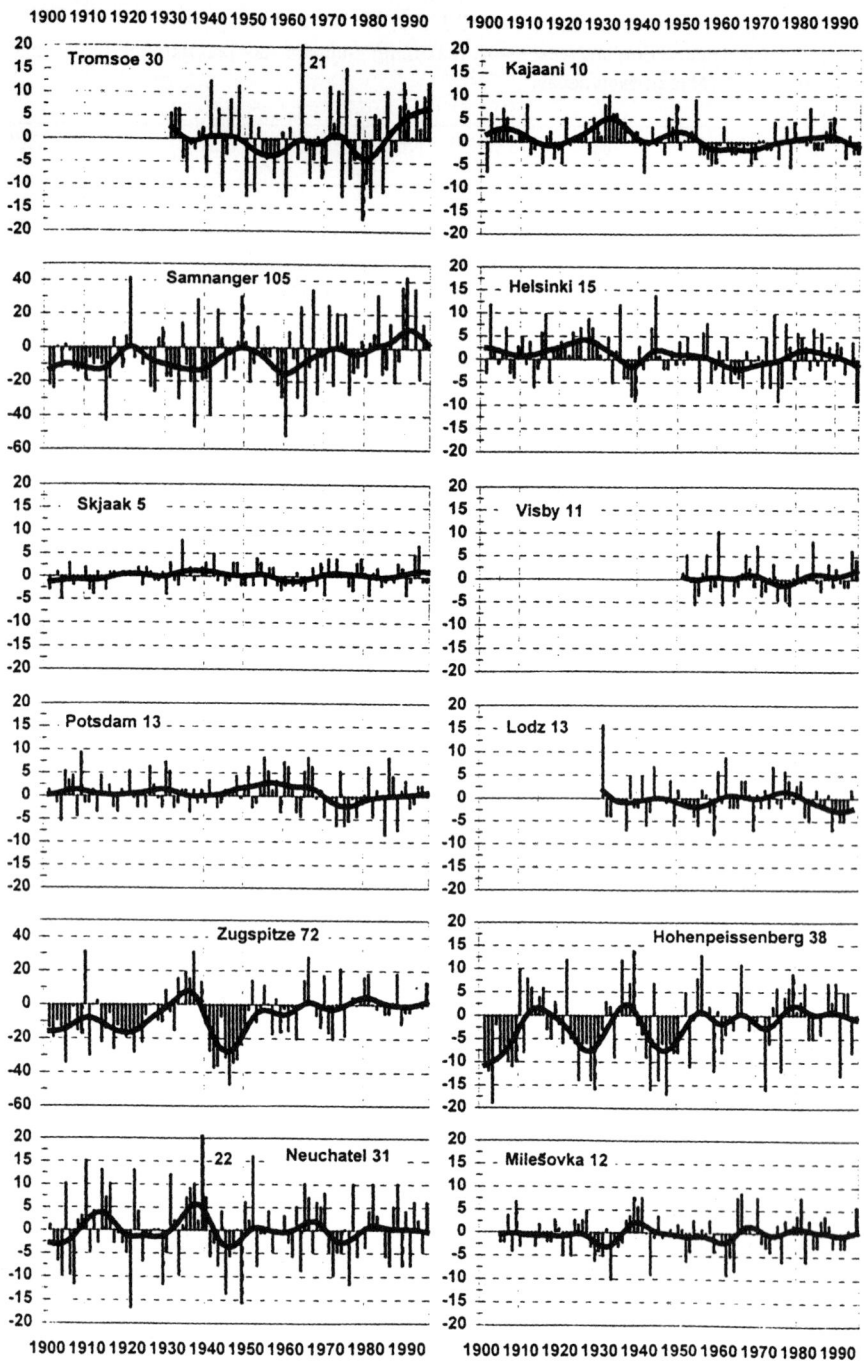

Figure 6. Annual anomalies of days with precipitation ≥ 10 mm for selected stations (corresponding mean in days after the station name), smoothed with a low-pass filter with Gaussian weighting coefficients equivalent to 10-year moving averages. Reference period: 1961-1990. Single columns exceeding 20 mm are indicated by their values.

Table IX
Linear trends in the anomalies of the numbers of precipitation days per year ≥10 mm (days/10 years)
at each station in 1910-1995, together with the statistical significance. Left column is the linear
trend tested with two-sided t-test. Right column is Sen's nonparametric estimate of slope (Gilbert,
198/) and the significance of the trend is evaluated with nonparametric Mann-Kendall test (Sneyers,
1990); (* 5%, ** 1%, *** 0.1%).

station	linear trend	Sen's estimate
Sodankylä (Fin)	-0.07	0.00
Kajaani (Fin)	-0.24	-0.19
Skjaak (Nor)	0.05	0.00
Samnanger (Nor)	1.88*	1.56
Helsinki (Fin)	-0.31	-0.28
Potsdam (Ger)	-0.06	0.00
Milešovka (CR)	0.19	0.16
Hohenpeissenberg (Ger)	0.21	0.27
Zugspitze (Ger)	1.59*	1.75*
Neuchatel (Swi)	-0.22	0.00

The paper by Beniston et al. (1994) has shown that in Switzerland there is a
strong interannual variability of precipitation, which does not allow any
significant trend to be established; however, Rebetez (1995) has demonstrated
that, particularly in summer, high monthly temperatures are correlated with low
monthly precipitation. This is confirmed by modelling studies using high-
resolution Regional Climate Models centered over the Alpine region (Beniston et
al., 1995; Marinucci et al., 1995; Rotach et al., 1997).

3.3. OTHER SELECTED EXTREMES

3.3.1. Geostrophic wind in the German Bight
In Central Europe, strong winds related to extratropical cyclones, causing widespread damage, have been observed in the 1990s (WMO, 1995). It was,
therefore, necessary to study whether observations showed any long-term
increase in strong winds.

Because observational records of wind speed and direction are highly biased
due to changes in observational instruments and methods as well as the
roughness of the environment, trends in extratropical storminess are best
assessed by using indices based on sea-level pressure measurements or fields.
Measurements of air pressure, even in the 19th century, can be assumed
relatively homogeneous, given correct reduction to mean sea level, gravity and

compensation for any change in observing hour. This is because the instrumentation has not changed much and air pressure is independent of surface characteristics. The geostrophic wind computed from the pressure readings of a few stations can thus be regarded as an estimate of the real wind. Any trend in the wind statistics will be reflected in the geostrophic wind statistics. By this method annual frequency distributions of daily wind can be obtained for long periods. However, small-scale and very short-term storms may not be detected by this method if daily, low-resolution (e.g. 150 km) pressure gradients are used.

To ascertain whether storminess is really changing, a systematic examination of the geostrophic wind in the German Bight, in the south-eastern part of the North Sea, has been carried out (Schmidt and v. Storch, 1993).

The different data bases of pressure, thoroughly checked and evaluated, were as follows (Behrens and Schmidt, pers. comm):
- daily pressure data, one and three readings per day, for three stations around the German Bight for the period 1878 to 1992,
- all available pressure data for the region, taken from synoptic reports, usually 10 to 19 per day, analysed for time steps of three hours for the period 1949-1995.
Pressure gradients and wind speeds estimated from the analysed maps may differ substantially from those estimated from the triangle of stations, because the maps can incorporate closed isobars and spatial changes of pressure gradient.

Figure 7 shows the annual percentage frequencies of geostrophic wind speed in the German Bight exceeding the stated threshold (20 m/s), based on both the station-triangle and the mapped pressure fields.

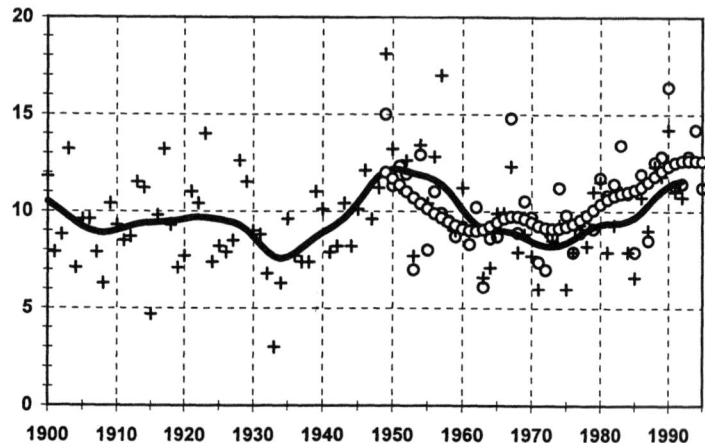

Figure 7. Annual percentage frequencies of geostrophic wind in excess of 20 m/s in the German Bight, smoothed with a low-pass filter with Gaussian weighting coefficients equivalent to 10-year moving averages, period 1901-1992 with one observation per day (crosses/black line) and period 1949-1995 with 8 observations per day (circles/line of circles), cf. the text for further explanations. Reference period: 1961-1990 (Behrens and Schmidt, pers. comm.).

The two methods give comparable results. Neither of the time series shows any indication of a long-term trend. Evaluations of the geostrophic wind for annual means or for higher wind speed thresholds also lead to the same conclusion.

The results presented here are in general agreement with analysis of the historical storm climate in north-western Europe (Alexandersson et al., 1998). The homogeneous storm indicators show no clear long-term trend, but a strong inter-decadal variability. The recent increase of storminess from around the 1960s falls within the observed variability and the present level is comparable with that at the end of the 19th century.

3.3.2. Thunderstorms

Thunderstorms belong to a class of weather phenomena in which much damage occurs. This is due on the one hand to direct lightning strikes (deaths, fires), on the other hand to the accompanying phenomena of thunderstorms, such as hailstorms, downpours or cloud-bursts and strong wind gusts. For those characteristics which are not directly measured at the weather stations (e.g. other than the total precipitation, wind speed), their observation is affected by the capability of the observer. Since the possible homogenisation of series of local extreme weather phenomena is owing to the local nature of their occurrence, considerably limited, we are compelled to rely on the quality of the original observations. Although some papers investigating the fluctuation of thunderstorms and related weather extremes in the USA do exist (see e.g. Changnon, 1985; Changnon, 1997), on the European scale climatological analyses have a more or less traditional descriptive orientation.

Three professional weather stations in the Czech Republic were selected for analysis of the above phenomena during 1946-1995, namely the mountain observatory Milešovka (for more information see Stekl and Podzimek, 1993), the typical urban station Prague-Karlov and the more or less rural station Brno-airport (see Figure 1). The objective of the study was thunderstorms (at the station, i.e. within 3 km from the station - THUN), hailstorms (HAIL) and extreme precipitation represented by precipitation days with sum ≥ 20.0 mm (PREC). PREC includes not only heavy precipitation during downpours or cloud-bursts but also days with persistent frontal rainfall. During 1946-1995 there have been no changes in the methodology of observations. As an example, the series of annual frequencies of occurrence of THUN, HAIL and PREC obtained are presented in Figure 8 for Milešovka. The series are expressed as anomalies from the indicated 1961-1990 means.

For the THUN series the negative linear trends are statistically significant (5% level) for all three stations (Milešovka -1.45 cases/10 yrs, Prague-Karlov -1.39, Brno-airport -1.51), while for HAIL for Prague-Karlov (-0.39) and Brno

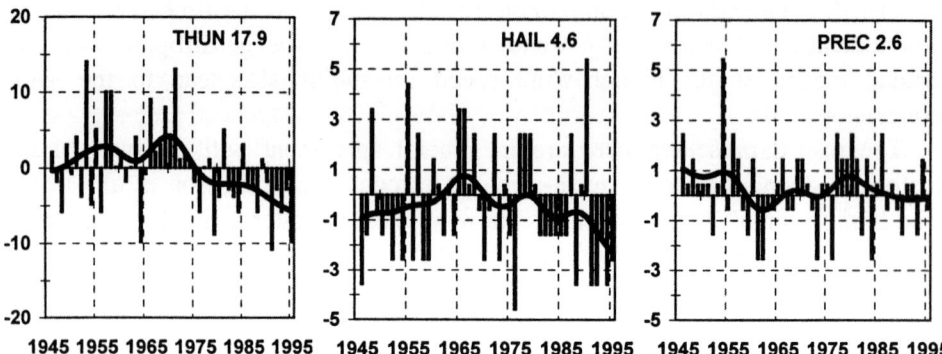

Figure 8. Annual deviations of frequencies of thunderstorms (THUN), hailstorms (HAIL) and heavy rainfall (PREC) at Milešovka (Czech Republic) in 1946-1995 (reference period 1961-1990, corresponding mean in days after the station name); smoothed with a low-pass filter with Gaussian weighting coefficients equivalent to 10-year moving averages.

airport (-0.17) and for PREC none of the negative trends is statistically significant. The decrease in annual frequencies of all three phenomena is related to the same tendency for the cyclonic weather types in which around 90 % of these phenomena occur (Brázdil et al., 1998). Simultaneously during 1946-1995 a positive linear trend in annual air temperature in the area of the Czech Republic was observed (0.15°C/10 yrs - Brázdil and Macková, 1998).

From the practical point of view, the impacts of thunderstorms and related weather extremes are important. These were studied in the region of south Moravia in 1957-1995 (area around 9000 km^2). As a source of information the regional newspaper Rovnost was used, in which reports about the damage, including picture documentation, appear. Out of the 135 records found, in 48.2 % damage due to downpours and thunderstorms was reported, in 37.8 % damage due to lightning, in 29.6 % due to hail and in 14.1 % due to gales. Cases with damage due to several factors overlap so the totals exceeds 100%. In south Moravia the mean annual frequency of occurrence of an extreme phenomenon with damage was 3.5 for THUN (i.e. 6.3 % of days with THUN), 1.0 for HAIL (i.e. 22.9 % of days with HAIL) and 1.7 for PREC (i.e. 10.8 % days with PREC) (Brázdil et al., 1998). This is consistent with Changnon (1997), who states that only 5 to 10 % of all thunderstorms and 10 to 25 % of all hailstorms produce damage.

3.4. CLIMATIC EXTREMES DERIVED FROM PROXY DATA

Because instrumental records are relatively short, documentary evidence about weather and climate as well as other proxy data are used to construct long time series. Proxy data, however, must first be transformed to quantities equivalent to modern instrumental data, to allow correct interpretation of the concatenated series. An example is given from the Swiss Mittelland, the region between the Jura mountains and the Alps, where most of the population lives.

Preparation of these data involved two steps. In the first one the monthly temperature anomalies were defined. The data fall into an instrumental sub-period that begins with the temperature series of Basel in 1755, and a pre-instrumental sub-period preceding that year. Within the instrumental sub-period, anomalies are defined as departures from the 1901-1960 average of more than 180 per cent of the standard deviation. The anomalies of the pre-instrumental sub-period were estimated from documentary proxy data (observations of ice and snow features, phenological and biological observations), from tree-ring data and from descriptive data in historical documents. The values are expressed in terms of an index of seven discrete levels: -3, -2, -1, 0, +1, +2, +3 (Pfister 1995; 1998).

In the second step the monthly anomalies were classified according to precipitation. For the instrumental sub-period (from 1864) the classification draws on an aggregate of four precipitation series from the main parts of the Mittelland and it is based on departures from the 1901-1960 average of this aggregate series. Months with a positive precipitation deviation of greater than 130 per cent of the standard deviation are designated 'wet', those with a negative precipitation deviation of more than 130 per cent of the standard deviation are designated 'dry'. Months within these thresholds are not designated. For the pre-instrumental sub-period prior to 1864 the classification again draws on an index of seven discrete levels. This index is based on the number of rainy days from daily weather observations and information on severe floods. This combined temperature-precipitation classification thus yields six types of anomalies: Warm-wet, warm-dry, warm; cold-wet, cold-dry and cold (Figure 9). Anomalies in both periods include only index levels of -3 and +3.

The following conclusions can be derived:
- The number of anomalies displays large fluctuations: In some decades (e.g. 1546/55, 1656/65, 1906/15) anomalies hardly occurred at all, whereas in others (e.g. 1676-1685) they were abundant.
- Cold and dry anomalies are associated with persistent blocking anticyclones over the North Sea or Scandinavia (negative mode of the North Atlantic Oscillation) and advection of cold and dry continental air masses. Months of this type frequently occurred during the last two Little Ice Age events (late sixteenth

to early seventeenth century) and 1810 to the late nineteenth century. They were rather rare between 1496 and 1565 and again from 1926 to 1995.
- The onset of the second Little Ice Age event stands out by virtue of a sudden 'cold' shift: Between 1556/65 and 1566/75 the number of cold anomalies increased sevenfold.
- The last decade from 1976/85 to 1986/1995 stands out on account of a 'warm' shift in which the number of warm anomalies rose to its maximum over the last 500 years. In conjunction with worldwide evidence for rising temperatures and glacial retreat (Nicholls et al., 1996), this might be seen as an indication of global warming.

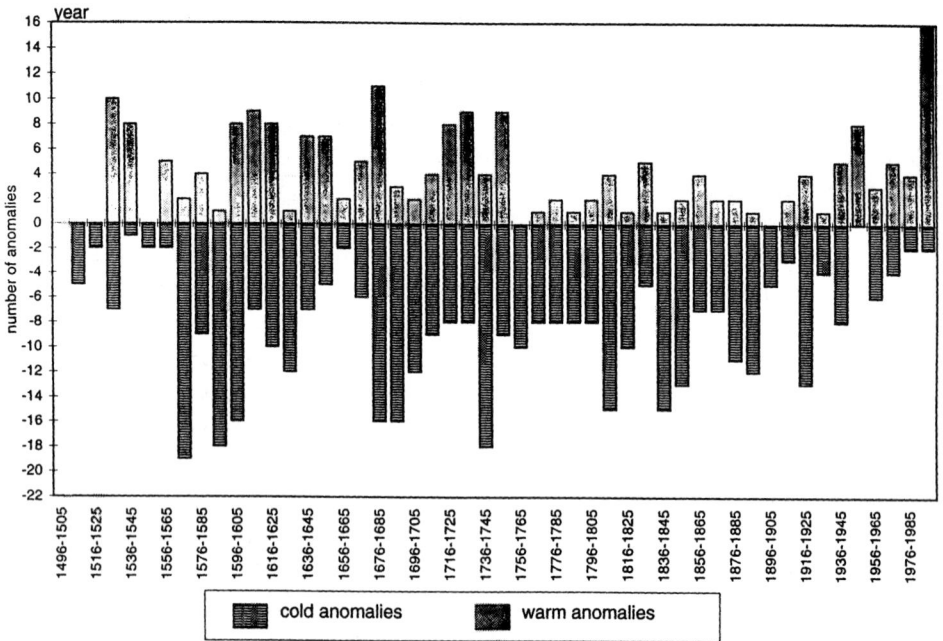

Figure 9 Number of monthly temperature and precipitation anomalies per decade in 1496-1995 in the Swiss Mittelland (cf. the text for further explanations).

4. Concluding Remarks

This paper has concentrated on some aspects of climatic extremes in Northern and Central Europe. Based on data for a few stations, results of the analysis of selected climatic elements were presented, mostly for the period 1901-1995. They include

- a decrease in the diurnal temperature range (Figure 3, Table III)
- a decrease in the number of frost days (Table IV)
- no major change in daily precipitation extremes (Figure 5, Table VI)
- no major change in the number of precipitation days ≥10 mm (Figure 6)
- no long-term intensification of strong winds (Figure 7) and
- a decrease in occurrences of thunderstorms and hails (Figure 8).

Some of the extreme elements are very sensitive to various inhomogeneities in the data. However, the quality of the daily maximum and minimum temperatures can be improved using the same methods that are widely used for the homogenization of mean temperatures. Precipitation extremes are less sensitive to changes in station location or instrumentation, except at mountain stations. However, the normal quality checking procedures should be applied effectively. Homogenized air pressure data, on the other hand, provide a good possibility for analysis of wind speed extremes and other circulation parameters.

A new area in the study of climatic extremes is the use of proxy climatic data. Their calibration with long-term instrumental records provides the necessary ability to create extra long records for extreme studies also covering the pre-instrumental period (Figure 9).

This study has revealed that extreme-related data sets are already partly ready or in preparation in many European countries. Regional action to collect the information on these data and subsequently, the data themselves, should be initiated in order to make the data available for global studies. Studies on climatic extremes in Europe and its regions need to extend beyond the traditional parameters of air temperature and precipitation. As discussed in this paper, the scope for weather and climate extremes to have impacts on nature and human society in densely populated Europe is very high. In this way Europe, with its dense network of stations and great potential of climatologists, can positively contribute to the third climate change assessment of the Intergovernmental Panel on Climate Change.

In addition to the purely observational-based studies on extremes, efforts need to be made, and are indeed currently taking place in a number of climate modelling centres, to investigate numerical model capabilities in simulating variability and extremes. Intuitively-reasonable variability can already be identified in high-resolution General Circulation Models. While models are still far from being accurate in terms of extremes, the prospects of using detailed Regional Climate Models to investigate extremes are improving. It should be stressed that extremes constitute the major stress factor on a number of environmental and associated socio-economic systems, and that if models are to be of use to the climate impacts community, they need to be capable of simulating current extreme event episodes, as well as the shifts in the extremes in relation to the changes in climate in the future. It is also in this sense that the joint efforts reflected in this paper will become increasingly important in the future, as a means of validating model simulations of extreme climatic events.

Acknowledgements

The authors would like to express their thanks as follows:
- H. Alexandersson, E. Førland, R. Heino and H. Tuomenvirta:
for financial support from the Nordic Council of Ministers to the REWARD-project (NMR-contract FS/HFj/X-93001)
- R. Brázdil:
for financial support from the Grant Agency of the Czech Republic (No. 205/95/0509 and 205/96/0527)
- C. Pfister:
for financial support from the Swiss National Science Foundation, Priority Programme Environment (No. 5001, 34888)
- M. Rebetez:
for the support of the Swiss Forest Investigation Programme
- G. Rosenhagen and S. Rösner:
for the contributions made by Jutta Herzog and Gerhard Müller-Westermeier.

In addition, all the authors express their thanks to the National Meteorological Services of their respective countries for providing the necessary data.

References

Alexandersson, H.: 1986, 'A homogeneity test applied to precipitation data', *J. Climatol. 6*, 661-675.

Alexandersson, H., Schmith, T., Iden, K., and Tuomenvirta, H.: 1998, 'Long-term trend variations of the storm climate over NW Europe', *Global Atmosphere-Ocean System* (accepted).

Bartels, H., Malitz, G., Asmus, S., Albrecht, F.M., Dietzer, B., Günther, T., and Ertel, H.: 1997, '*Starkniederschlagshöhen für Deutschland*', Deutscher Wetterdienst, Offenbach am Main.

Beniston, M., Rebetez, M., Giorgi, F., and Marinucci, M.R.: 1994, 'An analysis of regional climate change in Switzerland', *Theor. Appl. Climatol. 49*, 135 - 159.

Beniston, M., Ohmura, A., Rotach, M., Tschuck, P., Wild, M., and Marinucci, M.R.: 1995, '*Simulation of climate trends over the Alpine Region: Development of a physically-based modeling system for application to regional studies of current and future climate*', Final Scientific Report Nr.4031 - 33250 to the Swiss National Science Foundation, Bern, Switzerland, 198 pp.

Brázdil, R.: 1993, 'The homogeneity of air temperature measurements Prague-Klementinum with respect to the intensification of the urban heat island', *Zeszyty Naukowe Univ. Jagiell., Prace Geograficzne 95*, 67-74.

Brázdil, R., Machů, R. and Budíková, M.: 1994, 'Temporal and spatial changes in maxima and minima of air temperature in the Czech Republic in the period of 1951-1993', In: Brázdil, R. and Kolář, M. (Eds.), *Contemporary Climatology*, Brno, 93-102.

Brázdil, R. and Kolář, M. (Eds.): 1994, '*Contemporary Climatology*', Proceedings of the Meeting of the Commission on Climatology of the International Geographical Union, Brno, 620 pp.

Brázdil, R., Budíkova, M., Auer, I., Böhm, R., Cegnar, T., Faško, P., Lapin, P., Gajic-Čapka, M., Zaninovic, K., Koleva, E., Niedzwiedz, T., Ustrnul, Z., Szalai, S., and Weber, R.O.: 1996, 'Trends of maximum and minimum daily temperatures in Central and Southeastern Europe', *Int. J. Climatol. 16*, 765-782.

Brázdil, R. and Macková, J.: 1998, 'Series of annual air temperatures for the Czech Republic in 1828-1995 (in Czech)', *Meteorol. Zpr. 51*, 17-21.

Brázdil, R., Štěpánek, P., and Vais, T.: 1998, 'Temporal and spatial analyses of thunderstorms, hailstorms and extreme precipitation in the south part of Moravia in 1946-1995 (in Czech)', *Meteorol. Zpr. 51*, 45-52.

Böhm, R.: 1992, 'Lufttemperaturschwankungen in Österreich seit 1775', *Österr. Beitr. z. Met. u. Geophys. 5*, 96 pp.

Changnon, D.: 1997, 'Damaging storms in the United States: selection of quality data and monitoring indices', In: *Workshop on Indices and Indicators for Climate Extremes*, NCDC, NOAA, Asheville, 3-6 June 1997, 24 pp.

Changnon, S.A.: 1985, 'Secular variations in thunder-day frequencies in the twentieth century', *J. Geophys. Res. 90*, 6181-6194.

Easterling, D.R. and Peterson, T.C.: 1995, 'A new method for detecting undocumented discontinuities in climatological time series', *Int. J. Climatol. 15*, 369-377.

Easterling, D., Horton, B., Jones, P., Peterson, T., Karl, T., Parker, D., Salinger, J., Razuvayev, V., Plummer, N., Jamason, P., and Folland, C.: 1997, 'Maximum and minimum temperature trends for the globe', *Science 277*, 364-366.

ECSN: 1995, '*Climate of Europe - Recent variation, present state and future prospects (First European Climate Assessment)*', KNMI, De Bilt, 72 pp.

Førland, E.J., van Engelen, A., Ashcroft, J., Dahlström, B., Demaree, G., Frich, P., Hanssen-Bauer, I., Heino, R., Jonsson, T., Mietus, M., Müller-Westermeier, G., Pålsdottir, T., Tuomenvirta, H., and Vedin, H.: 1996a, 'Changes in 'normal' precipitation in the North Atlantic region (second edition)', *DNMI Report 7/96 Klima*, 27 pp.

Førland, E.J., Alexandersson, H., Frich, P., Hanssen-Bauer, I., Heino, R., Helminen, J., Jonsson, T., Nordli, P.Ø., Pålsdottir, T., Schmith, T., Tuomenvirta, H., and Tveito, O.E.: 1996b, 'REWARD, Progress report 01.01.1996 - 30.09.1996', *DNMI-Report 30/96 KLIMA*, 14 pp.

Førland, E. (coordinator), Alexandersson, H., Frich, P., Hanssen-Bauer, I., Heino, R., Helminen, J., Jonsson, T., Nordli, P.Ø., Pålsdottir, T., Schmith, T., Tuomenvirta, H., & Tveito, O.E.: 1998, 'REWARD - Relating Extreme Weather to Atmospheric circulation using a Regionalized Dataset - Final Report' *DNMI-Report 17/98 KLIMA*, 23 pp.

Frich, P. (coordinator), Alexandersson, H., Ashcroft, J., Dahlström, B., Demaree, G.R., Drebs A., van Engelen, A., Førland, E.J., Hanssen-Bauer, I., Heino, R., Jonsson, T., Keegan, L., Nordli, P.Ø., Schmith, T., Stefffensen, P., Tuomenvirta, H., and Tveito, O.E.: 1996, 'North Atlantic Climatological Dataset (NACD Version 1) - Final report', European Commission (DG-XII), Environment Programme, Contract (EV5V CT93-0277), *Danish Meteorological Institute, Scientific Report, 96-1*, 47 pp.

Gerstengarbe, F.-W. and Werner, P.C.: 1993, 'Extreme klimatologische Ereignisse an der Station Potsdam und an ausgewählten Stationen Europas', *Berichte des Deutschen Wetterdienstes Nr. 186*, Offenbach am Main.

Gilbert, R.O.: 1987, *'Statistical methods for environmental pollution monitoring'*, Van Nostrand Reinhold Co., New York, 320 pp.

Hanssen-Bauer, I. and Førland, E.J.: 1994, 'Homogenizing long Norwegian precipitation series', *J. Climate 7*, 1001-1013.

Hanssen-Bauer, I., Førland, E.J., Tveito, O.E., and Nordli, P.Ø.: 1997, 'Estimating regional precipitation trends - comparisons of two methods', *Nordic Hydrology 28*, 21-36.

Heikinheimo, P. (Ed.): 1995, 'International Conference on Past, Present and Future Climate. Proceedings of the SILMU conference held in Helsinki, Finland 22-25 August 1995'. *Publications of the Academy of Finland 6/95*, Helsinki, 490 pp.

Heino, R.: 1979. 'Urban effect on climatic elements in Finland.' *Geophysica 15*, p. 171-188.

Heino, R. (Ed.): 1994a, 'Climate variations in Europe. Proceedings of the European Workshop held in Kirkkonummi (Majvik), Finland 15-18 May 1994', *Publications of the Academy of Finland 3/94*, Helsinki, 386 pp.

Heino, R.: 1994b, 'Climate in Finland during the period of meteorological observations', *Finnish Meteorological Institute Contributions 12*, 209 pp.

Heino, R.: 1996, 'Data homogeneity and metadata', In: Obrebska-Starkel, B. & Niedzwiedz, T. (Eds.): *Proceedings of the International Conference on Climate Dynamics and the Global Change Perspective*, Cracow, Poland, 17-20 October 1995, p. 13-21.

Houghton, J.T., Meira Filho, L.G., Callander, B.A., Harris, N., Kattenberg, A., and Maskell, K.: 1996, *'Climate Change 1995 - The Science of Climate Change'*, Intergovernmental Panel on Climate Change, Cambridge Univ. Press, 572 pp.

Karl, T.R., Jones, P.D., Knight, R.W., Kukla, G., Plummer, N., Razuvayev, V., Gallo, K.P., Lindseay, J., Charlson, R.J., and Peterson, T.C.: 1993, 'A new perspective on recent global warming: Asymmetric trends of daily maximum and minimum temperature', *Bull. Amer. Meteor. Soc. 74*, 1007-1023.

Kattenberg, A., Giorgi, F., Grassl, H., Meehl, G., Mitchell, J., Stouffer, R., Tokioka, T., Weaver, A. and Wigley, T.: 1996, 'Climate models - Projections of future climate', In: Houghton, J.T., Meira Filho, L.G., Callander, B.A., Harris, N., Kattenberg, A., and Maskell, K. (Eds.): 1996, *Climate Change 1995 - The Science of Climate Change*, Intergovernmental Panel on Climate Change, Cambridge Univ. Press, 285-357.

Katz, R.W. and Brown, B.G.: 1992, 'Extreme events in a changing climate: Variability is more important than averages', *Climatic Change 21*, 289-302.

Marinucci, M. R., Giorgi, F., Beniston, M., Wild, M., Tschuck, P., and Bernasconi, A.: 1995, 'High resolution simulations of January and July climate over the Western Alpine region with a nested regional modeling system', *Theor. Appl. Climatol. 51*, 119-138.

Nicholls, H., Gruza, G., Jouzel, J., Karl, T., Ogallo, L., and Parker, D.: 1996, 'Observed climate variability and climate', In: Houghton, J.T., Meira Filho, L.G., Callander, B.A., Harris, N., Kattenberg, A. and Maskell, K. (Eds.): *Climate Change 1995 - The Science of Climate Change*, Intergovernmental Panel on Climate Change, Cambridge Univ. Press, 133-192.

Pfister, C.: 1995, 'Monthly temperature and precipitation in central Europe from 1525-1979: quantifying documentary evidence on weather and its effects', In: Bradley, R. S. and Jones, P. D. (Eds.), *Climate since A.D. 1500'*, Routledge, London, 118-142.

Pfister, C.: 1998, *'Wetternachhersage, 500 Jahre Klimavariationen und Naturkatastrophen 1496-1995'*, Bern (Haupt).

Rebetez, M.: 1995, 'Seasonal relationship between temperature, precipitation and snow cover in a mountainous region', *Theor. Appl. Climatol. 54*, 99-106.

Rebetez, M., Lugon, R., and Baeriswyl, P.-A.: 1997, 'Climatic change and debris flows in high mountain regions: the case study of the Ritigraben torrent (Swiss Alps)', *Climatic Change 36*, 371-389.

Rotach, M. W., Marinucci, M. R., Wild, M., Tschuck, P., Ohmura, A., and Beniston, M.: 1997, 'Nested regional simulations of climate change over the Alps for the scenario of a doubled greenhouse forcing', *Theor. Appl. Climatol.* (in press).

Šamaj, F., Valovič, Š., Brázdil, R., and Gulčíková, V.: 1982, 'Maximum daily totals of atmospheric precipitation in the CSSR (in Czech)', *Meteorol. Zpr. 35*, 129-135.

Šamaj, F., Valovič, Š., and Brázdil, R.: 1985, 'Daily precipitation sums with extraordinary intensity in the CSSR in 1901-1980 (in Slovak)', *Zbornik prác Slovenského hydrometeorologického ústavu 24, 9-112.*

Schmidt, H. and v. Storch, H.: 1993, 'German Bight storms analysed', *Nature 365*, 791.

Sneyers, R.: 1990, 'On the statistical analysis of series of observations', *Technical Note 143, WMO - No. 415*, 192 pp.

Von Storch, H., Guddal, J., Iden, K.A., Johnson, T., Perlnitz, J., Reistad, M., Ronde, de, J., Schmidt, H., and Zorita, E.: 1993, 'Changing statistics of storms in the North Atlantic', *MPI report, No. 116*, 19 pp.

Stekl, J. and Podzimek, J.: 1993, 'Old mountain meteorological station Milešovka (Donnersberg) in Central Europe', *Bull. Amer. Meteorol. Soc. 74*, 831-834.

Tuomenvirta, H. and Heino, R.: 1996, 'Climatic changes in Finland - recent findings', *Geophysica 32*, 61-75.

WMO: 1981, 'Selection of distribution types for extremes of precipitation', WMO-No. 560, *Operational Hydrology Rep.15*, World Meteorological Organization, Geneva, 64 pp.

WMO: 1995, *'The Global Climate System Review'*, WMO-No. 819, World Meteorological Organization, Geneva, 150 pp.

(Received 5 November 1997; in revised form 1 September 1998)

CHANGES IN CLIMATE EXTREMES OVER THE AUSTRALIAN REGION AND NEW ZEALAND DURING THE TWENTIETH CENTURY

NEIL PLUMMER[1], M. JAMES SALINGER[2], NEVILLE NICHOLLS[1], RAMASAMY SUPPIAH[3], KEVIN J. HENNESSY[3], ROBERT M. LEIGHTON[1], BLAIR TREWIN[4], CHER M. PAGE[3], and JANICE M. LOUGH[5]

[1]*Bureau of Meteorology, GPO Box 1289K, Melbourne, Victoria 3001, Australia*
[2]*National Institute of Water and Atmospheric Research, PO Box 109 695, Newmarket, Auckland, New Zealand*
[3]*CSIRO Division of Atmospheric Research, PMB No.1, Aspendale Victoria 3195, Australia*
[4]*The University of Melbourne, School of Earth Sciences, Parkville, Victoria 3052, Australia*
[5]*Australian Institute of Marine Science, PMB No.3, Townsville, Queensland 4810, Australia*

Abstract. Analyses of high quality data show that there have been some interesting recent changes in the incidence of some climate extremes in the Australian region and New Zealand.

For the Australia region:

- the percentage area of Australia experiencing extreme wet conditions has increased slightly while the area of extreme dryness has reduced slightly since 1910

- heavy rainfall has also increased in some areas during the same period, although not significantly

- the frequency of extreme warm days and nights has increased while extreme cool days and nights has decreased since 1961

- a decrease in the total number of tropical cyclones since the late 1960s is largely explained by variations in the El Niño Southern Oscillation phenomenon while the number of stronger cyclones has increased slightly

- extratropical cyclonic activity has decreased over much of the mid-latitude waters south of Australia since the mid-1960s but increased at higher latitudes further southwest

For New Zealand:

- warming from 1941 to 1990 has resulted in about 10 fewer days per year with temperatures less than 0°C and around 2 more days per year greater than 30°C in warmer locations

- the occurrence of moderate and severe drought decreased during the period 1951 to 1980 compared to the previous 30 year period

- changes in temperature extremes and drought frequency have been in response to changes in atmospheric circulation in the region

Climatic Change **42**: 183–202, 1999.
© 1999 *Kluwer Academic Publishers. Printed in the Netherlands.*

1. Introduction

There have been fewer studies of regional changes in climatic extremes compared to studies of changes in climatic means although some recent regional assessments have been made (e.g. Karl, 1996 and other papers in this volume). Reasons for this include a lack of adequate data, since non-climatic factors (e.g. changes in station location) often bias extremes more than climatic means, and problems related to choosing appropriate thresholds to define extremes (Nicholls, 1995). Also, analyses of extremes require data at daily or better resolution and these data have not routinely been assembled into climatological databases. Plummer et al. (1997) discuss some further issues, such as historical changes in instrumentation and observation practices, which have affected Australian data.

Salinger et al. (1996) provide a thorough investigation into observed changes in mean climate, including atmospheric circulation, over the Australian, New Zealand and South Pacific regions. Annually, surface air temperatures increased by 0.4 to 0.8°C throughout most of this region from 1951 to 1993 and this is consistent with post-1910 trends from observations of both surface ocean and air temperatures (Folland et al., 1997). Longer-term series suggested an increase of about 0.7°C since the start of the century, broadly consistent with that observed for the Southern Hemisphere (Salinger et al., 1996). Recent decades have seen a decrease in the diurnal temperature range (DTR), i.e. the difference between daytime maximum temperatures and overnight minimum temperatures, over much of Australia and the central southwest Pacific. Summer precipitation has increased over eastern Australia this century while decreases have been observed over parts of southwestern Australia in winter. Increases in precipitation have occurred to the northeast of the South Pacific Convergence Zone (SPCZ) while decreases have occurred to the southwest of the SPCZ. Salinger and Mullan (1996) describe changes in the atmospheric circulation regime of the New Zealand/South Pacific region, occurring around 1950 and 1976.

Many critical climate impacts are a consequence of rarer extreme events rather than a result of changes in mean values. The climatic indices investigated in this study are closely related to the incidence of floods, droughts, heat waves, frosts and strong winds and so their variations are of prime interest to the climate impacts community. High temperatures can exacerbate drought conditions and increase the likelihood of fires. Low temperatures are expressed through frosts and heavy snowfalls in some parts of the Australian and New Zealand region. This study is a collection of preliminary analyses to gain an insight into changes in a range of climatic extremes over the Australian and New Zealand region. Some of the indices suggested here may not be the most appropriate for impact studies and future work is expected to follow the recommendations made at (and since – refer other papers in this volume) the recent CLIVAR/GCOS/WMO Workshop on Indices and Indicators for climate extremes.

2. Data

While most Australian data used in this study have been extracted from the Australian Bureau of Meteorology's climate database, they have undergone further quality analysis as described in Plummer et al. (1997). Homogeneous daily rainfall data from 1910 to 1995 have been obtained for 379 stations (Lavery et al., 1997). Daily maximum and minimum temperatures for the period 1961 to 1995 were from a 48 station network where monthly mean data satisfied homogeneity checks (Plummer et al., 1995; Torok and Nicholls, 1996). Unfortunately, daily temperature data are not widely available in digital form prior to the late 1950s. Manually drawn analyses prepared by the Bureau's National Meteorological Centre for the Australian region have been examined for changes in tropical cyclone activity from 1969/70 to 1995/96 and for extratropical cyclonicity from 1965 to 1993. While there are concerns about using manual or numerical analyses for climate change studies (e.g. Nicholls, 1995; Karl et al., 1995), there are grounds to suggest that the changes found here are reasonable.

Forty-one New Zealand stations with homogeneous temperature records and at least 40 years of data were selected for analysis. These stations were carefully adjusted for site changes and other disturbances consistent with the techniques described in Rhoades and Salinger (1993). Drought frequency over New Zealand was assessed in terms of changes in soil moisture deficit from water balance modelling (Coulter, 1973).

Indices and periods of data used in this study are summarised in Table I.

Region	Climate Extreme Index	Period Examined
Australia	Percentage wet & dry	1910-95
Australia	Rainfall intensity	1910-95
Australia	Temperature extremes	1961-95
Australia (105-160°E)	Tropical Cyclones	1969/70-95/96
Australia (30-55°S, 80°E-180°)	Extratropical cyclonic activity	1965-93
New Zealand	Temperature extremes	1941-90 but some longer records
New Zealand	Drought frequency	1921-80 but 1901-80s for some

Table I. Summary of indices used in this study. Note that changes in the New Zealand temperature extremes index are inferred from changes in mean maximum and minimum temperatures over the period shown (refer 3.2.1). Changes in drought frequency over New Zealand are based on water balance modelling.

3. Observed Trends in Climate Extremes Indices

3.1. AUSTRALIAN REGION

Area-weighted averages for the indices based on rainfall and temperature data were derived for Australia and its four quadrants (separated by 26°S and 135°E) using a modification of the Thiessen Polygon method (Lavery et al., 1997). Trends were calculated from simple linear regression and, unless otherwise stated, their statistical significance was assessed using the Kendall-tau non-parametric test (Kendall and Gibbons, 1990). Changes in data coverage do not generally have a large influence on the continental scale trends presented here because one of the major criterion for station selection was homogeneous and continuous observations. However, changes in data coverage over the sparse Australian interior made it necessary for the removal of four of the 379 stations used in the analysis of rainfall intensity (refer 3.1.2).

3.1.1. *Percentage of Australia extremely wet and dry*
Figure 1 shows that the percentage of Australia experiencing extreme dry conditions (below decile 1 or 10th percentile of annual total) has decreased slightly since 1910. Annual percentile values (Moore and McCabe, 1993) were computed from data over the period 1910 to 1995 for individual stations and area weightings were derived from the Thiessen Polygon method. Due to widespread high rainfall totals in the mid-1970s, a small increase in the area experiencing very wet conditions (above decile 9 or 90th percentile) was found. However, like the Australian average rainfall series (Nicholls and Lavery, 1992; Lavery et al., 1997), time series in Fig. 1 are dominated by high interannual and interdecadal variability so there is little to suggest that the extents of dry or wet conditions have been appreciably different during the past few decades compared to earlier this century. Neither of the trends are statistically significant at the 95% confidence level.

3.1.2. *Rainfall intensity*
Studies on trends in heavy rainfall intensity in Australia during the past century (Yu and Neil, 1993; Lough, 1993, 1997; Nicholls and Kariko, 1993; Suppiah and Hennessy, 1996) reveal results which are affected by the use of different definitions for heavy rain events and different statistical methods (Nicholls, 1995). In this paper, heavy rainfall was defined as the 99th percentile of daily data as calculated from the 1910 to 1995 period. This is the 4th highest value per year and the highest value per season. To supplement the analysis of changes in the intensity of heavy rainfall, changes in the frequency of heavy rainfall are presented for a threshold of 25.4 mm/day (1 inch). Heavy rainfall indices were computed at individual stations. Since stations are irregularly spaced, each station was weighted according to the area it represents. Stations with more-distant

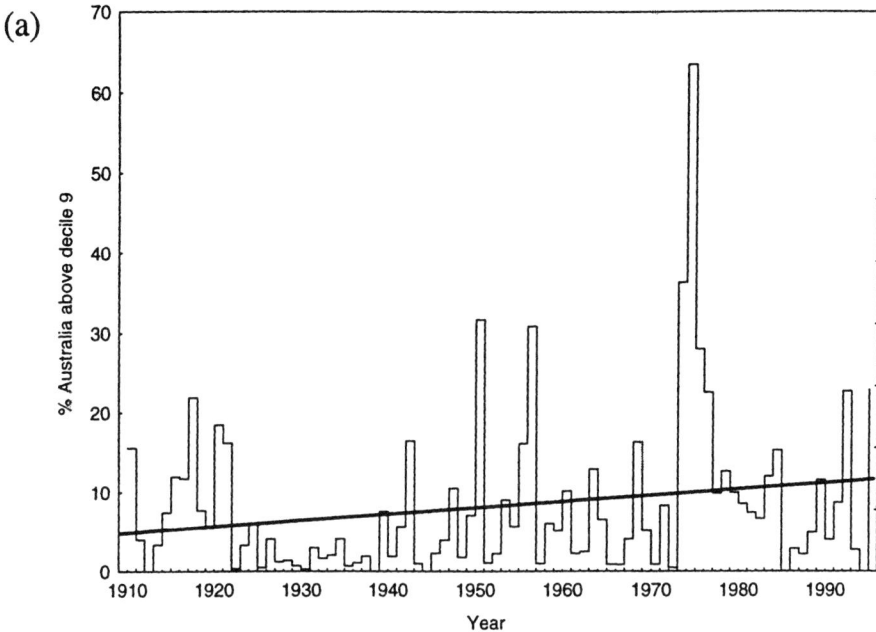

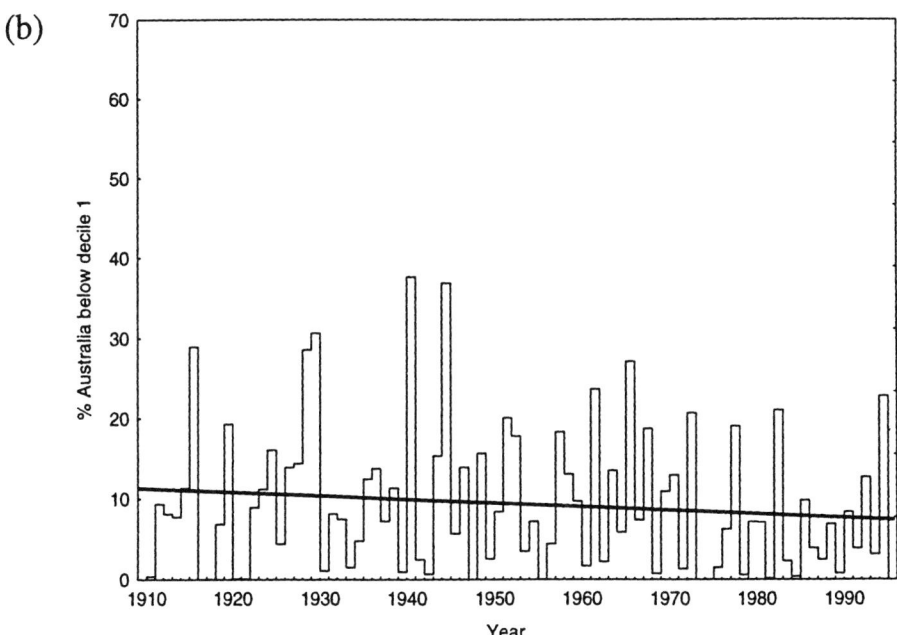

Figure 1. (a) Percentage of Australia experiencing extreme wet conditions (above decile 9, or 90th percentile, of annual total) and (b) extreme dry conditions (below decile 1, or 10th percentile) in each year from 1910 to 1995. Linear trend lines shown in bold.

neighbours represent larger areas and were given higher weight using a modification of the Thiessen Polygon method (Hennessy et al. 1998). Quadrant area-averages were then computed by summing weighted indices from each station. The significance of trends is determined from the non-parametric Kendall-tau test and the magnitude of trends is computed from linear regression analysis.

The number of stations passing the quality control processes (Hennessy et al. 1998) increased from 266 in 1910 to 370 in 1988, then declined to 236 in 1995. Most changes to the data network occur within areas already having good coverage, and so network variations have little effect on regional trends. Four stations (Giles, Forrest, Rabbit Flat and Rawtinna) in data-sparse parts of central and western Australia were introduced between 1931 and 1970, each having a noticeable effect on continental and regional trends. To eliminate such effects, these four stations were removed from the analysis.

The heaviest rainfall occurs in summer and autumn in northern Australia (Fig. 2a). Southern Australia generally has a uniform seasonal distribution of heavy rainfall but the southwest has a winter peak in less-intense rainfall. It is important to interpret seasonal and regional changes relative to this climatology.

The 99th percentile intensity has experienced negligible change over the period 1910-1995 for the continent overall. While increases in the 99th percentile occur in more quadrants and seasons than decreases (Fig. 2b), none of the changes are significant at the 5% level (although the 24% increase in the southeast in autumn is significant at the 6% level). Non-significant changes judged to be of hydrological importance include increases of about 20% in the northwest in winter and southwest in summer, and decreases of about 20% in the northeast in winter and in the southwest in autumn. In a related study looking at smaller spatial scales, Hennessy et al. (1998) found a significant 31% increase in the 99th percentile from 1910-1995 in New South Wales (in southeastern Australia) in autumn and a significant 13% decrease in southwest Western Australia in winter.

For Australia overall, there has been a non-significant 5% increase in the annual number of days with a rainfall total over 25.4 mm (Fig. 2d). Small increases have occurred in summer and autumn with decreases in winter and spring. Regionally and seasonally, increases in frequency are more prominent than decreases. None of these changes are significant, but the increases in the southeast in summer and autumn may be hydrologically important. For Australia, Groisman et al. (1998, this volume) found a non-significant increase in the number of summer days with totals over 50.8 mm (2 inches) of 1.1% per decade relative to the 1910-1996 mean, and a significant increase of 4.6% per decade was found for coastal regions of New South Wales and Victoria in southeastern Australia.

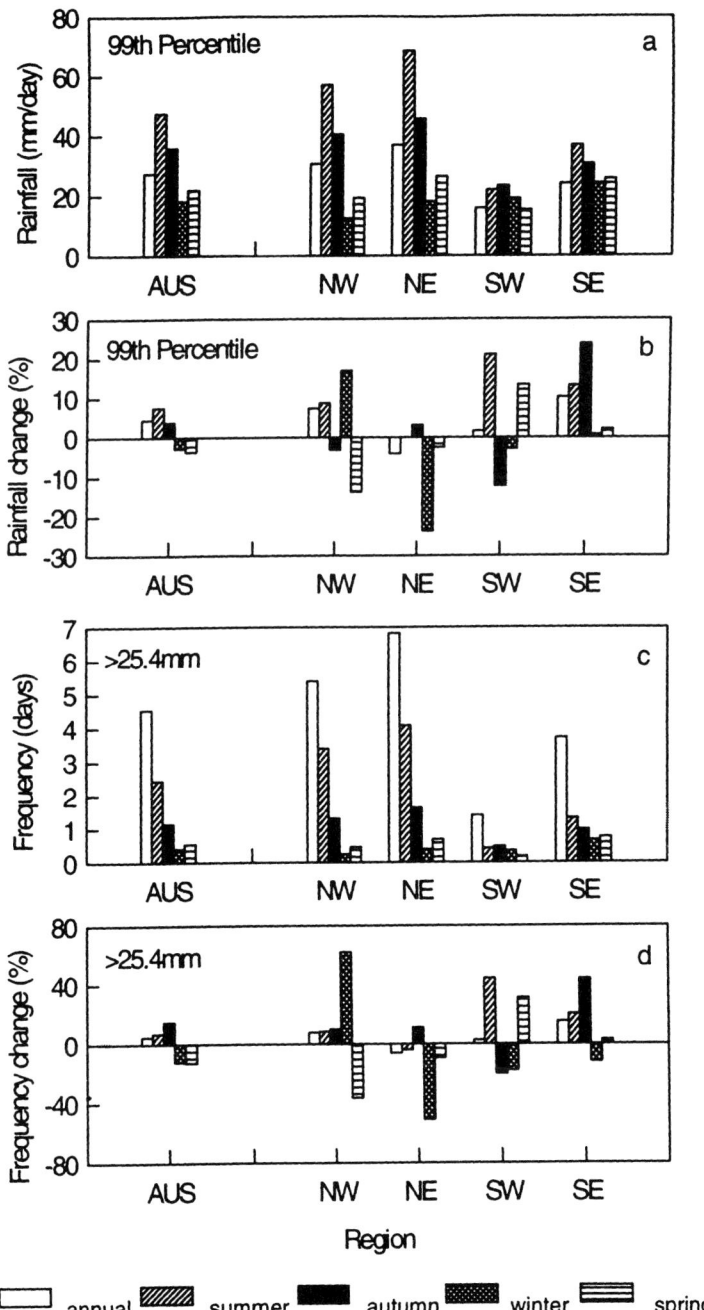

Figure 2. Australian (AUS) heavy rainfall indices. (a) 1910-1995 average 99th percentile intensity, (b) percentage changes in the 99th percentile, (c) 1910-1995 average frequency of days of at least 25.4 mm, and (d) percentage changes in days of at least 25.4 mm. Changes are expressed as a percentage of the 1910 regression value. The quadrants northwest (NW), northeast (NE), southwest (SW) and southeast (SE) are of similar size and are separated by latitude 26°S and longitude 135°E.

3.1.3. *Temperature extremes*

Warming has occurred over Australia during the second half of this century, particularly at night (Torok and Nicholls, 1996). Changes in the frequency of days and nights exceeding extreme warm threshold values and falling below extreme cool thresholds were analysed from 1961 to 1995. For each station, seasonal and annual time series of the percentage frequency of periods for which temperatures exceeded their (warm) 90th percentile threshold and fell below their (cool) 10th percentile were derived. Percentiles for individual stations were calculated from data over 1961 to 1990. Periods examined were single days/nights and three consecutive days/nights (i.e. occurrences of three consecutive days/nights where an extreme threshold was reached on every day/night). Percentages of periods were calculated, instead of counts of the period, to provide reasonable seasonal and annual values of the index when a (pre-defined) small number of missing observations were encountered. Similar to the rainfall analyses, these individual station time series were spatially averaged for Australia and each of its quadrants using the Thiessen Polygon approach.

There has been an increase in the frequency of warm days and nights (Fig. 3(a)) and a decrease in cool days and nights (Fig. 3(b)). The strongest trends were found for decreases in the frequency of cool nights (3% decrease over Australia annually and 5% in winter) while reductions in the number of cool days were relatively small. Since the stations used in this analysis were from small towns or remote locations, it is unlikely that trends are greatly affected by urbanisation. The most marked changes occurred over the northern quadrants, and daytime extremes have shown the greatest change in autumn. However, changes were weaker than trends in the number of days with temperatures exceeding the median. Trends in the frequencies of consecutive extremes (shown as the lower two series in both Fig. 3(a) and Fig. 3(b)) and more-extreme events (defined by the 95th and 5th percentile thresholds, not shown), although of similar sign, were also weaker than trends in the single and less-extreme (90th, 10th percentile) events, respectively. Of the eight annual time series in Fig. 3, the only changes not statistically significant at the 95% confidence level (at least) were the decreases in periods (both 1 and 3 days) with maximum temperature below the 10th percentile. Consistent with trends in the cooler extremes, there has been a significant decrease in the occurrences of minimum temperatures below 0°C and a trend towards an earlier date of last frost, over inland eastern Australia during this century (Stone et al., 1996).

3.1.4. *Tropical cyclones*

Trends in tropical cyclone activity in the Australian region (105-160°E) were examined from the 1969/70 season - considered to be the first reliable season due to the availability of satellite pictures - to 1995/96. Tropical cyclone numbers in the region are influenced by the El Niño Southern Oscillation (ENSO) phenomenon (Nicholls, 1992).

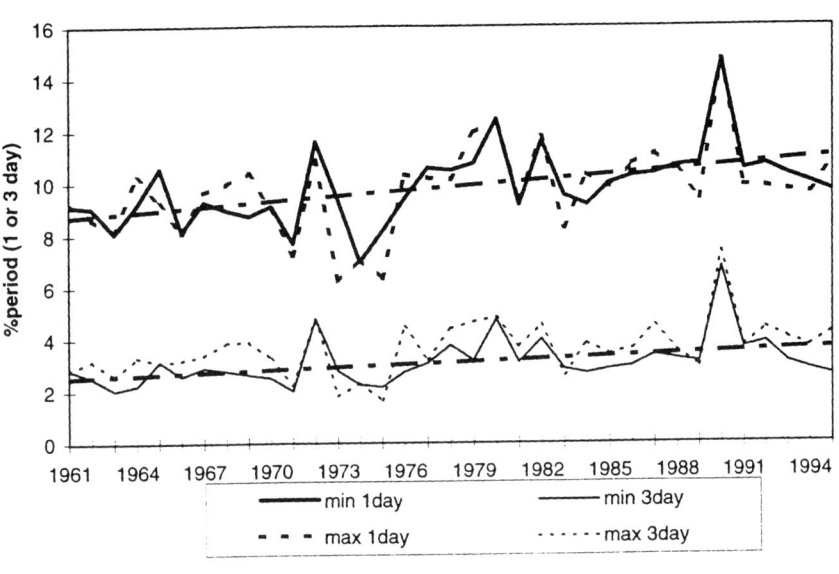

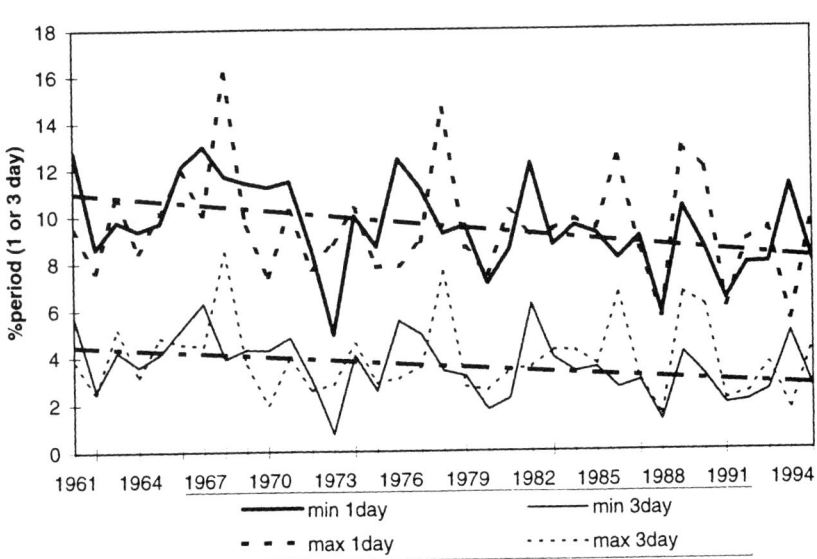

Figure 3. Percentage of periods where (a) warm - 90th percentile, and (b) cool - 10th percentile extreme temperatures reached over Australia. Periods are for both 1 day (bold lines - top two in each graph) and 3 days (lower two). Variations in periods of extreme maximum (broken lines) and minimum temperatures (solid) are shown. Linear trends (bold dashed-dot lines) are only shown for the minimum temperatures to give a comparison between 1 day and 3 day events.

Figure 4 shows that while the total number of cyclones has decreased, the number of stronger cyclones (minimum central pressure, MCP, $\leq$ 970hPa) has increased (Nicholls et al., 1998) although neither trend is statistically significant at the 95% confidence level. The total number of cyclone days has also increased slightly. Much of the decline in the weaker cyclones (MCP > 990hPa) occurred suddenly in the mid-1980s and this, with a concomitant change in relationship between cyclone numbers and the Southern Oscillation Index (SOI - a measure of the state of ENSO) around this time, suggest this decrease may be artificial. Improved understanding of weaker systems has resulted in them being less likely to be analysed as tropical cyclones (Nicholls et al., 1998). From regression analysis, the decrease in cyclones with MCP between 970hPa and 990hPa is largely due to a negative trend in the SOI over the past few decades.

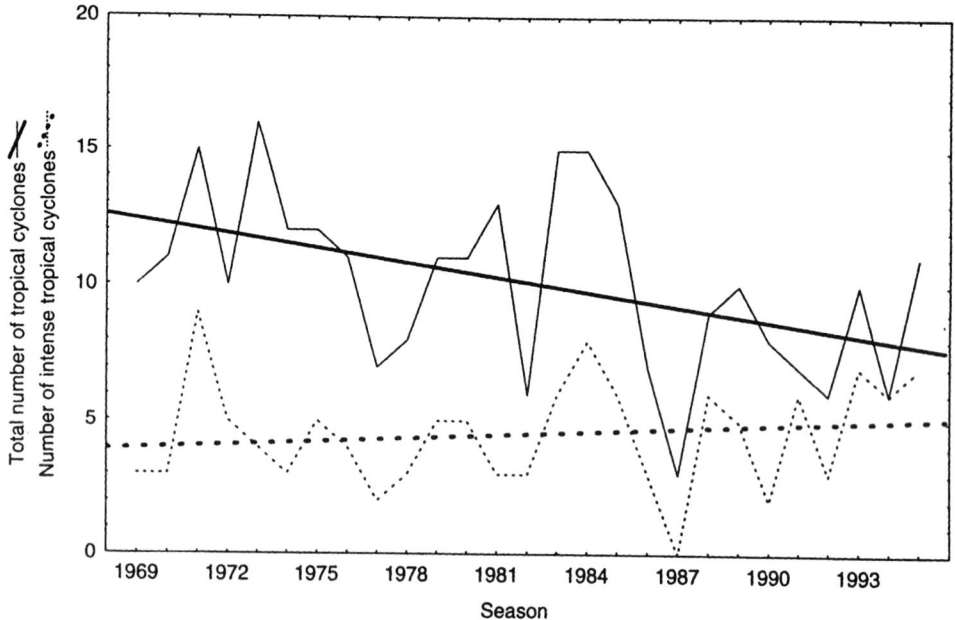

Figure 4. Total number of tropical cyclones (solid lines) and the number of intense systems (MCP $\leq$ 970hPa, dotted lines) in the Australian region (105-160°E). Linear trends are shown in bold. The abscissa gives the year in which the cyclone season starts (e.g. the 1969/70 season is plotted against "1969").

3.1.5. Extratropical cyclonic activity

Recent work found an increase in east-coast cyclones since the late 1950s (Hopkins and Holland, 1997). These systems are intense cyclonic systems that develop occasionally on the middle and southern parts of Australia's east coast. From updated analyses of mean sea level pressure (MSLP) charts (Leighton and Deslandes, 1991), changes in extratropical cyclonicity were examined over the Australian region (30-55°S, 80°E-180°) from 1965 to 1993. Cyclonicity is defined as the time (in hours) during which cyclone centres occupy a 5° lat/lon grid-box during a given period. An extratropical cyclone centre could be an identifiable closed circulation with or without an associated front or a wave low which developed into a closed circulation. Note that variations in this index do not necessarily suggest changes in the intensity of extratropical cyclones and a decrease may be associated with either fewer or faster moving systems. However, good correlations with other climatic variables over parts of southern Australia in recent decades (e.g. rainfall) suggest that this index may provide a useful indicator of changes in baroclinic activity in the extratropics. Some of the earlier years analysed may be suspect due to satellite pictures not being available.

A substantial area of decreasing cyclonicity extends northwest-southeast over waters south of Australia but increases have occurred further southwest (Fig. 5). Although the overall field significance of these changes have not been assessed (as, for example, in Wigley and Santer (1990)), locally statistically significant changes at the 95% confidence level (from a Students t-test) are shown shaded in Fig.5. This pattern of change is consistent with increases in winter MSLP (Allan and Haylock, 1993) which are linked to decreases in rainfall over southwestern Australia during the past three decades or so (Wright, 1974a, 1974b; Pittock, 1983; Nicholls and Lavery, 1992; Suppiah and Hennessy, 1998).

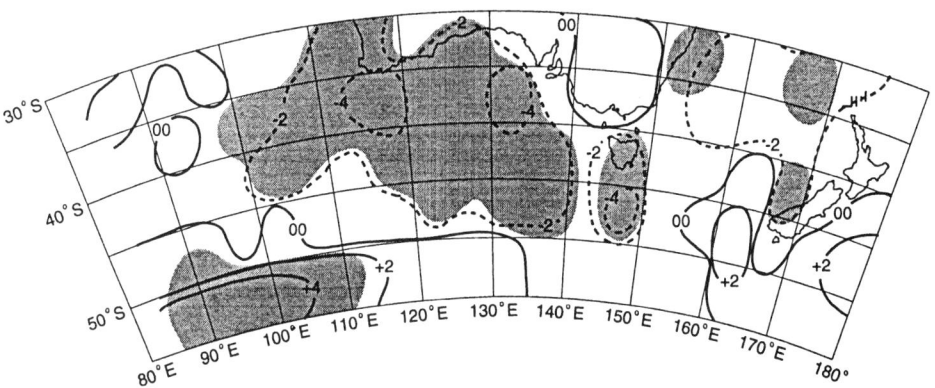

Figure 5. Annual variations in extratropical cyclonicity over the Australian region. Linear trends are in hours per year and shaded areas denote trends statistically significant at the 95% confidence level.

3.2. NEW ZEALAND

3.2.1. *Temperature extremes*

To examine the sensitivity of temperature extremes over New Zealand to changes in monthly and annual mean temperatures, temperature extremes were regressed against mean temperature data (Salinger, 1997). The New Zealand analysis was based on changes in specific temperature thresholds rather than changes in percentiles (as in 3.1.3). Temperate and sub-tropical grasses and crops are grown in New Zealand and the juxtaposition of the subtropical/temperate division is strongly determined by temperatures above and below certain thresholds. Additionally, the thermal regime in New Zealand is more moderate to that of Australia, with a relatively small difference in mean temperature between the warmest and coolest months. Therefore the sensitivity of temperature extremes to mean temperature changes is higher than for extremes in continental climates (Salinger, 1988). Regressions were performed for average maximum temperature compared with the number of days reaching 30°C or more and for average minimum temperature compared with the number of days of 0°C or below. In both cases, separate regression equations were calculated for monthly (combining all 12 months) and annual data although similar results were found for both. The regressions of monthly and annual numbers of days with average maximum and average minimum temperatures were then verified against the actual number of days. For days of 0°C or below, regressions were calculated for all 41 stations. For days of 30°C or above, only 18 stations were used since many had no (or very few) days above 30°C.

The relationship between changes in the number of cold days and changes in the average annual minimum temperature showed a robust response as shown by the large values for the percentage of variance explained (Table II). For much of the North Island a 1°C warming decreased the number of air frost days (i.e. temperatures below 0°C - ground frosts can occur when the air temperature is slightly above 0°C) by between 8 and 12 per year, and for inland areas by 20 days per year. In the South Island, the response was between 7 and 11 days in the north and west, 11 to 18 days in the east and south, and from 15 to 27 days per year for inland areas. The largest sensitivity of the number of air frost days to annual minimum temperature change was in the inland areas - where the highest incidence occurs. In most cases regression relationships were strong, explaining between 45 and 70 percent of the variance. The New Zealand average for all sites analysed was 14 fewer air frost days per year for a 1°C increase in average annual minimum temperature.

The change in the length of the frost free period (i.e. the number of days between the last spring frost, and first autumn frost) was investigated at 24 locations. Cold sites with a large range in minimum temperature, such as inland South Island areas, showed a change in frost-free period of about 15-25 days per °C change in average annual minimum temperature. Warmer sites with a small annual temperature range exhibited a larger sensitivity. Throughout the North Island the length of the frost-free period increased by at least 30 days for every 1°C increase in average minimum temperature. Time series of trends at two sites, Ruakura and Rotorua, which have not had site changes are shown in Fig. 6. These show a clear reduction (statistically significant above the 99% confidence level) in annual frequency of air frosts, particularly during the 1950s.

Presently, warmer locations only average between 1 and 9 days with

(a) Specific sites

	<0°C		>30°C	
	No. Days	Variance explained	No. Days	Variance explained
North Island				
Ruakura	-8	49	1	47
Gisborne	-5	34	3	17
Hastings	-11	62	2	11
Palmerston North	-8	48	-	-
South Island				
Nelson	-7	20	-	-
Winchmore	-18	71	2	12
Ophir	-22	74	2	8
Invercargill	-18	53	-	-

(b) Districts

	<0°C	
	No. Days	Variance explained
Inland North Island	-20	67
Rest of North Island	-8 to -12	33 to 63
North and west of South Island	-7 to -11	20 to 44
Inland South Island	-15 to -27	46 to 74
Eastern South Island	-11 to -18	46 to 71
New Zealand	-14	

Table II. Changes in the annual number of days below 0°C per °C change in average annual minimum temperature, and above 30°C per °C change in average annual maximum temperature for (a) specific sites and (b) districts in New Zealand.

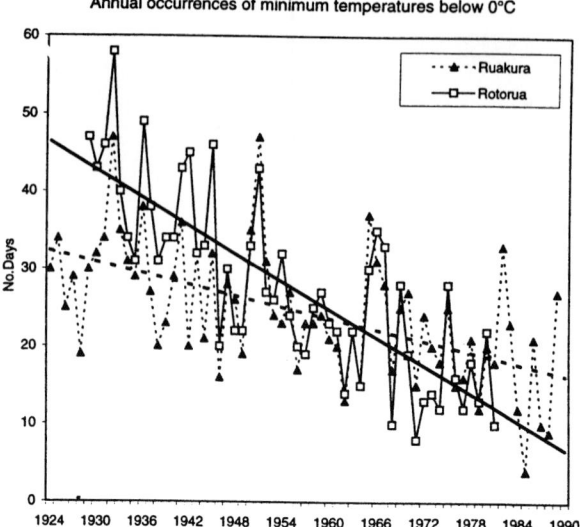

Figure 6. Annual frequency of days below 0°C (air frosts) at two New Zealand sites (Ruakura and Rotorua). Linear trends are shown in bold.

temperatures above 30°C. Therefore, the results of similar analyses were much weaker than those for temperatures below 0°C (the regression variance explained ranged from 8 to 47 percent; Table II). Typically the response was for an increase in the number of days above 30°C by between 1 and 4 days for every 1°C increase in average maximum temperature. The average for those sites that showed a response was 2.4 days/°C. Time series for two sites, Hastings and Ashburton, are shown in Fig. 7. These sites have experienced a non-statistically significant increase in the frequency of days above 30°C since 1950.

New Zealand average mean air temperatures have increased by 0.7°C over the period 1941 to 1990 (Salinger, 1997). On this basis, there would be a trend, on average, to 10 fewer days (ranging from 4 to 19 depending on the site) with temperatures less than 0°C. In the east of the North Island, and inland and eastern areas of the South Island, this temperature trend equates to an increase in incidence of days above 30°C of around 2 days from 1941 to 1990.

3.2.2. Drought frequency

Using water balance modelling, an index of drought or dry periods is provided by summations of the number of days within a year at wilting point (ND). This assumes that water is lost from the soil at the potential evapotranspiration rate, until all the readily available water capacity is exhausted. At this point the model assumes that evapotranspiration ceases until rainfall on any one day exceeds the mean daily evapotranspiration for the calendar month in which the day falls. When the readily available water capacity is exhausted, the soil is assumed to be at wilting point, and plant growth is assumed to stop. The number of days at wilting point is thus an index of agricultural drought.

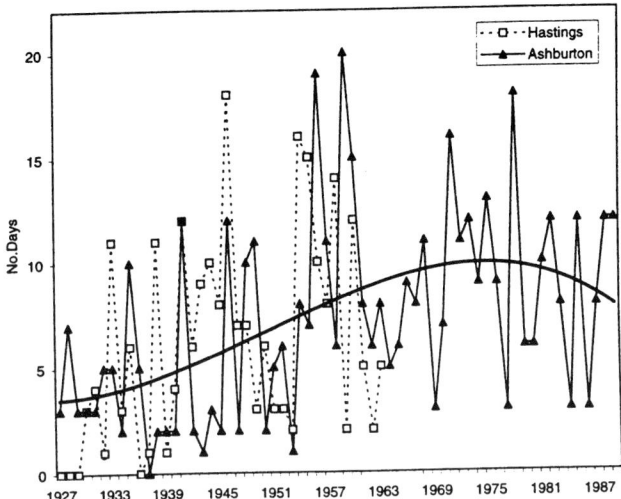

Figure 7.   Annual frequency of days above 30°C at two New Zealand sites (Hastings and Ashburton). Linear trend, for Ashburton only, is shown in bold.

This model, developed by Coulter (1973), has been successfully applied and tested in the New Zealand situation for many years.   Porteous et al. (1994) calibrated the performance of this model for pasture sites with two weekly measurements of soil moisture from neutron probe measurements over a two year period at four sites.   This study found that predicted deficit regressions against measured deficit explained between 88 and 97% of the variance in model runs. They concluded that this model can be applied with confidence to many pastoral agricultural applications such as drought assessments.   Barringer et al. (1995) also validated this model in predicting wilting point deficits with four years of neutron probe data and concluded that this model accurately predicted the gross trends in soil moisture deficit, giving reasonable estimates of the timing and magnitude of deficits.   New Zealand has mountainous orography, giving strong precipitation gradients, and young soils of varying depths.   This gives a rather complex mixture of soils with different field capacities.   Evapotranspiration shows small variability compared with rainfall, and the rainfall gradients.   Because of these factors, the water balance model has been successfully applied in New Zealand.   For this study, the readily available water capacity was set at 75 mm as this is representative of many New Zealand soils.

Wilting point (ND) summations were performed for the period 1921 to 1980 for stations from representative climate regions (Salinger, 1997).   ND summations with return periods of 1 in 20 years and 1 in 10 years were classified as 'severe' and 'moderate' droughts, respectively.   These return periods are useful in identifying the inconstancy in the frequency of extreme events.   The summations were compared between periods 1921-1950 and 1951-1980, because the westerly circulation over New Zealand had weakened in the latter period.

'Moderate' and 'severe' droughts from water balance return period analysis are listed in Table III. The results suggest a trend to fewer serious droughts in many New Zealand districts for the 1951-80 period compared with the previous 30 years. From matched pairs Students t-tests, this decrease was found to be statistically significant above the 99% confidence level. Reductions were particularly apparent in the north and east of the North Island and in the south of the South Island. These changes were associated with a change in atmospheric circulation with more east to northeast airflow in the latter period compared to south to southwest airflow from 1921 to 1950 (Salinger and Mullan, 1996). More frequent airflow from the east and northeast gives higher rainfall in the drier, more drought prone lower rainfall areas of New Zealand. Examples of time series used to derive drought frequency are shown in Fig. 8.

	MODERATE		SEVERE	
	1921 - 50	1951 - 80	1921 - 50	1951 - 80
Northern North Island	3.4	1.2	0.8	0.6
Western North Island	2.5	2.3	1.0	1.0
Eastern North Island	3.2	1.8	1.7	0.8
Western South Island	2.3	3.0	1.0	0.8
Northern South Island	3.5	1.5	1.5	1.0
Eastern South Island	2.0	1.5	1.5	0.5
Southern South Island	4.7	2.0	2.3	0.2

Table III. Moderate and severe drought frequency for districts of New Zealand. Values are frequencies as defined by the 1 in 10 (moderate) and 1 in 20 (severe) year return period for moderate (severe) events.

4. Discussion and Conclusions

This study provides evidence that there have been some interesting changes in climate extremes over the Australian and New Zealand region during the twentieth century. However, several data series span only a few decades and, for some indices, the El Niño Southern Oscillation phenomenon contributes much of the observed variability. Further work is required to investigate the influence of this phenomenon on these observed changes. Perhaps the most apparent and consistent change in the region has been a decrease in the number of cold nights in recent decades. Although regionally there have been some exceptions, there is little evidence to suggest a trend towards drier conditions over Australia or New Zealand. Nevertheless, further work is necessary to provide optimum information on variations in climatic extremes. In the future, indices of direct relevance to climate sensitive sectors need to be developed, as do indices for more problematic data types such as occurrences of strong winds and thunderstorms. Further, a large

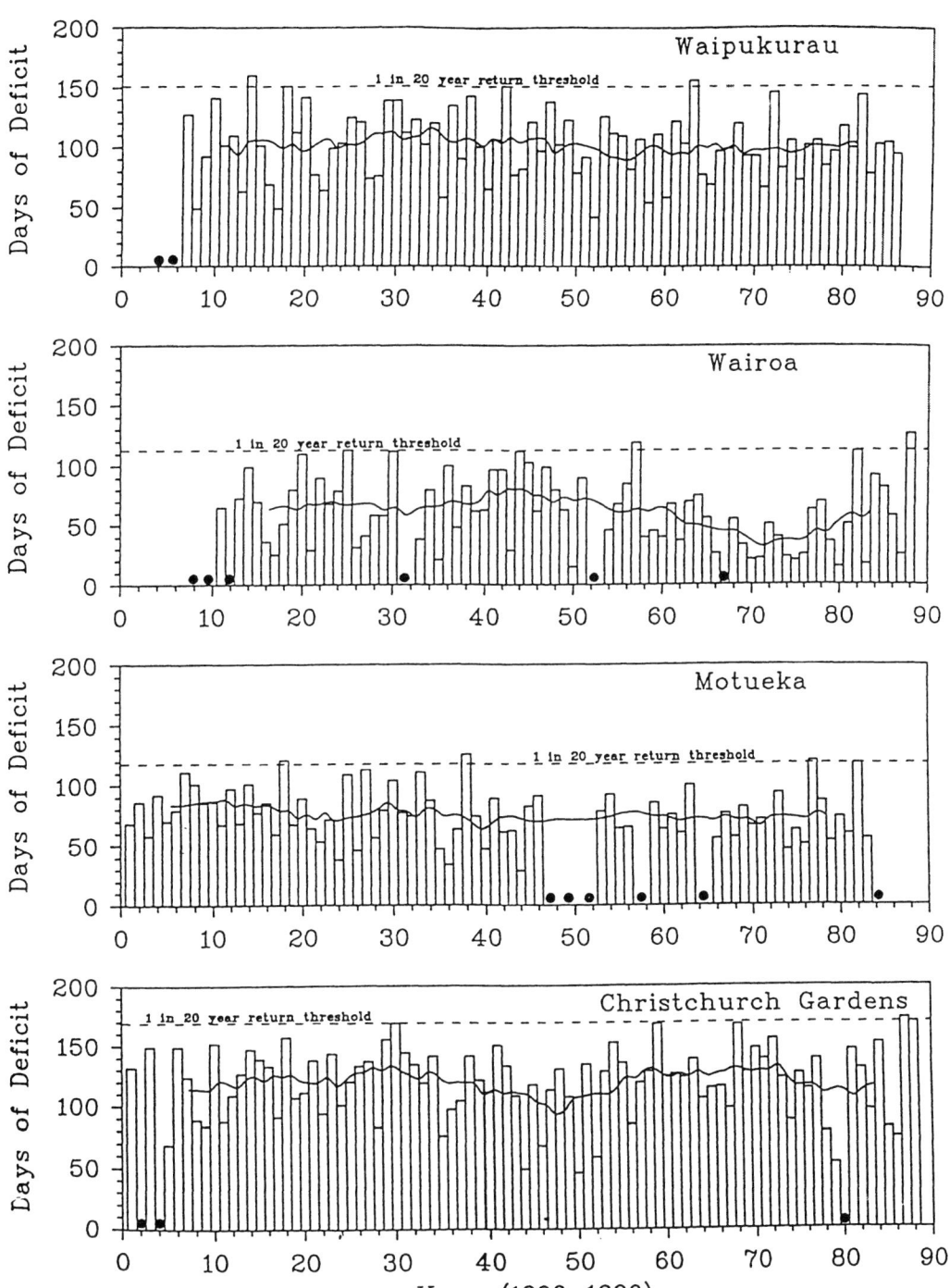

Figure 8. Annual number of days of deficit (ND, explained in text), for soils with an available water capacity of 75 mm, for locations in eastern areas of New Zealand. A dot represents missing data.

volume of daily and hourly historical data are still to be digitised and archived.

Clearly, future long-term climate monitoring will require adherence to a number of guiding principles so that data homogeneity and continuity are maintained (Karl et al., 1995). These should ensure that influences on the climate record caused by changes in instruments, observing practices, observation locations, sampling rates, etc. are known prior to implementing such changes.

Acknowledgments

The authors would like to thank Simon Torok (CSIRO Land and Water, Australia) for providing information on data quality for some of the Australian stations used in the temperature analyses. Beth Lavery (Bureau of Meteorology) contributed to analysis of the Australian data and Brett Mullan and Alan Porteous (both NIWA) provided some analyses of the New Zealand data. Mary Voice, Harvey Stern, Dean Collins (all Bureau of Meteorology), Tahl Keston (Monash University, Australia), Kevin Walsh and Graeme Pearman (both CSIRO Division of Atmospheric Research) provided helpful comments on this or earlier drafts of the paper. Research towards the New Zealand component of the work was funded by a New Zealand Foundation for Research, Science and Technology contract CO1621, and is a contribution to the international CLIVAR programme.

References

Allan, R. J. and Haylock, M. R.: 1993, 'Circulation features associated with the winter rainfall decrease in southwestern Australia', *J. Clim.* **6**, 1356-1367.

Barringer, J. R. F., Porteous, A., Salinger M. J., and Trangmar, B. B.: 1995, 'Estimating spatial patterns of wilting point deficit using a water balance model and a geographic information system', *Journal of Hydrology (New Zealand)* **34(1)**, 42-59.

Coulter, J. D.: 1973, 'A water balance assessment of New Zealand rainfall', *Journal of Hydrology (New Zealand)* **12**, 83-91.

Folland, C., Salinger, M. J., and Rayner, N.: 1997, 'A comparison of annual South Pacific island and ocean surface temperatures', *Weather and Climate: Journal of the Meteorological Society of New Zealand (Inc.)* **17**, 23-42.

Groisman, P., Karl, T., Easterling, D., Knight, R., Jamason, P., Hennessy, K., Suppiah, R., Page, C., Wibig, J., Fortuniak, K., Razuvaev, V., Douglas, A., Førland, E., and Zhai, P.: 1998, 'Changes in the probability of extreme precipitation: important indicators of climate change', *Climatic Change (in press)*.

Hennessy, K. J., Suppiah, R., and Page, C. M.: 1998, 'Australian rainfall changes, 1910-1995', *Aust. Met. Mag. (in press)*.

Hopkins, L. C. and Holland, G. J.: 1997, 'Australian heavy-rain days and associated east-coast cyclones: 1958-1992', *J. Clim.* **10**, 621-635.

Karl, T. R.: 1996, 'Indices of climate change for the United States', *Bull. Amer. Meteor. Soc.* **77**, 279-292.

Karl T. R., Derr, V. E., Easterling, D. R., Folland, C. K., Hofmann, D. J., Levitus, S., Nicholls, N., Parker, D. E., and Withee, G. W.: 1995, 'Critical issues for long-term climate monitoring', *Clim. Change* **31**, 185-221.

Kendall, M.G. and Gibbons, J.D. 1990. *Rank Correlation Methods, 5th ed.* Edward Arnold, London, UK.

Lavery, B. M., Joung, G., and Nicholls, N.: 1997, 'An extended high-quality historical rainfall dataset for Australia', *Aust. Met. Mag.* **46**, 27-38.

Leighton, R. M. and Deslandes, R.: 1991, 'Monthly anticyclonicity and cyclonicity in the Australian region: averages for January, April, July and October', *Aust. Met. Mag.* **39**, 149-154.

Lough, J. M.: 1993, 'Variations of some seasonal rainfall characteristics in Queensland, Australia: 1921-1987', *Internat. J. Climatol.* **13**, 391-409.

Lough, J. M.: 1997, 'Regional indices of change: Queensland, Australia', In: *CLIVAR/GCOS/WMO Workshop on indices and indicators for climate extremes*, Asheville, NC, USA, 3-7 June 1997.

Moore, D. S. and McCabe, G. P.: 1993, *Introduction to the practice of statistics, 2nd ed*, W. H. Freeman and Company, USA, 854pp.

Nicholls, N.: 1992, 'Recent performance of a method for forecasting Australian seasonal tropical cyclone activity', *Aust. Met. Mag.* **40**, 105-110.

Nicholls, N.: 1995, 'Long-term climate monitoring and extreme events', *Clim. Change* **31**, 231-245.

Nicholls, N. and Kariko, A.: 1993, 'East Australian rainfall events: interannual variations, trends, and relationships with the southern oscillation', *J. Clim.* **6**, 1141-1152.

Nicholls, N., Landsea, C., and Gill, J.: 1998, 'Recent trends in Australian region tropical cyclone activity', *Meteorology and Atmospheric Physics* **65**, 197-205.

Nicholls, N. and Lavery, B. M.: 1992, 'Australian rainfall trends during the twentieth century', *Internat. J. Climatol.* **12**, 153-163.

Pittock, A. B.: 1983, 'Recent climatic change in Australia: Implications for a CO_2 warmed earth', *Clim. Change* **5**, 321-340.

Plummer, N., Lin Z., and Torok, S.: 1995, 'Trends in the diurnal temperature range over Australia since 1951', *Atmospheric Research* **37**, 79-86.

Plummer, N., Trewin, B. C., Hicks, R., Nicholls, N., Torok, S. J., Lavery, B. M., and Leighton, R.

M.: 1997, 'Australian data for documenting changes in climate extremes', In: *CLIVAR/GCOS/WMO Workshop on indices and indicators for climate extremes*, Asheville, NC, USA, 3-7 June 1997.

Porteous, A. S., Basher, R. E., and Salinger, M. J.: 1994, 'Calibration and performance of the single-layer soil water balance model for pasture sites', *New Zealand Journal of Agricultural Research* **37**, 107-118.

Rhoades, D. A. and Salinger, M. J.: 1993, 'Adjustment of temperature and rainfall records for site changes', *Internat. J. Climatol.* **13**, 899-913.

Salinger, M. J.: 1988, 'Climatic warming: Impact on the New Zealand growing season and implications for temperate Australia', In: *Greenhouse: Planning for climate change (G. I. Pearman, ed.)*, CSIRO Publications, Melbourne, 564-578.

Salinger, M. J.: 1997, 'Indices and indicators of changes in Extreme Events in the South Pacific: Preliminary Results from New Zealand', In: *CLIVAR/GCOS/WMO Workshop on indices and indicators for climate extremes*, Asheville, NC, USA, 3-7 June 1997.

Salinger, M.J., Allan, R., Bindoff, N., Hannah, J., Lavery, B.M., Lin, Z., Lindesay, J., Nicholls, N., Plummer, N. and Torok, S.J.: 1996, 'Observed variability and change in climate and sea-level in Australia, New Zealand and the South Pacific', In: *Greenhouse - Coping with climate change* (Bouma, W.J., Pearman, G.I. and Manning, M.R. eds), CSIRO Publishing, Collingwood, Australia.

Salinger, M. J. and Mullan, A. B.: 1996, 'CLIMPACTS 95/96: Variability of monthly temperature and rainfall patterns in the historical record', *NIWA Report AK96051*, 21pp.

Stone, R., Nicholls, N., and Hammer, G.: 1996, 'Frost in northeast Australia: trends and influences of phases of the Southern Oscillation', *J. Clim.* **9,** 1896-1909.

Suppiah, R. and Hennessy, K. J.: 1996, 'Trends in the frequency and intensity of heavy rainfall in tropical Australia and links with the Southern Oscillation', *Aust. Met. Mag.* **45**, 1-17.

Suppiah, R. and Hennessy, K. J.: 1998, 'Trends in seasonal rainfall, heavy-rain days, and number of dry days in Australia, 1910-1990', *Int. J. Climatol. (in press)*.

Torok, S. and Nicholls, N.: 1996, 'A historical annual temperature data set for Australia', *Aust. Met. Mag.* **45**, 251-260.

Wigley, T.M.L. and Santer, B.D.: 1990, 'Statistical comparison of spatial fields in model validation, perturbation and predictability experiments', *J.geophys.Res.* **95**, 851-865.

Wright, P. B.: 1974a, 'Seasonal rainfall in southwestern Australia and the general circulation', *Mon. Wea. Rev.* **102**, 219-232.

Wright, P. B.: 1974b, 'Temporal variations in seasonal rainfalls in southwestern Australia', *Mon. Wea. Rev.* **102**, 233-243.

Yu, B. and Neil, T. D.: 1993, 'Long-term variations in regional rainfall in the south-west of Western Australia and the difference between average and high intensity rainfalls', *Internat. J. Climatol.* **13**, 77-88.

(Received 5 November 1997; in revised form 18 December 1998)

CHANGES OF CLIMATE EXTREMES IN CHINA

PANMAO ZHAI[1], ANJIAN SUN[1], FUMIN REN[1],
XIAONIN LIU[2], BO GAO[1] and QIANG ZHANG[1]

[1]*National Climate Center, Beijing, China,*
[2]*National Meteorological Center, Beijing, China*

Abstract. Changes in China's temperature and precipitation extremes have been studied by using observational data after 1950. The results reveal that mean minimum temperature has increased significantly in China during the past 40 years, especially in the winter in northern China. Meanwhile, nation-wide cold wave activity has weakened and the frequency of cold days in northern China has been reduced significantly. Mean maximum temperatures display no statistically significant trend for China as a whole. However, decreasing summer mean maximum temperatures are obvious in eastern China, where the number of hot days has been reduced. Seasonal 1-day extreme maximum temperatures mainly reflect decreasing trends, while seasonal 1-day extreme minimum temperatures are increasing.

A statistically significant reduction of much above normal rain days in China has been detected. Contrarily, an increasing trend was detected in much above normal of precipitation intensity (precipitation/number of precipitation days) during the past 45 years.

1. Introduction

China is strongly influenced by the East Asian monsoon. During the winter-half year (Oct.-Mar.), the climate is mainly cold and dry. Cold waves, strong winds and low temperatures are the major extreme climate events. During summer period, the rain belt moves gradually from the south to the north. The climate in eastern China is hot and humid. Large precipitation variability is one of the major features of the East Asian summer monsoon climate. In some years, rainfalls are abundant enough to cause flood, while in other years they are too deficit to support agriculture. Drought and flood are outstanding climate issues of concern to both the Chinese government and general public.

Previous studies on climate change have focused on mean temperature and total precipitation, yet climate extremes can result in severe social economic disasters. We have less information about how they varied or changed. Using several indicators of climate extremes, this study focuses on their changes using China's meteorological records. Changes are assessed through an examination of linear trends in climate extremes.

Climatic Change **42**: 203–218, 1999.
© 1999 *Kluwer Academic Publishers. Printed in the Netherlands.*

2.   Data Source and Methods

2.1. DATA SOURCE

Several long-term meteorological data sets were used in this study. Monthly statistics were derived from a meteorological data set that consists of 369 Chinese observation stations. They have very good spatial coverage over mainland China. The data period is from 1951 through 1995. It includes variables such as monthly precipitation totals, mean maximum and minimum temperatures, monthly extreme maximum and minimum temperatures, and other variables. Monthly precipitation totals served as cross-check for the daily precipitation for many stations. Possible biases caused by station relocations or by the influence of urban heat island were minimized by examining changes of station latitudes and longitudes, and the population of the observation cities. Random errors were also controlled by analyzing the spatial consistency of anomalies as by (Zhai and Ren, 1997). After rejecting 121 stations with poor data quality, 248 stations were used in this study.

Investigation of changes of extreme precipitation requires short-time scale data. A primary Chinese data set was used, which contains daily precipitation and temperature observations. This data set includes data from 196 Chinese stations during the period from 1951 through 1995.

Precipitation data from many stations were found incomplete. Moreover, the spatial coverage is not adequate in western China. Therefore, another daily precipitation data set with 321 stations was added. After removal of the duplicates, 361 stations remained. Karl et al. (1995) indicated that time dependent missing daily data can bias precipitation trend analyses if assumed to zero or at the daily average of the month. For handling missing data, the following procedure was used: if there were more than 20 days of precipitation data missing during a year, the statistics of that year were assumed missing. Stations were rejected with more than 5 years of random missing or 3 years of consecutive missing data. Exceptions were made for very limited station data in the Tibetan Plateau. Missing data was more serious here, but it usually happened in dry season. This should not be expected to introduce important biases to trends of our study. Stations with significant relocations or in the regions with high population density network were also rejected. To ensure the quality of daily precipitation data, annual precipitation totals for all the stations were generated and were cross-checked with those reported from other data sets. Through the above considerations, 296 stations of daily precipitation in China were finally selected in this study. Small amounts of random missing data should not introduce important biases in trends especially since data are used to generate the indicators in large regions that consist of many stations. All the stations selected in the study are plotted in Figure 1.

Due to the greater spatial correlation of temperature anomalies, daily temperature data during 1951-1995 from a fixed surface network with 149 well distributed stations were used to study nation-wide cold wave activity. They were also used to investigate changes of the number of cold and hot days.

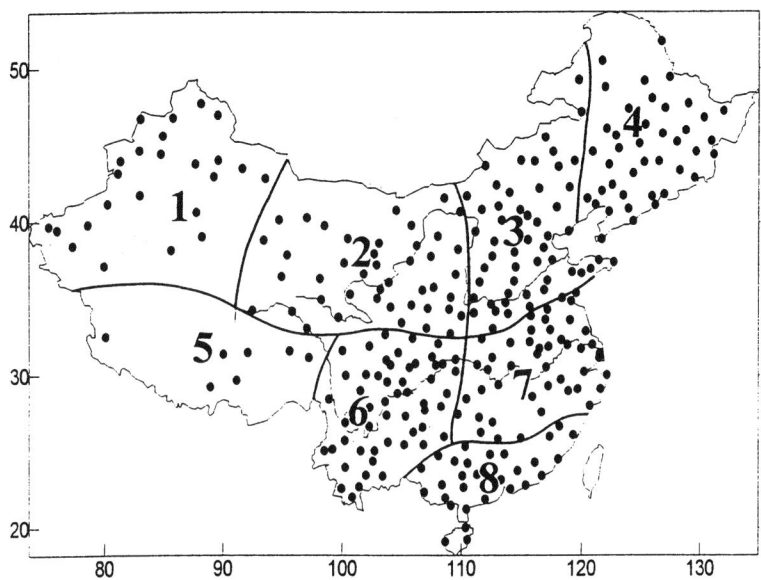

Figure 1. Selected daily precipitation stations (dots) and separate regions in China: 1 - western Northwest China, 2 - eastern Northwest China, 3 - North China; 4 - Northeast China, 5 - Tibet, 6 - Southwest China, 7 - East China, 8 - South China. Northern China consists of region 1, 2, 3, 4, Eastern China consists of region 3, 4, 7, 8.

2.2. METHODS

Regions discussed in this paper are shown in Figure 1. Regional climate extreme indices were generated by arithmetic means inside each province and then weighted by area to give national trends.

Seasonal averages were generated by the arithmetic means of three consecutive months, but seasonal 1-day extreme maximum and minimum temperatures were derived from the highest and lowest extremes in the three months. The four seasons were defined as spring (MAM), summer (JJA), autumn (SON) and winter (DJF).

Time series related to extreme precipitation consisted of:

a. total annual precipitation and the number of days with measurable precipitation ≥ 0.1 mm/day;

b. annual 1-day and 3-day maximum rainfalls;

c. 1-day precipitation totals exceeding 10 mm, 50 mm and 100 mm/day;

d. the annual mean intensity of precipitation defined as the annual total precipitation divided by annual number of rain days;

e. on an annual basis the longest consecutive number of rain days (daily precipitation $\geq$0.1 mm)

To better understand changes of extreme precipitation, a scheme was applied similar to Karl et al. (1996). On an annual basis the highest and lowest 4 years were considered as the upper and lower deciles for the 45 years' of data. The time series of proportion of the area affected by these extremes over all of China for the 8 different regions (shown in Figure 1) were derived for trend analysis. This method aggregates information of climate change from all the stations inside a large region and overcomes the difficulties of detecting long-term trends with spatial variability.

A rainfall index (RI) was calculated to define regional droughts and floods. In this paper:

$$RI = [(\sum_{i=1}^{m} P_i / \sum_{i=1}^{m} R_i) + k / m] \times 100\%$$

Where, P is rainfall anomaly in Jun.-Aug. each year, R the mean total rainfall in Jun.-Aug. for the 30-year normal period during 1961-1990. m is the number of stations in the region, and k is the number of stations with positive summer (JJA) rainfall anomalies.

A flood (drought) is defined when RI is higher than 25% (or lower than -25%). If RI is higher than 35%(or lower than -35%), severe flood (severe drought) is defined.

A hot day is defined when daily maximum air temperature is equal to or higher than 35°C.

A cold day can be defined by several thresholds, such as the days of daily minimum air temperature equal to or below 0°C, -10°C, -20°C, -30°C. Daily temperatures lower than -30°C are very few in China. Therefore, cold days of daily minimum air temperature equal to or lower than -20°C are used in this paper to define cold days.

China's nation-wide cold wave index is defined when at least 1/3 of 96 selected stations in northern China and 1/4 of 53 selected stations in southern China had a temperature drop of more than 10°C in a day, and temperature anomalies were lower than -5°C.

A nine-point binomial filter and a linear regression model were used to illustrate climate variations and changes. The statistical significance of linear trends different from zero was tested using the Kendall-tau statistic, the 0.05 significance level is used throughout this analysis.

3. Results

3. 1. TEMPERATURE EXTREMES

3.1.1. *Mean maximum and minimum air temperatures*

For China as a whole, there exists a slightly increasing, but not statistically significant in annual mean maximum temperature during 1951-1990 (shown in table I). The greatest increase was found in winter, about 0.1 °C/decade, while no obvious trends were detected during autumn and summer. Trends of maximum mean temperature in each specific region reveal some cooling trends in eastern China during summer (Zhai and Ren, 1997).

Table I

Trends for China's national maximum and minimum mean temperatures （0.1°C/decade）, R^2 for variance explained by the trend. Statistically significant trends at the 0.05 significance level are underlined as tested by Kendall-τ test.

	Mean Maximum Temperature		Mean Minimum Temperature	
	Trend	R^2	Trend	R^2
Spring	0.27	-	1.79	24
Summer	-0.06	-	0.01	-
Autumn	0.00	-	1.53	18
Winter	1.44	-	4.17	34
Year	0.30	-	1.75	45

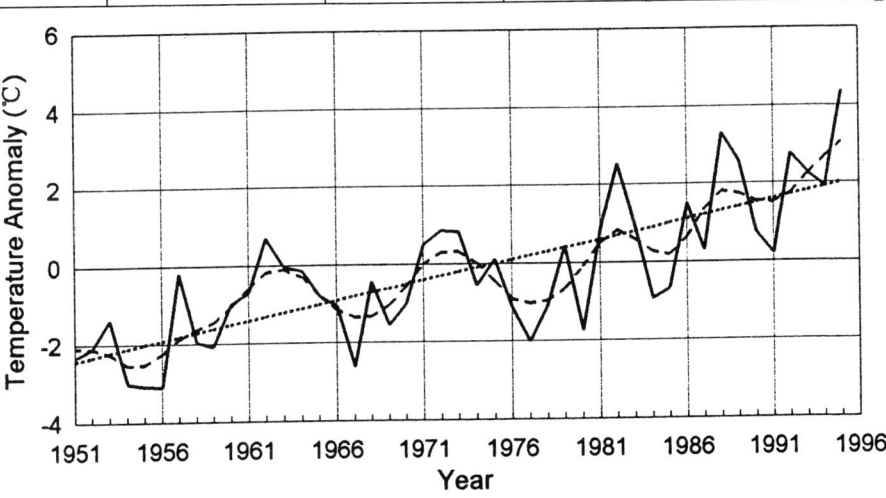

Figure 2. Variations of mean minimum temperature in northern China (dashed lines are filtered using a 9-point binomial smoothing)

Annual mean minimum temperatures in China exhibit a statistically significant increasing trend, which is about 0.2°C/decade during 1951-1990. The largest trend was found in winter with a warming rate of 0.4°C/decade (table II). Generally, annual mean minimum air temperatures display increasing trends all over China, with the greatest increase in the north and smaller increases in the south (Zhai and Ren, 1997). In northern China (shown in Fig. 2), annual mean minimum air temperatures were consistently lower than normal during 1950s and 1960s, while after the 1970s they were only 2 years lower than normal. The trend for increasing mean minimum temperatures in northern China is statistically significant. During winter beginning in the 1980s, there is a strong warming of the minimum air temperature.

3.1.2. *Extreme maximum and minimum temperatures*

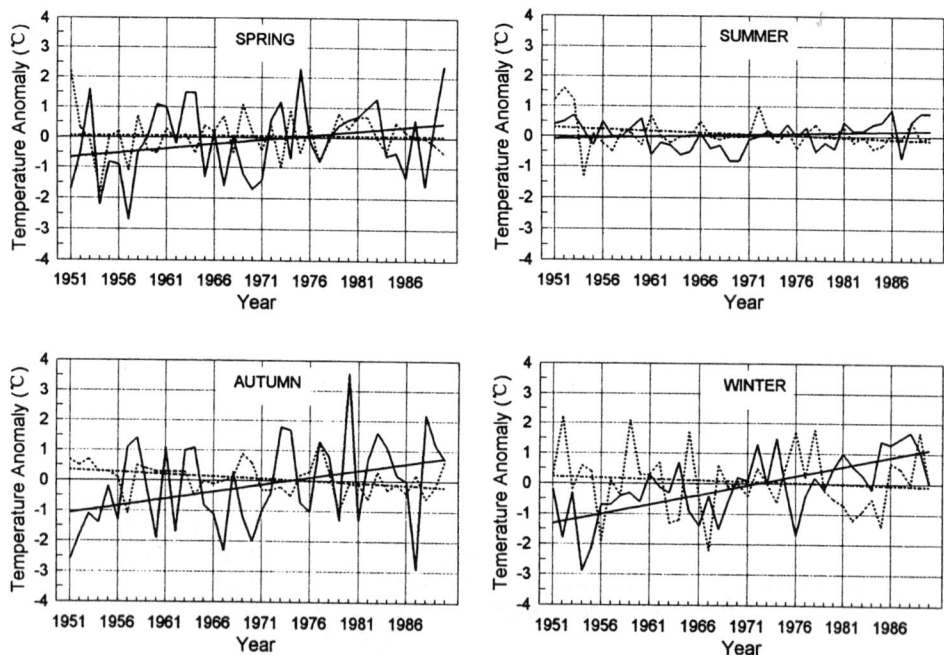

Figure 3. Variations in China's seasonal 1-day extreme maximum (dashed lines) and minimum (solid lines) temperatures during 1951-1990.

Fig. 3 contains data related to China's seasonal 1-day extreme temperatures during the past 40 years. For extreme maximum temperature, decreasing trends were found in summer and autumn. Extreme maximum temperatures dropped 0.6°C during the past 40 years during summer, but these were not statistically significant. Increasing trends in the extreme minimum temperatures are evident in all seasons, but the magnitude of the trends was not the same. The trends at the 0.05 significance level were detected in winter and autumn. The most significant increase was found in winter with a 2.5℃ increase during the past 40 years (1951-90). The increases of extreme minimum temperatures during the past four decades are 1.8℃ and 1.1℃ in autumn and spring, respectively.

Fig. 4 depicts variations of trends in monthly extreme temperatures in China during the past 40 years on a monthly basis. For extreme maximum temperature, decreasing trends are evident in Feb.-Apr., in June- July and in September. For extreme minimum temperature, decreasing trends were detected only in July, while the largest increasing trend was found in January. The increasing trends in winter are greater than in summer months. Noticeably, the increasing trend for extreme minimum temperature in each month is greater than that for extreme maximum temperature. The range of temperature extremes in China is decreasing.

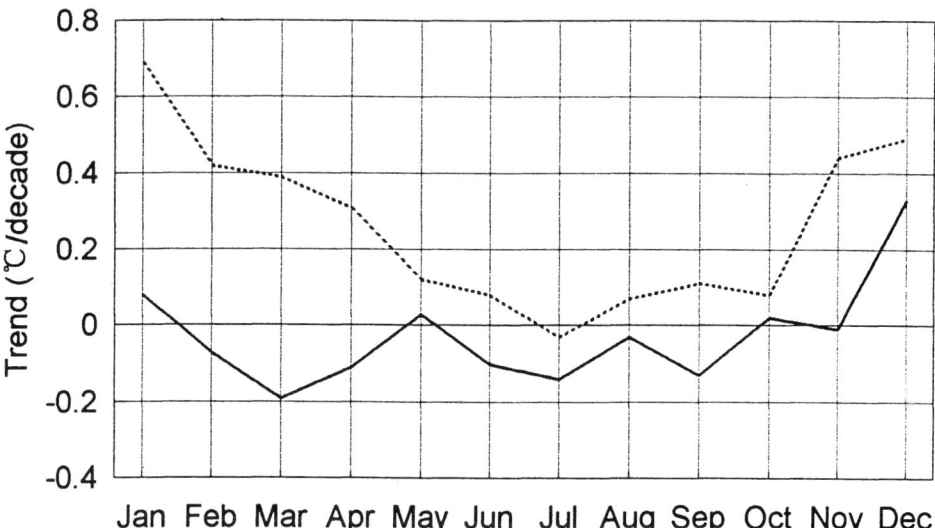

Figure 4. Monthly trends of extreme maximum (solid) and minimum (dashed) temperatures during 1951-1990.

3.1.3. *Number of hot days*

The trends of the annual number of hot days are similar in North China, the Mid-lower reaches of the Yangtze River and regions to the south. They were higher than normal from 1950s to mid-1970s, and thereafter the numbers of hot days were less than normal. There is a slight decreasing trend after the 1970s in Northeast China, but an increasing trend since the 1970s in South China. These trends of annual hot days are basically consistent with the trend for summer mean maximum temperature in the corresponding regions. In eastern China, there were more frequent hot days during 1950s and 1960s when 70% of years experienced greater-than-normal number of hot days. Since the 1970s, however, in 68% of the years the numbers of hot days were less than normal.

3.1.4. *Number of cold days*

Fig. 5 reveals a decreasing statistically significant trend in the number of cold days in northern China. During 1981-1995, there were 13 years of negative anomalies related to the number of cold days, while during 1950s and 1960s there were only 4 years of negative anomalies. The largest negative anomaly during 1951-1980 was -4.6 days, but in 1995 the anomaly reached -14.5 days. There were 9 years of negative anomalies in the most recent 15 years period. This reduction in the number of cold days is related to the remarkable warming since the 1980s. In northern China there has been 10 successive warm winters during the 1986-1995 period.

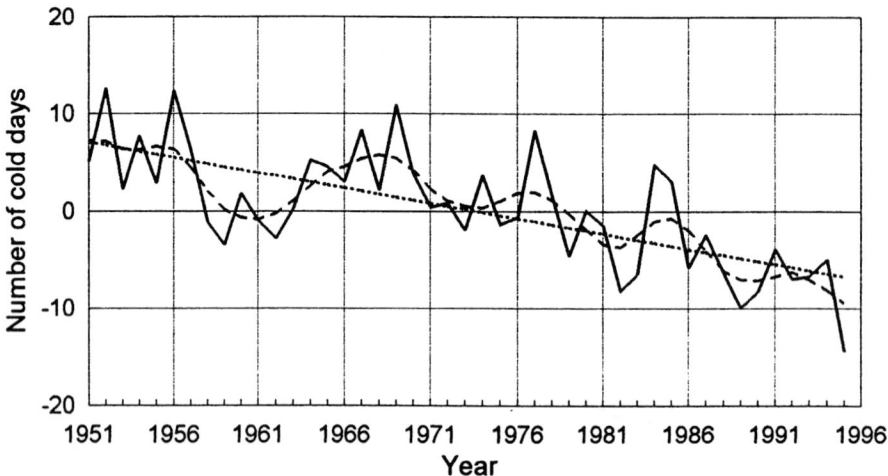

Figure 5. Variations in the number of cold days in northern China (dashed lines are 9-point binomial filtered).

3.1.5. *Cold waves*

Annually, on average, there are 1.8 nation-wide cold waves in China. The frequency of cold waves reached their highest frequency during 1950s, and since then they have decreased gradually. Cold wave activity weakened pronouncedly during 1980s and early 1990s, when the annual average number was only 0.9 per year. The decreasing trend in the number of nation-wide cold waves is statistically significant at 0.05 level. This suggests winter warming since 1980s in mid-high latitudes is reflected by the weakened cold wave activity in China.

3.2. PRECIPITATION EXTREMES

3.2.1. *Annual rainfalls and rain days*

Figure 6 displays that no obvious increasing or decreasing trend but large inter-annual variations are apparent in the proportion of China affected by extreme annual precipitation totals. In 1964, a severe wet year in many parts of China, about 22% of China was covered by extreme annual precipitation amount. Meanwhile, only about 4% of the country had much less than normal annual precipitation (figure not shown). During late 1950s through early 1960s, and since 1980s, the nation often experienced above average areal extent extremely excessive annual precipitation. In 1970s, the areal extent of extreme heavy annual precipitation was relatively low. Table II indicates that a trend toward increasing spatial coverage of much more than normal amount of precipitation is very obvious in western Northwest China. The other regions, such as Northeast China, Southwest China and North China all exhibit decreasing trends. Only in North China is a decreasing trend statistically significant (0.05 significance level).

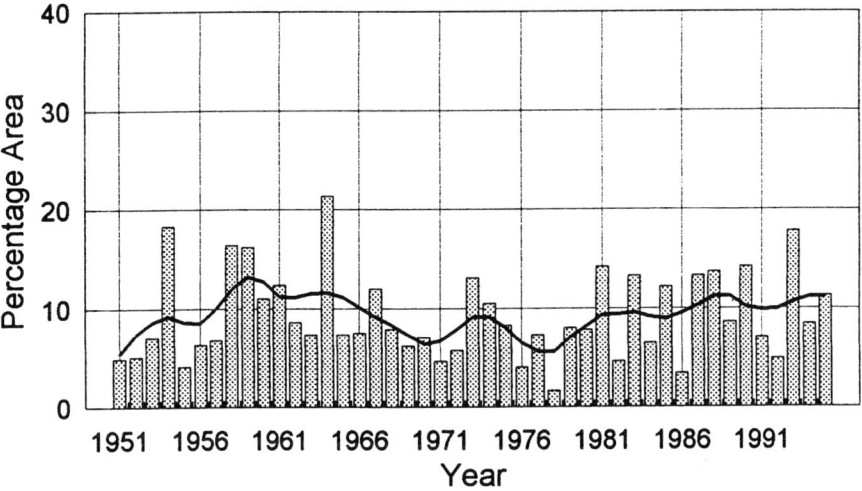

Figure 6. Percent area of mainland China affected by extreme precipitation (bars are for interannual variations, and the data are smoothed by a 9-point binomial filter).

The proportion of the country covered by a much greater than normal number of rain days (Figure 7) decreased significantly during 1951-1995. The decreasing trends are very evident in all the regions in China, except western Northwest China (Table II).

Table II.

Linear trends (1%/decade.) for various indicators of precipitation. The underlined values are statistically significant at 0.05 level.

Indicators	NE	WNW	ENW	N	E	SW	S	TIBET	CHINA
Annual precipitation	-1.6	2.3	-0.6	-3.2	0.0	-1.0	0.5	2.1	-0.9
rain days	-5.8	-0.4	-2.5	-5.8	-4.7	-3.1	-3.7	-2.0	-3.9
1-day max. Rainfall	-0.1	1.9	-0.6	-1.5	-0.9	0.5	0.1	-0.9	-0.3
3-day max. Rainfall	-0.6	2.0	-0.7	-1.8	-1.3	0.4	0.3	-1.5	-0.5
Rainfalls with ≥10 mm daily	-1.0	2.9	-0.3	-2.1	0.4	-0.7	0.9	-1.4	-0.4
Number of raindays for ≥10 mm/day	-3.3	0.6	-1.1	-1.8	-0.7	-2.2	-1.8	-3.9	-1.7
Rainfalls with ≥50 mm daily	-0.9	-	0.4	-0.9	-0.1	0.4	0.7	-	-0.2
Number of raindays for ≥50 mm/day	-4.1	-	-3.9	-3.1	-1.2	-1.4	0.1	-	-2.3
Rainfalls with ≥100 mm daily	-0.4	-	0.7	-1.4	-0.3	1.0	0.2	-	-0.1
Number of raindays for ≥100 mm/day	-4.7	-	-1.1	-5.6	-3.4	-2.7	-2.5	-	-3.6
Duration of longest consecutive raindays	-4.1	-3.4	-1.7	-3.3	-2.7	-2.8	-1.8	-3.6	-2.9
Intensity of rainfall	1.5	0.7	-0.3	2.3	3.3	1.0	3.1	-0.8	1.5

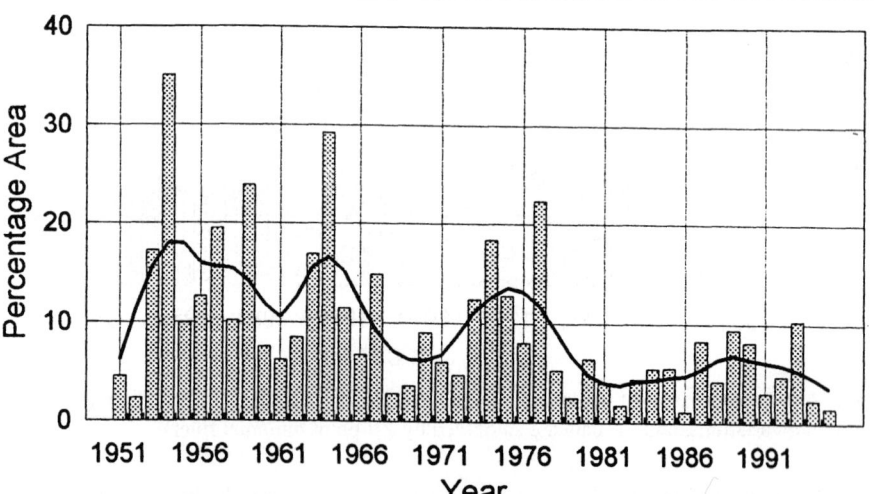

Figure 7. Percent area of mainland China affected by much more than normal number of rain days (bars are for inter-annual variation, and curve is a 9-point binomial filter) .

The big difference between the trends in the extreme amount of annual precipitation and the number of rain days indicates that rainfall intensity is changing.

3.2.2. *1-day and 3-day maximum rainfalls*

Generally, there were no statistically significant trends in 1-day and 3-day maximum rainfall extremes for China as a whole. Regionally, statistically significant decreasing trend in both 1-day and 3-day rainfall extremes were revealed in North China, while obvious increasing trends were observed in western Northwest China.

3.2.3. *Amount and days of high intensity rainfall*

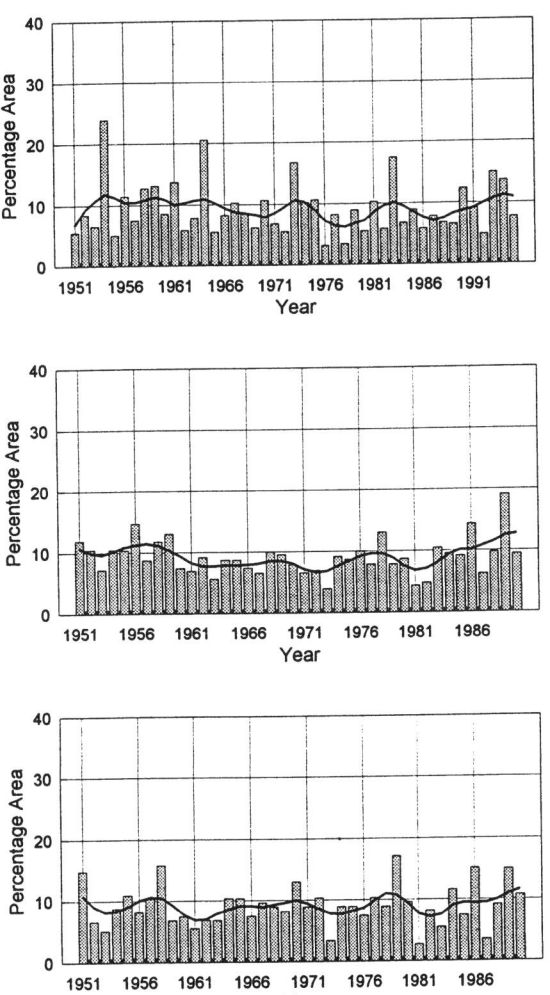

Figure 8. (a) Percent area of mainland China in the upper ten percentile of rainfall totals derived

from daily precipitation greater than 10 mm (bars are for inter-annual variation, and curve is 9-point binomial filtered), (b). Same as Figure 8a except for daily precipitation greater than 50 mm, (c). Same as Figure 8a except for daily precipitation greater than 100 mm.

The interdecadal variations are obvious but there were no significant trends in proportion of China affected by much above normal precipitation derived from daily precipitation greater than various thresholds (≥10 mm, 50 mm and 100 mm rainfalls daily) (shown in Fig.8, table II).

The year 1954 was noticed for the severe floods and this marks the peaks in fig. 8, while the extremely severe drought year, 1978, was a minima for all the time series regardless of their threshold. During late 1950s to early 1960s, large areas of China with much more than the normal amount of precipitation derived from heavy rains is very high. After the late 1970s, there is an increasing frequency of extreme amounts of heavy rains (≥50 mm and 100 mm rainfalls daily).

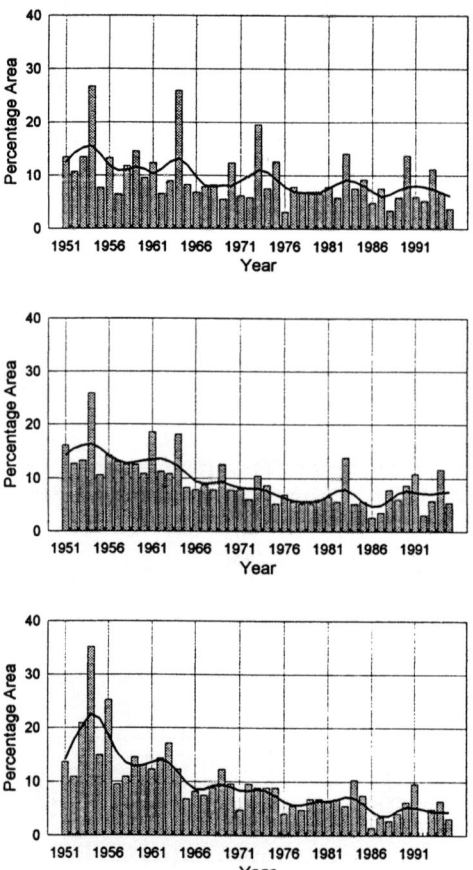

Figure 9. Percent area of China affected by the number of rain days that were in the upper ten percentile for three different thresholds (≥10mm, 50 mm and 100 mm per day).

The increasing trend in western Northwest China for the percent area of much more than normal rainfalls with a daily threshold above 10 mm is statistically significant, while a decreasing trend in North China is also very evident.

Decreasing trends are very evident in the time series related to the area of China affected by much greater than normal number of high intensity rainfall days for each of the three thresholds (Fig. 9). Fig. 9 also suggests a tendency forward more sharply decreasing trends for higher intensity precipitation. Regionally, statistically significant decreasing trends in the number of days with daily rainfall above 100 mm are apparent in much of China, such as South China, Southwest China, East China, and North China, and Northeast China. The largest decreasing trend was found in North China. For rainfall days with daily rainfall above 50 mm, statistically significant increasing trends are in Northeast China, eastern Northwest China, and North China. As for days with daily precipitation greater than 10 mm, decreasing trends are significant only in Northeast China and Southwest China.

3.2.4. *Rainfall intensities*

Figure 10 clearly displays a statistically significant positive trend in the area of China affected by much greater than normal intensity of precipitation. Regionally, significant increasing trends are mainly evident in eastern China, such as South China, East China, and North China, where the climate is typically influenced by East Asia monsoon. In the Northwest, the areal coverage of much greater than normal annual precipitation increases significantly, but rainfall intensity shows no obvious trend.

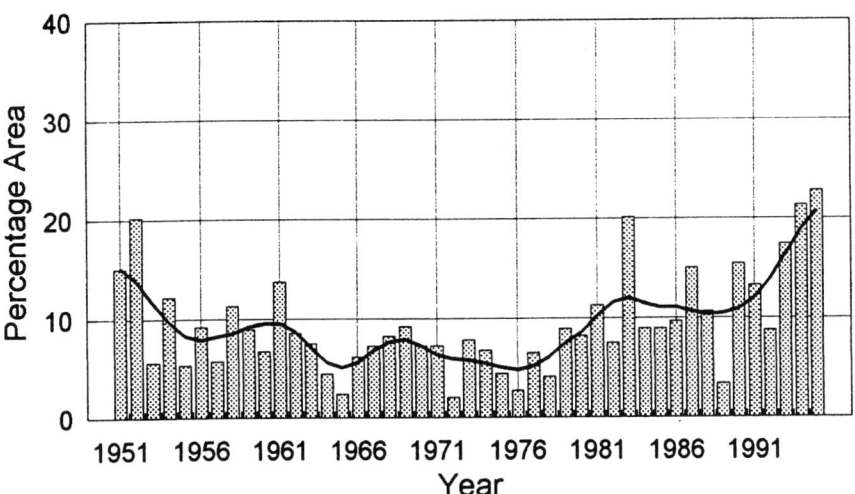

Figure 10. Percent area of mainland China in the upper ten percentile as related to annul mean rainfall intensity (bars are for inter-annual variation, and curve is 9-point binomial filtered).

3.2.5. *The longest consecutive rain days*

A statistically significant decreasing trend is evident in the percent of the country with much more than the normal duration consecutive rain days. This trend is more obvious after the 1960s. The regions of western Northwest China, Northeast China, eastern Northwest China, East China and North China all exhibit significant decreasing trends. Interestingly, in eastern China, the larger decreasing trends are at the higher latitudes (Table II). The chances for occurring consecutive day precipitation are decreasing.

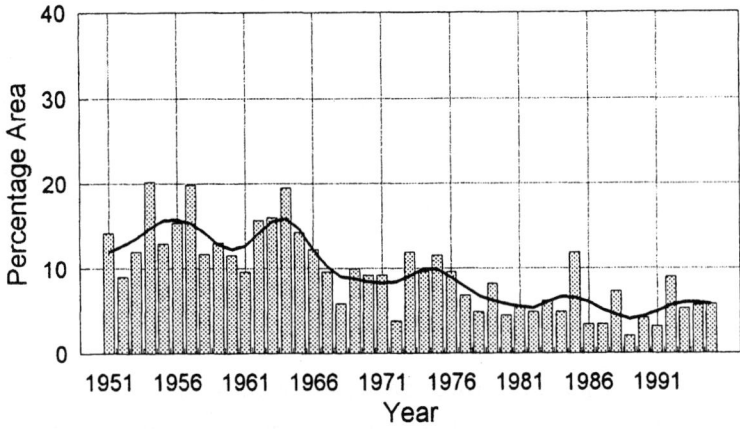

Figure 11. Percent area of mainland China affected by the upper ten percentile of consecutive rain days (bars are for inter-annual variation, and smooth curve is a 9-point binomial filter).

Significant trends related to increasing areal coverage of much higher than normal precipitation intensity and decreasing coverage of the extreme wet spells is consistent with the notion of increasing trends of both droughts and floods in China. We specifically checked for such trends.

3.2.6. *Droughts and floods in eastern China*

The occurrence of droughts and floods in eastern China displays obviously decadal variation (Table III).

Table III

Decadal mean number of drought or/and flood in eastern China.

Decade	Drought	Flood	Drought + flood
1951-1959	0.6	0.4	1.0
1960-1969	0.5	0.8	1.3
1970-1979	0.4	0.5	0.9
1980-1989	0.7	0.3	1.0
1990-1996	0.1	1.4	1.5

For eastern China as a whole, the mean decadal occurrence of droughts and floods reflects a slow steady increase since the 1970s, and reached its peak in the 1990s.

Droughts were most frequent in 1980s, and floods were most frequent in early 1990s. Droughts and floods occurred more frequently after the 1980s (Table III).

4. Conclusions and Discussion

For China as a whole, the mean minimum temperature increased significantly during the past 40 years. Generally, increasing trends were found throughout China, but dominant warming trends were detected at the higher latitudes. Consistently, fewer cold days in northern China and nationally less cold wave activity are apparent after the 1980s.

Although no obvious trend was detected in the annual mean maximum temperatures during the past 40 years for China as a whole, a cooling trend was very much evident in much of eastern China during summer time, especially in North and East China. Meanwhile, the number of hot days in eastern China reduced significantly.

China's extreme maximum temperature shows no obvious trends. However, extreme minimum temperature displays significant increases. This reflects the fact that the range of extreme temperatures tends to be smaller during the past 40 years.

For China as a whole, there were no significant trends in much greater than normal amounts of annual precipitation, in 1-day and 3-day maximum rainfalls, and in rainfalls derived from daily precipitation greater than 10 mm, 50 mm and 100 mm. However, decreasing trends are significant in the number of annual rain days that were much above normal, and in rain days with daily precipitation greater than 10 mm, 50 mm and 100 mm. These decreases, however, are accompanied with significant increase in the trend related to the proportion of China affected by extremely high rainfall intensities.

Western Northwest China is the region that has experienced obvious increasing trends in precipitation. Zhai and Zhou (1997) revealed that its summer precipitation (the main rainy season) is largely related to spring snow cover retreat in East Asia. In western Northwest China, the increasing trend in extreme annual precipitation is accompanied with an increase of daily precipitation exceeding 10 mm. As Groisman et al. (1994) indicated the spring retreat of snow cover extent may be related to continental warming. The increase in high intensity precipitation in western China is possible a response to the warming.

On the other hand, North China is a region that has experienced a drying trend. It displays a significant decrease in the percentage of area with excessive annual precipitation (table II). This is accompanied by a decreasing trend in the percentage of North China with much more than the normal number of rain days. More frequent high intensity rainfall is also very obvious.

References

Groisman, P. Ya., T. R. Karl and R. W. Knight, 1994, *Observed impact of snow cover on the heat balance and rise of continental spring temperatures, Science,* 263, 198-200.

Karl,T. R., R. W. Knight, D. R. Easterling and R. G. Quayle, 1996, *Indices of climate Change for the United States, Bull. Amer. Meteor. Soc.,* 77, 279-292.

Karl, T. R., R. W. Knight and N. Plummer, 1995, *Trends in the high-frequency climate variability in the twentieth century, Nature,* 377, 217-220.

Zhai, P. M. and Q. F. Zhou, 1997, *Northern Hemisphere's snow cover and China's summer precipitation, Quarterly J. of Applied Meteorology,* 8(2),230-235.

Zhai, P. M. and F. M. Ren, 1997, *Change of maximum and minimum temperatures during the Past 40 years in China, Acta Meteorogica Sinica,* 54(4), 418-429.

(Received 5 November 1997; in revised form 1 December 1998)

INDICATORS OF CLIMATE CHANGE FOR
THE RUSSIAN FEDERATION

G. GRUZA and E. RANKOVA

Institute for Global Climate and Ecology (IGCE), 20-b Glebovskay St., Moscow, 107258, Russia

V. RAZUVAEV and O. BULYGINA

*Russian Research Institute of Hydrometeorological Information - World Data Center
(RIHMI-WDC), 6 Korolev St., Obninsk, Kaluga Reg., 249020, Russia*

Abstract. Observed climate changes over the Russian Federation (RF) territory are considered. Several indicators based on monthly mean temperature and precipitation station data are used to quantify regional climate changes. Some of these are the components of two aggregated indices of climate change, suggested by Karl et al. (1996): the Climate Extremes Index (CEI) and the Greenhouse Climate Response Index (GCRI). For the RF territory as a whole, and for its western part, the "Russian Permafrost Free (RPF) territory" in particular, changes in surface air temperature are investigated, together with changes in precipitation and drought indices, and also the fraction of the Russian territory experiencing climatic anomalies below and/or above certain specified percentiles. Composite indices CEI-3 and GCRI-3 based on three parameters (air temperature, precipitation and drought indices) are examined, as well as the Climate Anomaly Index (CAI), known in Russia as Bagrov's coefficient of "anomaly".

It is shown, that over the area of the RPF as a whole, air temperature and the occurrence of drought has increased somewhat during the 20th century, while precipitation has decreased; these changes were non-uniform in space. The linear trend accounts for only a small fraction of the total variability, but the role of climate variations on decadal scales seems more substantial. The CEI, determined as the percentage of the area experiencing extreme anomalies (with a 10% or less frequency of occurrence) of either sign, increased for mean annual temperature, decreased for total precipitation and increased slightly for the occurrence of drought conditions; the aggregated index based on all three of these quantities increased slightly. There was also an increase in the GCRI-3 index, which is indicative of an agreement between the observed climate changes and the changes owing to the greenhouse effect as predicted by climatic models.

The observed climate changes are too small to enable us confidently to reject a hypothesis that they are a reflection of the natural variability of climatic parameters within the context of a stationary climate. However, there is no doubt about the reality and importance of the observed changes.

1. Introduction

The most recent accomplishments of Working Group I of the Intergovernmental Panel on Climate Change (IPCC) (1996) provide a comprehensive assessment of the climatic changes that have occurred over the past century. The structure of global climatic change appears to be rather complicated, and the question of whether the climate has become more variable and/or extreme over the 20th century remains unanswered. At the same time, more detailed data on climate

Climatic Change **42**: 219–242, 1999.
© 1999 *Kluwer Academic Publishers. Printed in the Netherlands.*

change in regions of special interest are required for assessment of the impact of climate change on the environment and on the economic activities, these are dependent on changes in the frequency of occurrence of extreme events (rather than merely on changes in climatic means). To study the variability in extreme events, it will be necessary to develop data sets on a daily time scale.

In this paper preliminary estimates of climatic changes within the Russian Federation (RF) territory as a whole, and for its western part, the "Russian Permafrost Free (RPF) territory" will be presented, based on available monthly data. We shall also examine an index of the "anomality" (i.e., the area-averaged value of the local root mean squared normalized anomalies) for indications of long-term trends. We made similar estimates in the past (Gruza et al., 1980, 1982) based on gridded data obtained as a result of visual analysis. The area averaged root mean squared (RMS) values of the gradient of temperature and temperature anomalies were used as a measure of the spatial variability. Intra-annual standard deviations of the monthly values were used as a measure of the temporal variability. The percentages of the area covered by anomalies above and below prescribed percentiles were used as a measure of the prevalence of "extremes", as well as Bagrov's coefficient of "anomality" (Bagrov et al., 1966). Some increases in the level of climatic variability and the anomality from the 1970's onward were noted.

In Gruza et al. (1985) a specified set of parameters relating to the characteristics of outliers in two-dimensional random fields was offered as a means of characterizing the large-scale features of climatic anomalies. Included are the number of outliers contained by a given field, their individual areas and intensities. Also included are the integral properties in terms of the total area coverage and integral intensity of anomalies exceeding specified threshold levels.

Some specific questions concerning the statistical methods for analyzing climatic time series and the resulting interpretations were discussed in Khlebnikova et al. (1988). Based on a careful objective analysis of the temperature field, Khlebnikova (1989) was able to reproduce the results of Gruza et al. (1985). But, she concluded that the observed increase in the area coverage of severe temperature anomalies might be due simply to the trend in the mean value and does not necessarily reflect an amplification of the temperature variability. An analysis of climate changes over the USA over the past century was carried out in Karl et al. (1996). Two aggregated indices were suggested there: the Climate Extremes Index (CEI) and the Greenhouse Climate Response Index (GCRI). The practical importance of these indices was demonstrated in Easterling et al. (1995).

A reliable analysis of this kind for the RF territory (including RPF) is available now by using monthly data only, where a sufficient number of observing stations have been in operation for the past 90-95 years (Figure 2). For considering climatic change over the full Russian territory it would be necessary to confine the investigation to the past 50 years.

2. Data

Datasets of monthly-observations, assembled and updated at the Institute for Global Climate and Ecology (IGCE), were used as a basis for the estimates being presented in this study. Monthly values of mean surface air temperature and precipitation totals observed at 455 meteorological stations over the Former Soviet Union (FSU) territory during the period of 1886 to the present were used. All available stations in the region whose records extend from 1951 to the present with fewer than 5 missing years are included. There are 169 temperature and 202 precipitation stations whose records include at least 80 years for monthly data observed (after 1901) contained within the region of study (Figures 2, 5). There are 111 temperature and 128 precipitation stations for the RF, of which 72 temperature and 91 precipitation stations are located in RPF.

Homogeneity of the data considered is assured by the uniformity of observing practices used throughout the country and by the corrections designed to compensate for temporal disruptions (change of a measuring instrument, stations replacements etc.). The corrections were introduced directly into monthly data. Stations located in urban areas with substantially increasing population were excluded from the analysis.

To analyze variations in the space-time distribution of extreme temperatures and precipitation at individual Russian stations for the last few decades, daily temperature (mean, minimum and maximum) and atmospheric precipitation data for the last WMO reference period 1961-1990 from 41 stations were used (Figure 13). A daily data set has been prepared at the RIHMI-WDC (Obninsk, Russia) based on a subset of stations from the previously compiled data set of Razuvaev et al. (1993). For the purpose of the present work, time-series for 41 stations have been updated up to 1995 and subjected to additional quality control.

3. Indicators of Regional Climate Changes

Area-average climatic parameters over the region of interest are first used to characterize the regional climate variability. The area averaging was performed here in two stages. First, averages in 5° latitude by 5° longitude boxes were calculated from the stations located within each box. Then, area-weighted averaging was performed, based on the mean value of each box.

The linear trend was estimated by the method of least squares, applied to the period 1901-1995 in all cases. In the figures that contain time series plots, smoothed versions (based on a 9-point binomial filter) are included, together with four selected statistics based on the 95-year record: the mean [mean], the standard deviation [std], the linear trend coefficient [b(tr) in dimensional units per 100 years] and the percentage of the total variance explained by the linear trend [d(tr) in %], which provides a measure of the significance of the estimated trend.

In Table 1 the probabilities of exceeding the observed d(tr)'s under the null hypothesis are presented as Prob{d(tr)|H$_o$ > d(tr)}. Here d(tr) means observed value, while d(tr)|H$_o$ is obtained from a stationary sequence of N independent random numbers. These probabilities are based on an F-ratio distribution with 1 and N degrees of freedom applied to the d(tr) being estimated. In practice, N should be set equal to the number of independent observations in the time series under investigation. In the case of climatic time series consisting of annual values, N tends to be smaller than the number of years by a factor of 1.5-2.0 because of autocorrelation. It is evident from Table 1 that trends, which account for as much as 12% of the variance are generally significant at the 95% level or above. Annual values of all three calendar periods (the year as a whole, and the cold and warm periods of year), are calculated as arithmetic averages of the appropriate monthly anomalies, namely: December through November, October through April and May through September, respectively.

Table I

The probabilities of exceeding the observed d(tr)'s under the null hypothesis
Prob(d(tr)|H$_o$ > d(tr))

d(tr), %	Number of observations				
	20	40	60	80	100
2.5	0.50555	0.32984	0.22759	0.16128	0.11615
5.0	0.34329	0.16544	0.08590	0.04617	0.02533
7.5	0.24265	0.08727	0.03423	0.01397	0.00583
10.0	0.17436	0.04682	0.01384	0.00427	0.00135
12.5	0.12621	0.02522	0.00559	0.00129	0.00031
15.0	0.09158	0.01355	0.00223	0.00039	0.00007
17.5	0.06642	0.00723	0.00088	0.00011	0.00001
20.0	0.04804	0.00381	0.00034	0.00003	0.00000
22.5	0.03459	0.00199	0.00013	0.00001	0.00000
25.0	0.02477	0.00102	0.00005	0.00000	0.00000

Percentiles and temporal empirical cumulative probabilities of the observed anomalies (so-called "percentile ranks") were estimated individually for each station based on the same period 1901-1995. The percentage of the area with extreme anomalies was calculated in terms of the spatial cumulative distribution function given to the year of concern. For example, to obtain the percent of the regional area with extreme temperature (below tenth percentile) in 1995, the following steps were completed. First, the observed temperature anomaly field was recalculated in terms of the local percentile ranks for each station considered. These percentile ranks were checked with 0.1 and two sequential ranks were selected, immediately below and immediately above 0.1. Eventually, both corresponding area percentages were estimated (with percentile ranks below each of the selected thresholds separately) and interpolated into tenth-point.

4. Results

4.1. SURFACE AIR TEMPERATURE

Time series of seasonal mean surface air temperature anomalies, averaged over the RPF territory are presented in Figure 1 for each of three periods: the year as a whole, and the cold and warm periods.

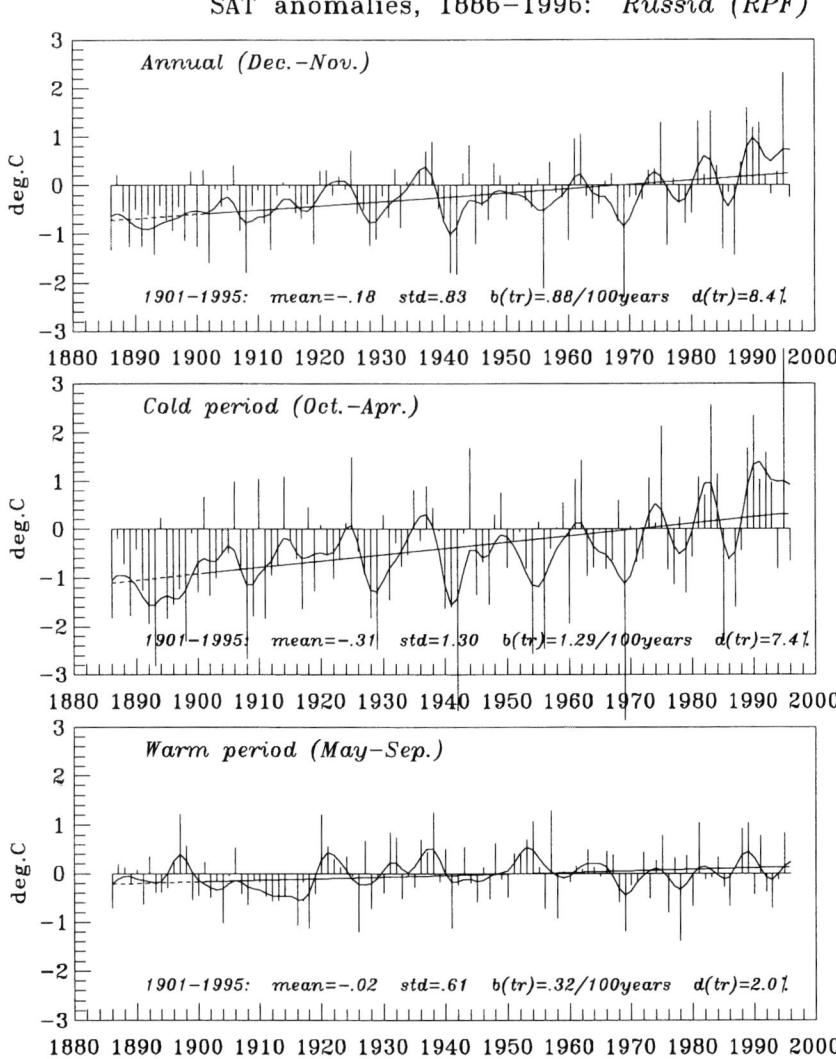

Figure 1. Departure from the 1961-1990 mean of the area-average surface air temperature over the RPF territory, annual (Dec. through Nov.) and for the cold (Oct. through Apr.) and warm (May through Sep.) seasons. Smooth curves in these and subsequent plots are based on a 9-point binomial filtering of the annual values. The long-term statistics are based on 1901-1995 for this and subsequent figures.

All three series reveal a positive linear trend (warming), which is stronger in the cold season and weaker in the warm season. The significance of the trends, as measured by fraction of the total variance that they explain, is largest for mean annual temperature, which exhibits less interannual variability than cold season temperature. It is noteworthy that the mean anomaly for the period 1901-1995 is negative (i.e., this period as a whole was colder than the base period 1961-1990, which was used for defining the normal). Notably, the linear trends -- which amount to increases of about 0.9, 1.3 and 0.3 °C per century for annual mean, cold season and warm season temperature, respectively, -- exceed the IPCC-95 trend estimates of the global and Northern Hemisphere temperature trends during the past century. However, as follows from estimates of d(tr), even these rather substantial trends are not large enough to provide a formal basis for rejecting the null hypothesis concerning its cause.

The linear trends only rather poorly reflect the structure of the long-term variability of area-average temperature in the region under consideration. A more substantial role is played by fluctuations on decadal scales, as is clearly evident in the plots and confirmed by spectral analysis. It is evident from Figure 2 that the warming is not uniform in space. It appears to be most substantial in the latitude belt 50-55N. A few stations in the Northern Caucasus and near the Black Sea exhibited negative trends over the century. The absence of stations with long records within the Asian territory of the country is notable: hence the focus was made on the western part of Russian territory (RPF), which is better sampled. The boundary of the RPF is indicated here and in Figures 5, 8 by a double line.

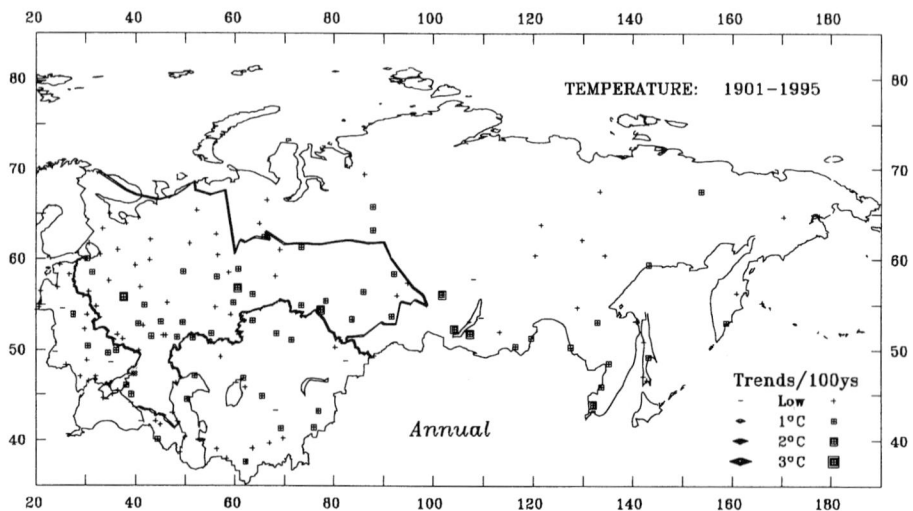

Figure 2. Surface air temperature trends (1901-1995) are indicated by the number (up to 3) of concentric squares/diamonds centered on the sites of stations considered. Squares and plus signs represent warming and diamonds and minus signs, cooling.

Time series of the percentage of the area of the RPF covered by temperature anomalies failing within the most extreme tenth and twentieth percentiles on both sides of the local frequency distribution are shown in Figure 3. The corresponding median time series is indicated by the smoothed curve in the middle of the figure. The areas under and above this curve correspond to the fractional area covered by negative and positive anomalies, respectively. [Recall that the percentage of the area affected by anomalies occurring once in a ten-year period is used as the basis for defining a climate extremes index in Karl et al. (1995, 1996).]

The tendencies in the changes in the percentages of the areas (covered by negative and positive anomalies of the level considered) are consistent with the upward trend in mean temperature and appear to be a more sensitive indicator of the observed warming. Although not shown here, the total area covered by extreme temperature anomalies of either sign increased slightly over the century being somewhat more in area for the cold season and less for the warm season (with d(tr) less than 7%).

Also, although the percentage of the area covered by anomalies below or above some threshold value may be of practical value as a climate change indicator, it can vary owing to changes in other parameters. For example, Khlebnikova (1988, 1989) showed that a trend in mean temperature leads to asymmetric changes of extreme values probabilities in tails of a normal distribution. As a result, a linear trend of either sign in the mean values results in an increase in the total area covered by extreme anomalies near the beginning and the end of the record relative to the area coverage near the middle of the record. For the case in point, we suspect that the rather sharp upward trend in mean temperature from the 1970's onward may be responsible for the increase in the area covered by extreme (mostly positive) temperature anomalies toward the end of the record. The role of other parameters should be analyzed in the future.

Thus, to investigate a structure of climate change and variations over the past century it is necessary first carefully analyze changes in means, variances and elements of the covariance's matrix, at least for temperature, whose distribution is close to normal.

4.2. PRECIPITATION

The variation of area-average precipitation over RPF territory is shown in Figure 4. Annual mean precipitation, and precipitation for both warm and cold seasons exhibited weak downward trends which are characterized by a high degree of spatial heterogeneity (Figure 5, 5[1]). Time series of the percentage of the area of the RPF covered by extreme anomalies (Figure 6) and their statistics are in agreement with the conclusion about a decrease of precipitation over Russian territory in general. Cold season precipitation accounts for most of the decrease in the annual mean.

We believe there are not enough reliable station data to permit an accurate estimate of precipitation trends for much of the Asian part of Russia over this period; only 37 of the 180 stations considered have at least 80 years of data during the period 1901-1995 in this region. This lack of data could explain the ambiguous estimates of long-period precipitation trends, which have been published and suggest further study is needed.

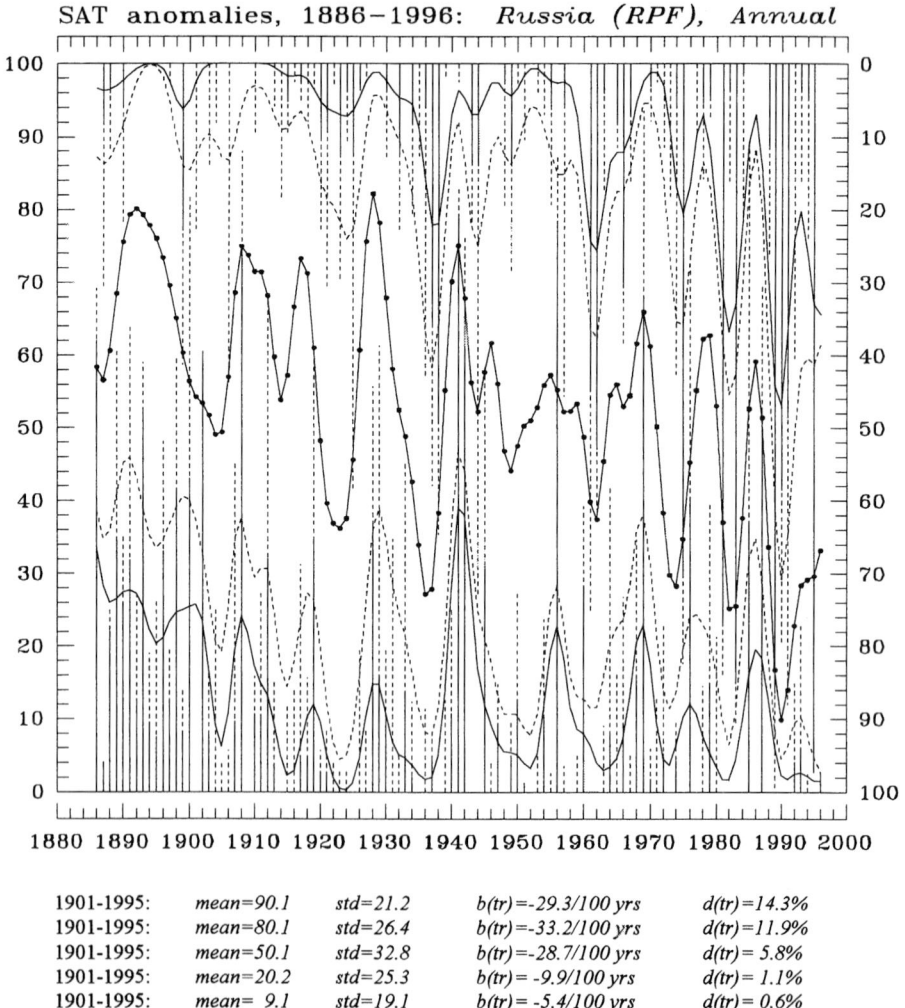

SAT anomalies, 1886–1996: *Russia (RPF), Annual*

1901-1995:	mean=90.1	std=21.2	b(tr)=-29.3/100 yrs	d(tr)=14.3%
1901-1995:	mean=80.1	std=26.4	b(tr)=-33.2/100 yrs	d(tr)=11.9%
1901-1995:	mean=50.1	std=32.8	b(tr)=-28.7/100 yrs	d(tr)= 5.8%
1901-1995:	mean=20.2	std=25.3	b(tr)= -9.9/100 yrs	d(tr)= 1.1%
1901-1995:	mean= 9.1	std=19.1	b(tr)= -5.4/100 yrs	d(tr)= 0.6%

Figure 3. Percentage of the area with the mean temperature falling in the lower (left scale) and upper (right scale) tenth percentiles of the local cumulative frequency distributions based on the century-long period of records. Dashed curves correspond to the twentieth percentiles and middle curve, the fiftieth percentile. The long-term statistics are all expressed in terms of the fractional areas falling in the lower percentiles.

Here we present additionally Figure 5[1] to show a pattern of precipitation trends as based on full available network (455 stations) but for the period of 1951-1995, when missing data are practically negligible. The Figure shows decreases during the last 50 years in many regions of Asian part of the FSU.

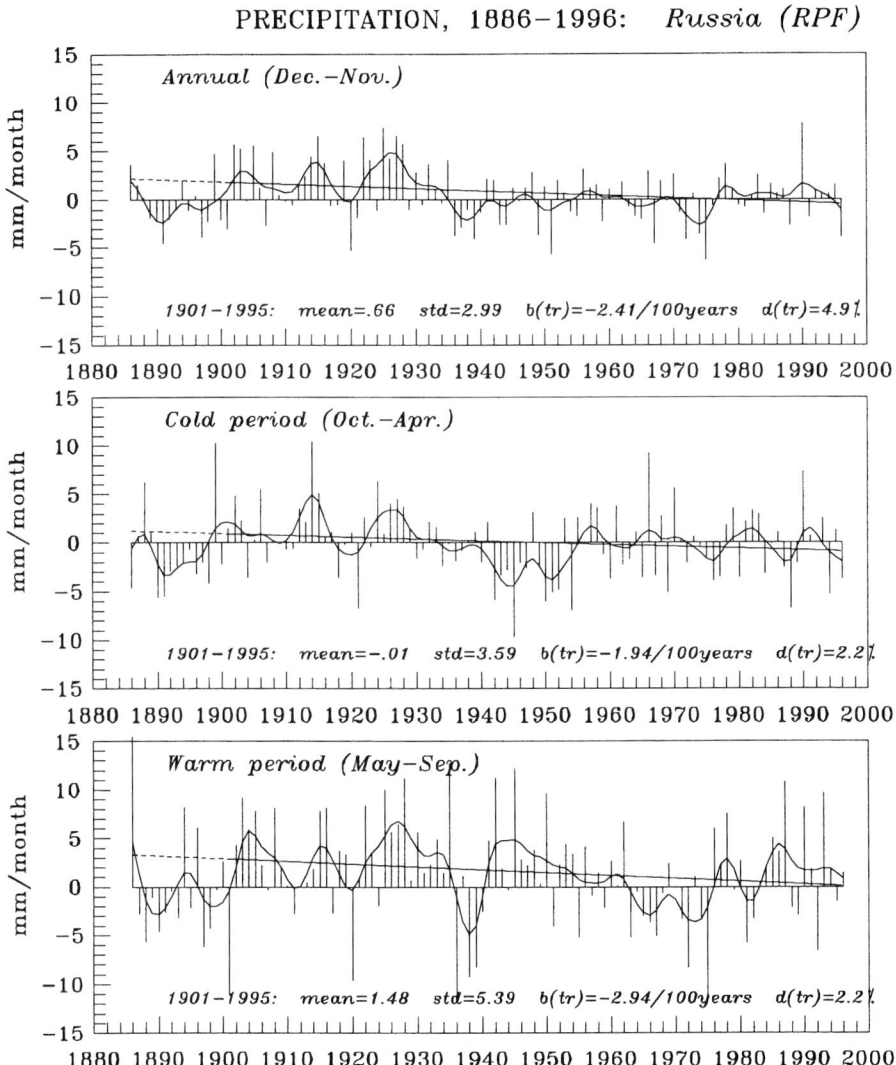

Figure 4. As in Figure 1 except for precipitation totals (in mm/month).

4.3. DROUGHT INDEX

An index defined in Pedj (1975) has been widely used in Russia in the past two decades for documenting droughts and moisture surplus conditions. It represents the difference between standardized anomalies of surface air temperature and atmospheric precipitation over the area of interest. While this index is inferior in a physical sense to Palmer's index (used in U.S.), it still enables us to draw some conclusions about changes in the frequency of occurrence of drought conditions over the past century.

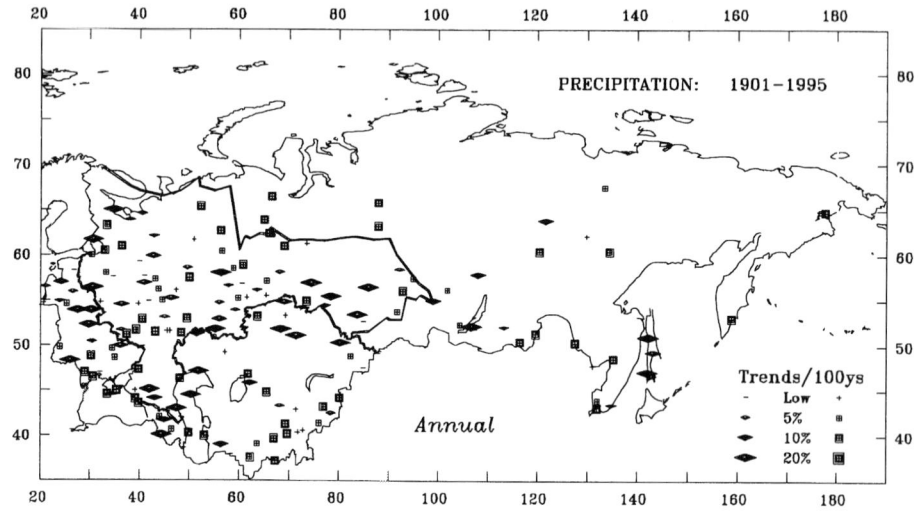

Figure 5. As in Figure 2 except for precipitation totals expressed as percentages of the 1961-1990 means. Squares and plus signs represent increases and diamonds and minus signs represent decreases.

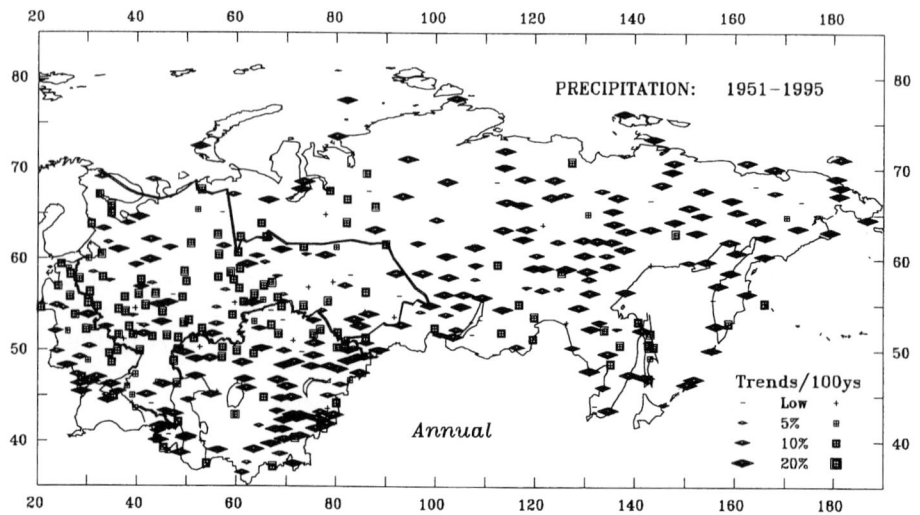

Figure 5[1]. As in Figure 5 except for the period of 1951-1995.

Drought index time series for the year as a whole and for cold and warm seasons are shown in Figure 7. There is an obvious tendency for an increase in the frequency of drought in all seasons. The trends are more substantial than those of temperature and precipitation, probably because trends in temperature and precipitation both contribute to an increased incidence of drought.

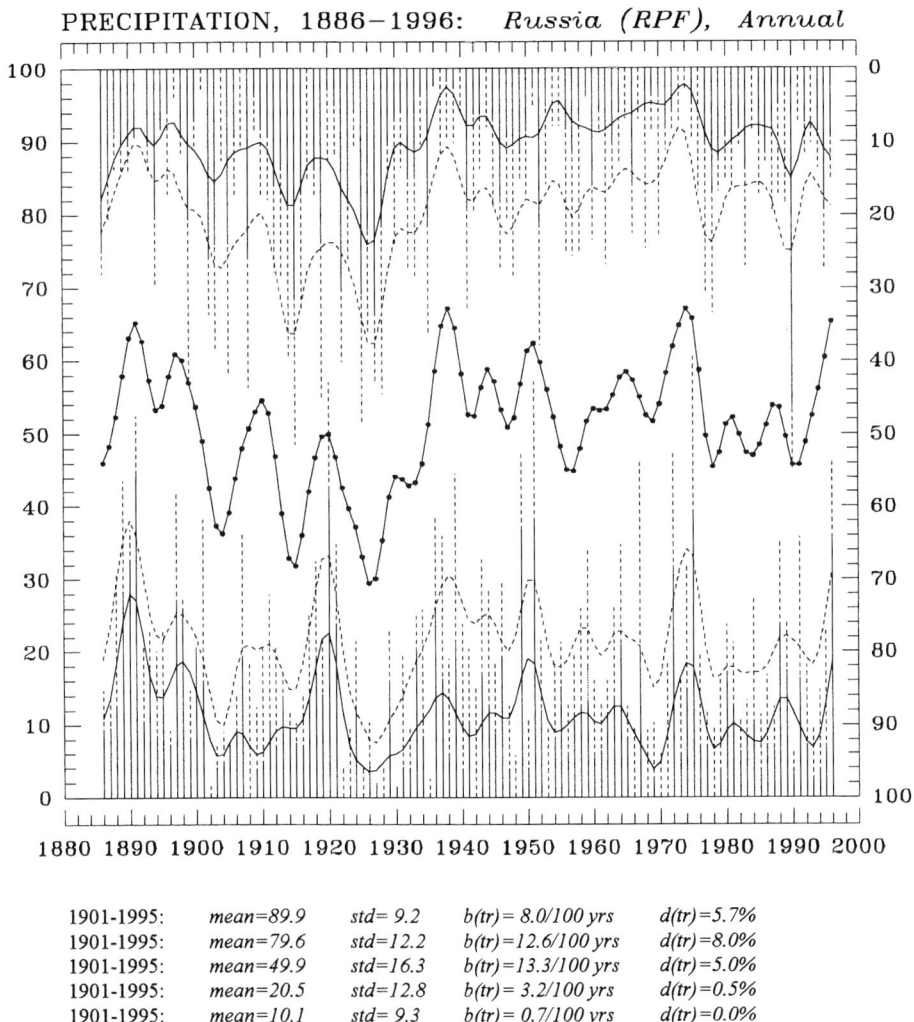

PRECIPITATION, 1886–1996: *Russia (RPF), Annual*

1901-1995:	*mean=89.9*	*std= 9.2*	*b(tr)= 8.0/100 yrs*	*d(tr)=5.7%*
1901-1995:	*mean=79.6*	*std=12.2*	*b(tr)=12.6/100 yrs*	*d(tr)=8.0%*
1901-1995:	*mean=49.9*	*std=16.3*	*b(tr)=13.3/100 yrs*	*d(tr)=5.0%*
1901-1995:	*mean=20.5*	*std=12.8*	*b(tr)= 3.2/100 yrs*	*d(tr)=0.5%*
1901-1995:	*mean=10.1*	*std= 9.3*	*b(tr)= 0.7/100 yrs*	*d(tr)=0.0%*

Figure 6. As in Figure 3 except for annual precipitation percentiles.

The spatial pattern of drought index trends (Figure 8) appears to be more uniform than that for temperature and precipitation separately. The northeast European part of territory considered is noteworthy due to a tendency towards decreasing drought occurrence that may be attributed to a local upward trend in precipitation. Time series of the area (in % of RPF) covered by extreme positive and negative anomalies of drought index (Figure 9) are in agreement with above conclusions.

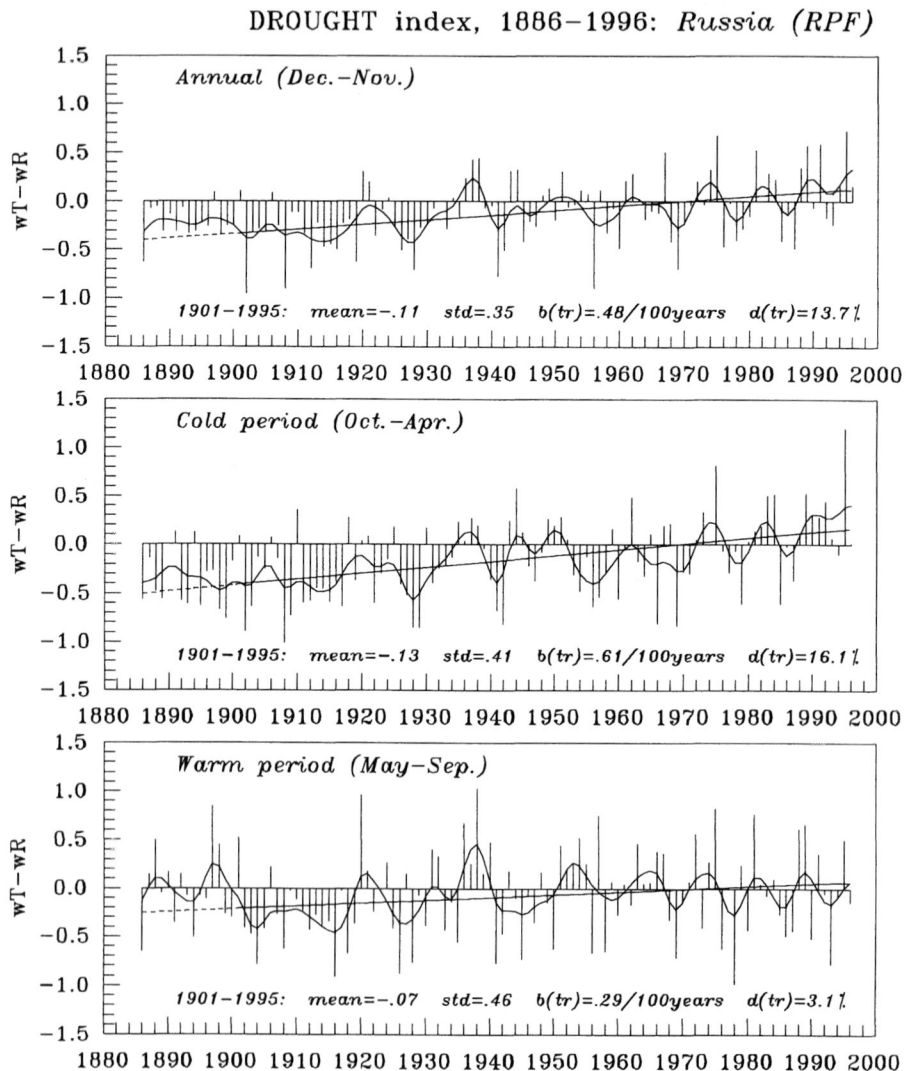

Figure 7. As in Figure 1 except for drought index (in dimensionless units).

However the total area with extreme drought index anomalies (that is a sum of the areas with extreme drought conditions and those with excessive moisture) varied only slightly during the century, being very similar in the warm and cold periods of the year.

4.4. CLIMATE EXTREMES INDEX (CEI-3)

The Climate Extremes Index in the present work is defined as the arithmetic average of the proportions of the considered area experiencing the extreme anomalies (i.e., values falling in the outermost 10% in either tail of the local frequency distribution) for temperature and precipitation totals and for the drought index.

The Climate Extremes Index for the U.S. discussed in Karl et al. (1996) also includes the percentage of the U.S. with a much greater than normal contribution of daily precipitation events exceeding 2 inches, and the same of the number of days with and without precipitation. The notation used here (CEI-3) is used to denote that our CEI index is based on only three indicators instead of five.

Changes in the climate extremes index CEI-3 (Figure 10) over the past century in the RPF territory are practically nonexistent. A small positive linear trend is present in the annual mean data, but it explains only one percent of total interannual variability. The cold season exhibits a negative trend, but it is not significant, and neither is the warm season trend.

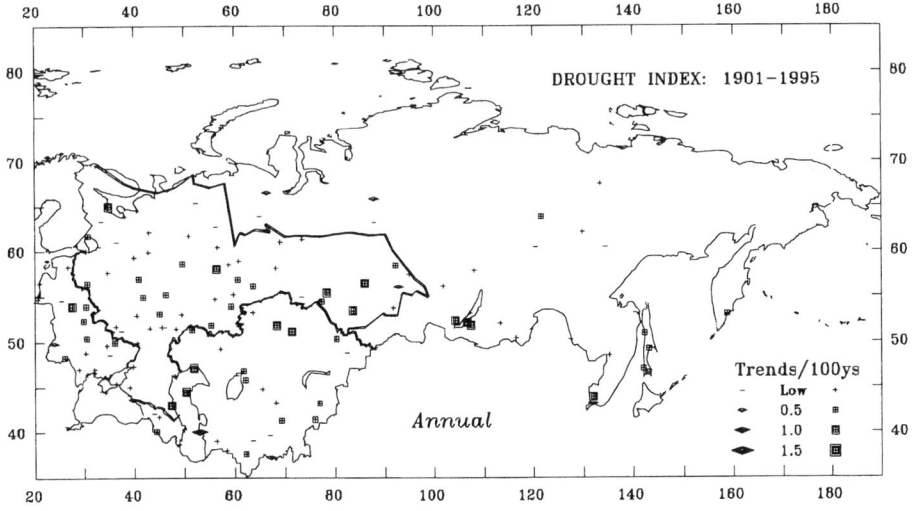

Figure 8. As in Figure 2 except for drought index. Squares and plus signs represent increases in prevalence of drought and diamonds and minus signs indicate decreases.

4.5. GREENHOUSE CLIMATE RESPONSE INDEX (GCRI-3)

Following Karl et al. (1995) we have attempted to develop a Greenhouse Climate Response Index (GCRI-3), that reflects effects of greenhouse gas warming. The three indicators used in the index are variables in which climate models predict large changes: namely, an increase in annual mean temperature (1), an increase in cold season precipitation (2), and an increase in the drought index during the warm season (3). As in the case of the previous index, the suffix 3 is used to emphasize that the index is comprised of three rather than five components. (The area percentages with much-above normal proportion of precipitation resulting from extreme daily precipitation events and the same with much-below normal diurnal temperature differences are not considered here.)

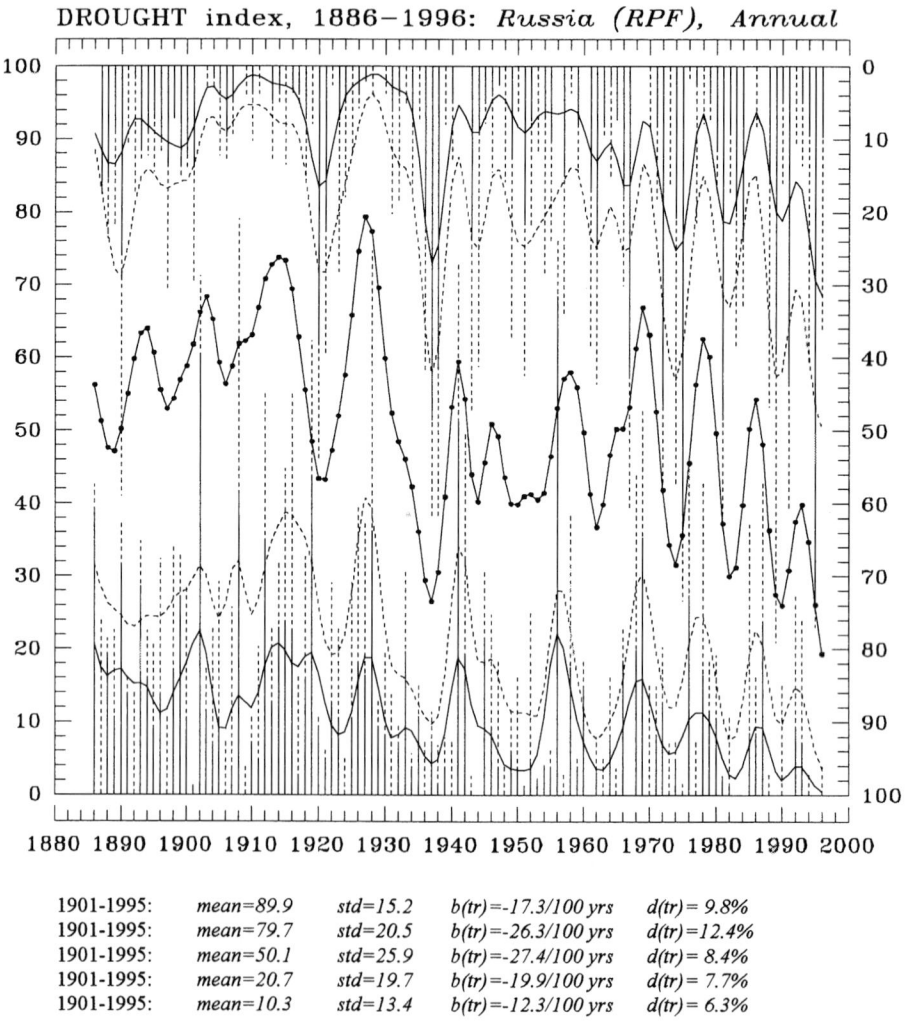

DROUGHT index, 1886−1996: *Russia (RPF), Annual*

1901-1995:	mean=89.9	std=15.2	b(tr)=-17.3/100 yrs	d(tr)= 9.8%
1901-1995:	mean=79.7	std=20.5	b(tr)=-26.3/100 yrs	d(tr)=12.4%
1901-1995:	mean=50.1	std=25.9	b(tr)=-27.4/100 yrs	d(tr)= 8.4%
1901-1995:	mean=20.7	std=19.7	b(tr)=-19.9/100 yrs	d(tr)= 7.7%
1901-1995:	mean=10.3	std=13.4	b(tr)=-12.3/100 yrs	d(tr)= 6.3%

Figure 9. As in Figure 3 except for percentiles of the annual drought index.

Time series of GCRI-3 shown in Figure 11 are for the RPF territory. There is definitely some agreement between observed climate changes and the changes predicted by climatic models, particularly during the past 50 years. Whereas, periods with appreciable upward trends in this index are observed from the late 1800s through the early 1900s.

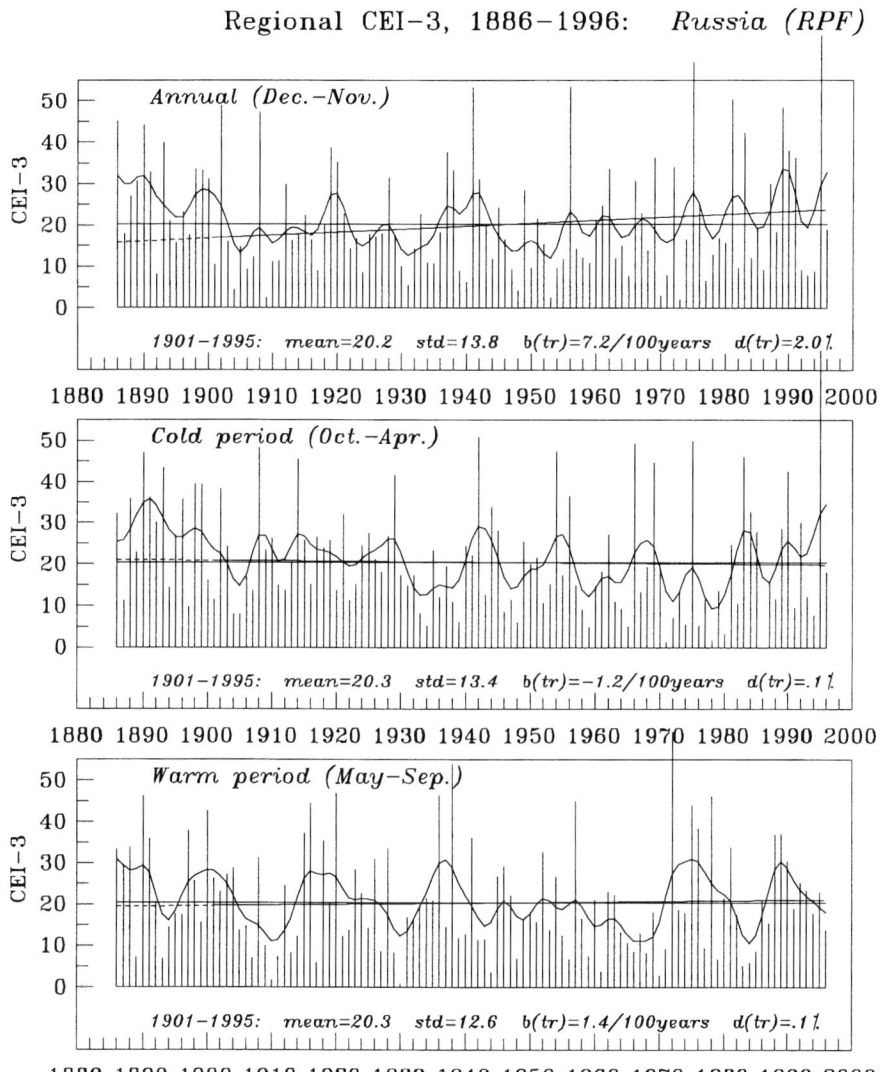

Figure 10. Annual Regional Climate Extremes Index (CEI-3) for year as a whole (Dec. through Nov.) and for cold (Oct. through Apr.) and warm (May through Sep.) seasons.

4.6. CLIMATE ANOMALY INDEX (CAI)

Bagrov (1966) defined a "coefficient of anomality" (as described in section 1) to characterize the spatial variability of the temperature field over Northern Hemisphere. We use this parameter as the basis for our Climate Anomaly Index

(CAI), whose interpretation can be understood as follows. Let us represent the variables chosen to describe the current state of the climate system, or some of its part as components of climate state vector in phase space. The current value of the CAI (Bagrov's coefficient of "anomaly") at any given time is the Euclidean distance in the chosen phase space between the point describing the current climate state and the point representing the time-mean state (the normal). The larger the CAI, the farther removed is the point representing the instantaneous climate state from the centroid of the points in phase space.

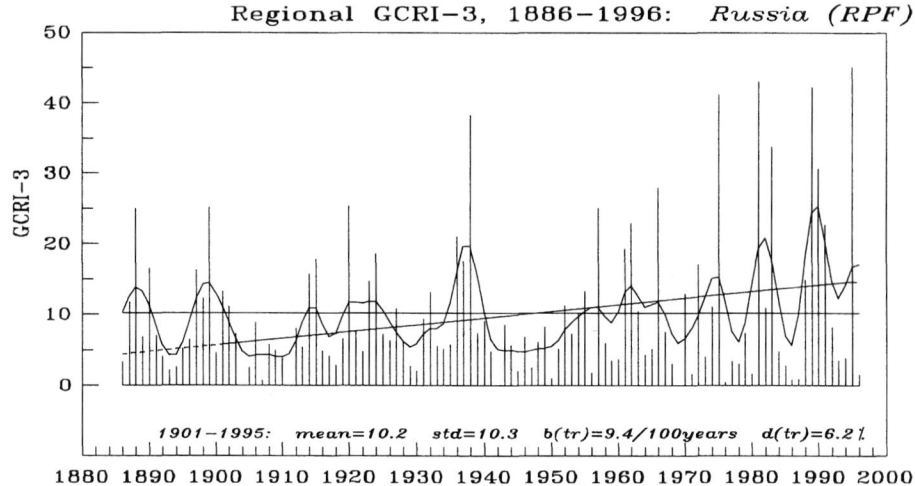

Figure 11. Annual Regional Greenhouse Climate Response Index (GCRI-3) based on three greenhouse climate response indicators: annual temperature, cold season precipitation and the warm season drought index.

Variations in the CAI in the surface air temperature field over the 20th century are shown in Figure 12 for the RPF territory. As in previous figures, there is an upward trend superimposed on substantial decade-to-decade fluctuations. It is worth examining climate anomaly indices not only in reference to the temperature field, but also to fields of other climatic parameters (in particular, precipitation and the drought index).

4.7. CHANGES IN DAILY EXTREMES

The total number of days for winter (December-February) and summer (June-August) seasons, when daily air temperatures and precipitation were found to be higher or lower than some of the fixed limiting values, was used in this case as an index of climatic extremes. To this end, daily values for the time period of interest (1961-1995 for temperature; 1966-1995 for precipitation) were sorted by magnitude and 5% and 95% percentiles were defined for each station on a seasonal basis (Figure 13). All values falling below the 5th percentile and above the 95th percentile were considered as daily extremes. The number of days, N, when daily temperatures or precipitation were within the above-mentioned intervals, was determined for the two seasons (winter and summer) of each year. Linear trends in the number of days were calculated for each station for the time period of interest.

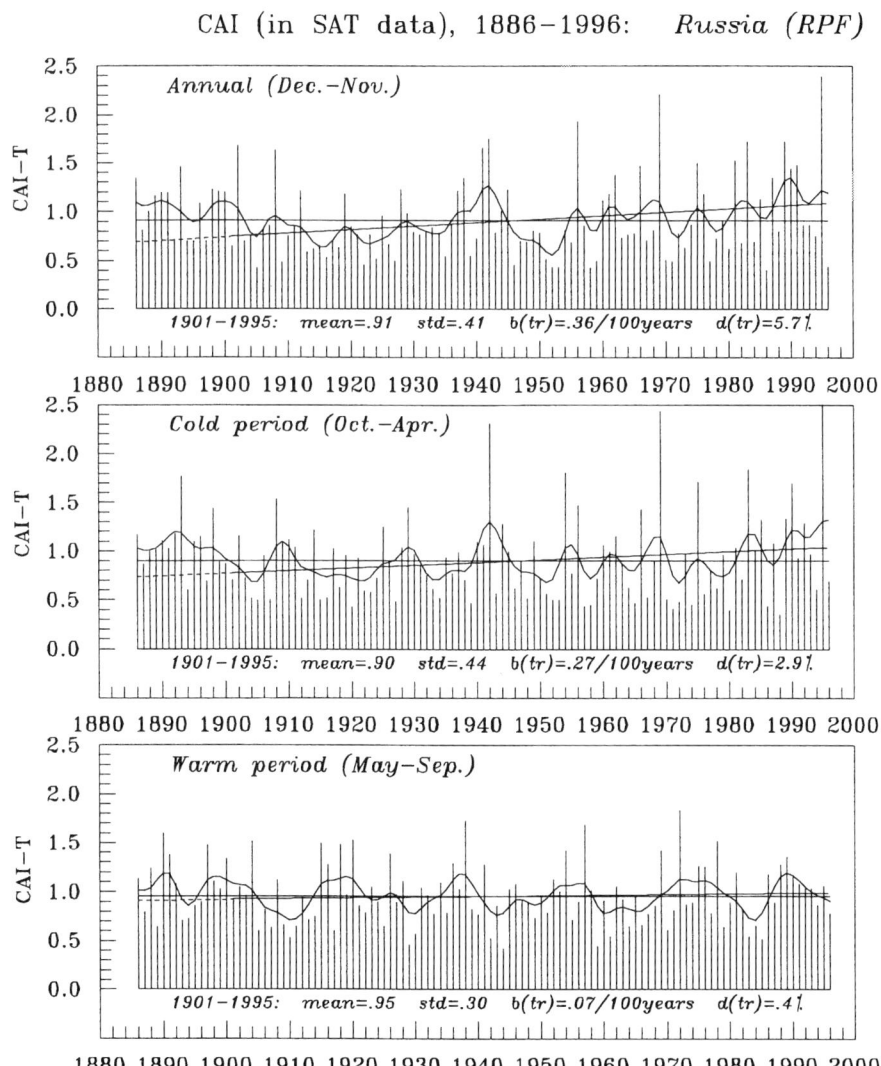

Figure 12. Annual Regional Climate Anomaly Index (CAI) based on standardized temperature anomalies, annual (Dec. through Nov.) and for the cold (Oct. through Apr.) and warm (May through Sep.) periods.

Changes in daily temperature extremes were studied for the period of 1961-1995. The period 1966-1995 was used to analyze the variability of extremes in daily precipitation, since the change of a measuring instrument in 1964-1965 disrupted the homogeneity of precipitation time-series and procedure for adjustment of daily precipitation is not yet prepared.

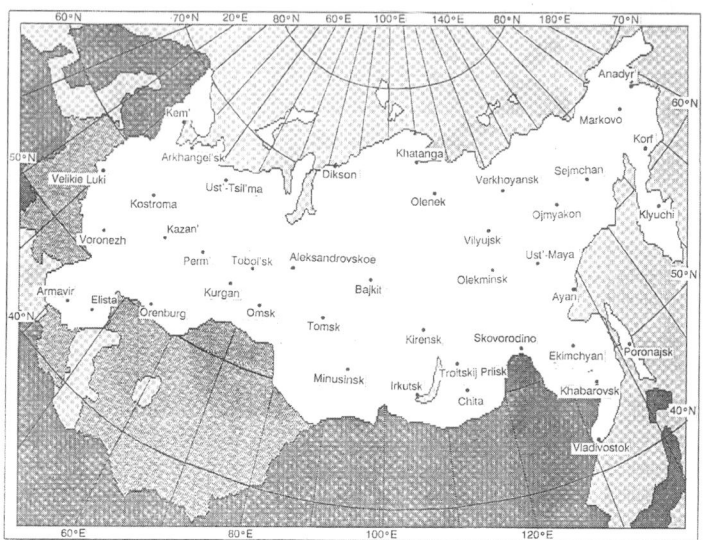

Figure 13. Russian stations used in the analysis of extreme events in daily temperature and precipitation.

4.7.1. *Winter (December-February)*

Temperature. Variability in the index of daily temperature extremes for 1961-1995 is characterized by linear trend coefficients obtained from the series of N values when the maximum air temperature, Tmax, is in the five percentile wide interval 95%-100% (coefficient β_1) and when the minimum air temperature, Tmin, is in the interval 0%-5% (coefficient β_2). Figure 14 shows a spatial distribution of coefficients β_1, β_2 and coefficients of the linear trend in the time-series of mean seasonal air temperatures, Tmean (β_T). Most of the stations under consideration experienced fewer days with abnormally low temperatures. The largest negative linear trend coefficients (β_2) were obtained in the southern part of Russia and in the eastern part of Yakutia (in Minusinsk $\beta_2 = -2.0$ n/10yr, in Skovorodino $\beta_2 = -2.3$ n/10yr, in Omsk $\beta_2 = -1.8$ n/10yr, in Seimchan $\beta_2 = -1.7$ n/10yr). The number of days with maximum temperatures higher than the 95% percentile has increased in most of European Russia (except for southern and southeastern regions) and Western Siberia. Negative values of linear trend coefficients (β_1) were obtained in the south of European Russia (the values vary from -0.2 to -1.0). The increasing number of extremes of maximum daily air temperatures is also observed in the east of the country (in Ust-Maya $\beta_1 = 2.1$

n/10yr). These changes in the index of daily temperature extremes contribute to changes in mean seasonal temperatures. Positive linear trend coefficients in mean seasonal temperature series have been obtained for almost the whole of Russian area, except for Chukotka, Kamchatka and Krasnodar.

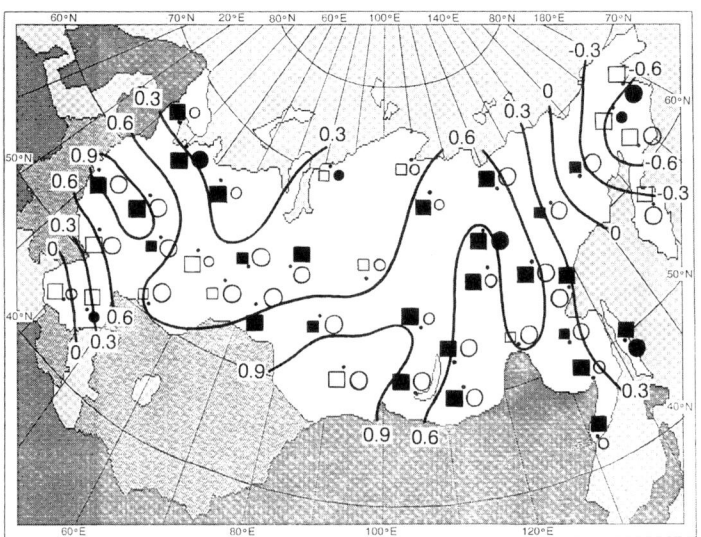

Figure 14. Linear trend coefficients in the time-series of seasonal temperature (β_T), number of days with abnormally high (β_1) and low (β_2) air temperatures in Winter (DJF). Time period 1961-1995:
β_T —— β_1: $\square$<-0.5 $\square$[-0.5,0] $\blacksquare$[0,0.5] $\blacksquare$>0.5 β_2: $\bigcirc$<-0.5 $\bigcirc$[-0.5,0] $\bullet$[0,0.5] $\bullet$>0.5

Precipitation. The increasing number of days with precipitation exceeding the 95% percentile was found at the stations of European Russia to the north of $55°$ N and in the center of Siberia (Figure 15), as well as at the station of Chukotka (Anadyr $\beta_N = 0.6$ n/10yr). This character of variability in extreme precipitation is consistent with the tendencies observed in variations of seasonal (winter) precipitation (Figure 15). Positive linear trend coefficients (β_R) were obtained over the European territory (in Kostroma β_R =9.8 mm/10yr; 10% significance level) and in the center of Siberia. The largest value of the coefficient β_R, which is equal to 9.6 mm/10yr and significant at the 1% level, was obtained for station Omsk. Decreasing trends in precipitation amount were found for the most other stations (Figure 15). For example, at stations Elista, Seimchan, Vladivostok, and Voronezh, the coefficient β_R was found to be equal to -10.1, -10.6, -9.5, and -8.9 mm/10yr, respectively.

4.7.2. Summer (June-August)

Temperature. When considering the index of daily temperature extremes in summer it was found (Figure 16) that the number of days with maximum temperatures exceeding 95% percentile increased at the stations in the east of

European Russia, in the center of Siberia, in Yakutia, and in the east of the country. At the same time, the negative values of linear trend coefficients in the number of days with abnormally low temperatures (β_2) were obtained for most of the stations. Only for some of the stations in the northeast of the country were positive values of β_2 found.

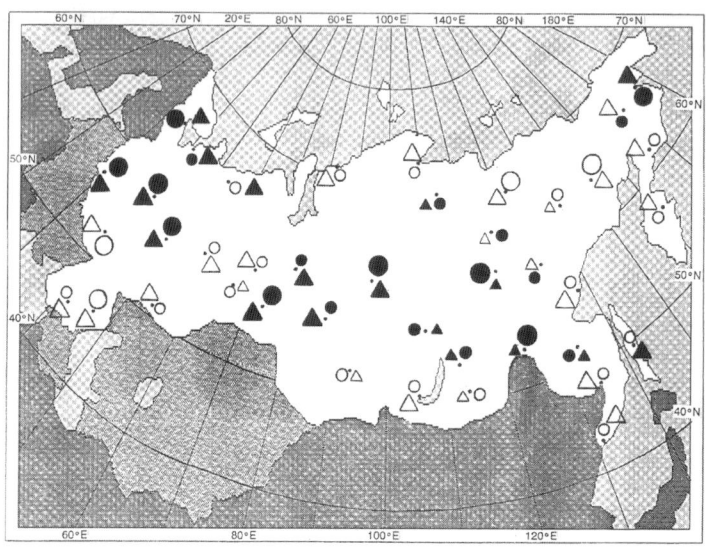

Figure 15. Linear trend coefficients in the time-series of seasonal sums of atmospheric precipitation (β_R) and the number of days with heavy precipitation (β_N) in Winter (DJF). Time period 1966-95:
β_R: $\triangle<-0.5$ $\triangle[-0.5,0]$ $\blacktriangle[0,0.5]$ $\blacktriangle>0.5$ β_N: $O<-0.5$ $o[-0.5,0]$ $\bullet[0,0.5]$ $\bullet>0.5$

Overall, it may be assumed that the mean seasonal (summer) temperature rise over most of Russia in the period 1961-1995 is related to the growing number of days with abnormally high temperatures and, to a smaller extent, to fewer days with very low temperatures.

Precipitation. There is a tendency for the increasing number of days with heavy precipitation in the period 1966-1995. Negative linear trend coefficients (β_N) in the time-series of the number of days with heavy precipitation were only observed at individual stations in central Siberia and in Magadan region (Figure 17). A coincidence of the tendency signs in seasonal precipitation and in the number of days with heavy precipitation, which is quite noticeable for winter season, is also found in summer precipitation data.

The increase in summer precipitation is observed at all stations under consideration that are located over European Russia. The largest linear trend coefficients were obtained for Voronezh (β_R=21.6 mm/10yr), Armavir (β_R =20.5 mm/10yr), and Velikie Luki (β_R =17.4 mm/10yr). Summer precipitation is observed to increase in the south of Siberia, Lake Baikal region, and Yakutia.

Negative linear trend coefficients in summer precipitation were generally obtained in the regions experiencing fewer days with heavy precipitation. However, at some stations with insignificant increases in the number of days with extreme precipitation, negative linear trend coefficients in the summer precipitation were also obtained (in Vladivostok β_R = -6.0 mm/10yr and β_N = 0.3 n/10yr; in Arkhangelsk β_R = -7.2 mm/10yr and β_N = 0.2 n/10yr; in Kem-Port β_R = -11.3 mm/10yr and β_N = 0.3 n/10yr).

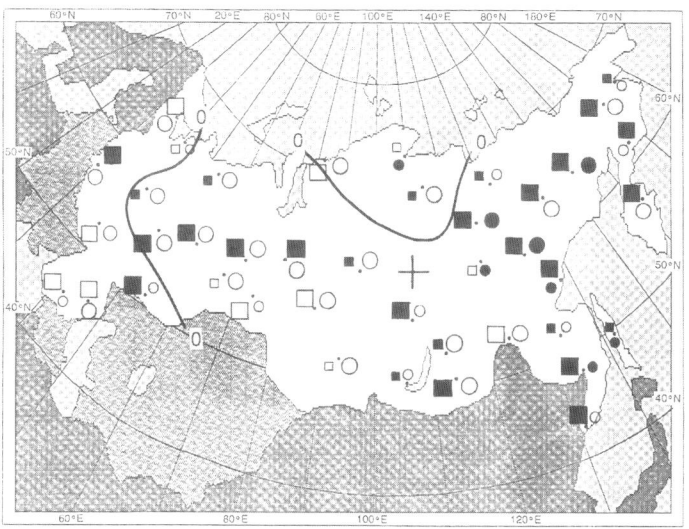

Figure 16. Same as Figure14 except for Summer (JJA).

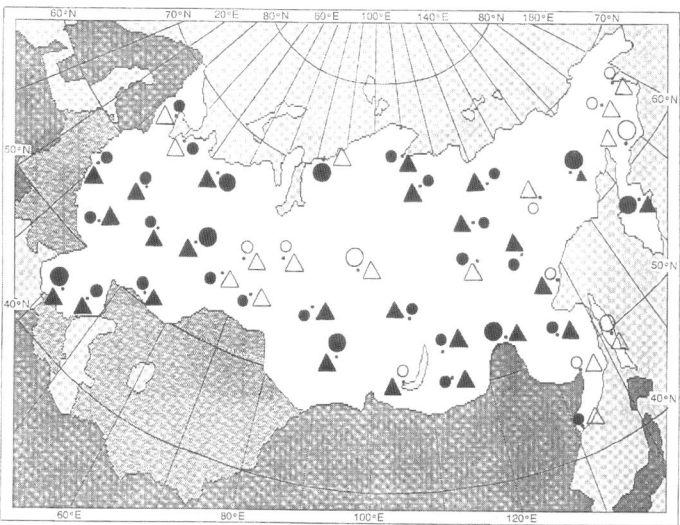

Figure 17. Same as Figure15 except for Summer (JJA).

5. Discussion and Conclusions

We have shown that over the course of the 20th century the RPF territory as a whole has experienced some increases in the air temperature and in the areal coverage of drought, together with decreases in precipitation; these changes were non-uniform in space. The linear trend captures only a small fraction of the total variability of area-average temperature in the region considered. A more substantial role is played by fluctuations on decadal scales. The climate extremes index, determined as the percentage of the area experiencing extreme (with 10% or less chance occurrence) anomalies of either sign increased for the annual mean temperature, decreased for total precipitation and weakly increased for drought conditions. The aggregated index CEI-3, based on the three indicators examined in this study, exhibits a slight increase in climate extremes within the RPF territory. There was also an increase in GCRI-3 that indicates some degree of agreement between observed climate changes and changes owing to greenhouse effect as simulated by climate models. As for variations of extreme events in daily air temperatures and precipitation, preliminary analysis of data from 41 Russian stations indicates that the number of days with abnormally low daily air temperatures is decreasing in both winter and summer seasons for most of the stations in question. The number of days with heavy precipitation is observed to increase for the stations of European Russia in summer period.

The estimated climate changes are too small to confidently allow us to reject the hypothesis as they could merely be a reflection of sampling fluctuations of the natural variability of climatic parameters in a stationary climate. Nevertheless there is no doubt regarding the reality of the observed changes, which may be large enough to be of some practical importance. It is necessary to carry out a more detailed analysis of the structure of climate variability, in order to determine which statistical parameters are responsible for the observed changes in the climate indicators and indices used in this analysis.

Acknowledgments

The authors thank Dr. J. M. Wallace for stimulating discussions and helpful editorial assistance and Mr. D. Kaiser for help in preparation of the English version of this paper. This work was supported by Projects RBRF96-0564883, RBRF97-0565102 and RBRF97-0564150.

References

Bagrov, N. A. and Mjakisheva, N. P., 1966: Some characteristics of monthly air temperatures. Trudy MMC, vyp. 9 (in Russian).

Easterling, W. E. and Kates, R. W., 1995: Indexes of leading climate indicators for impact assessment, in *Long-term Climate Monitoring by the Global Climate Observing System*, Karl, T. R. (ed.), Kluwer Academic Publishers, Dordrecht, The Netherlands, pp.493-518.

Gruza, G.V., 1982: Climate variability as estimated from atmospheric observations. Invited paper. Report of the WMO / ICSU Study Conference on physical basis for climate prediction on seasonal, annual and decadal time scales (Leningrad, 13-17 Sept., 1982), WMO, WCP-47, p. 25-50.

Gruza, G.V., 1991: Some problems concerning climate change detection using statistical methods. Report of the CC1 Working group on climate change detection (First session), Geneva, WMO/TD No. 466.

Gruza, G.V. and Rankova, E.Ya., 1980: Structure and variability of observed climate. Air temperature over the Northern Hemisphere. - Leningrad: Hydrometeoizdat, 72 p., (in Russian, summary in English).

Gruza, G. V. and Rankova, E.Ya., 1994: Climate variability with emphasis on extreme climatic events and circulation / synoptic regimes. Proc. 18th Annual Climate Diagnostics Workshop, Boulder, Co., Nov. 1-5 1993, 98-101

Gruza, G.V., Rankova, E. Ya. and Rocheva, E.V., 1989: Analysis of global data on variations in surface air temperature over the period of instrumental observations, Meteorology and Hydrology, No. 1, p.22-31, (in Russian, summary in English).

Gruza, G. V., Rankova, E. Ya. and Rocheva, E.V., 1994: Comparison of 1931-1960 and 1961-1990 air temperature normals for the territory of the former USSR. Proc. 18th Annual Climate Diagnostics Workshop, Boulder, Co., Nov. 1-5 1993, 276-279

Gruza, G. V., Rankova, E. Ya. and Semenyuk, E.A., 1985: Climatic variability in characteristics of large-scale anomalies of air temperature. Meteorology and Hydrology, No. 6, p. 21-28 (in Russian, summary in English).

IPCC, 1996: Climate Change 1995, The Science of Climate Change, Contribution of Working Group 1 to the Second Assessment Report of the Intergovernmental Panel on Climate Change, J. T. Houghton, L.G. Meira Filho and B. A. Callander, Eds., Cambridge University Press, 572 pp.

Karl, T. R., Knight, R. W., Easterling, D. R. and Quayle, R. G, 1995: Trends in U.S. Climate during the Twentieth Century, Consequences 1, 1, 3-12.

Karl, T. R., Knight, R. W., Easterling, D. R. and Quayle, R. G., 1996: Indices of climate change for the United States. Bull. Amer. Meteor. Soc., 77, 279 - 292.

Khlebnikova, E. I., 1989: On variability of the spatial characteristics of anomalies for the field of monthly mean surface air temperature. - Proceedings of MGO, N 525, pp.52-62 (in Russian).

Khlebnikova, E. I., Salle, I. A. and Sibir, E. E., 1988: On using spatial crossing-level characteristics of meteorological fields for analysis of climate changes. - Proceedings of MGO, No. 516, p. 110-120 (in Russian).

Khlebnikova, E.I. and Salle, I. A., 1989: On evaluating crossing-level characteristics and their sensitivity for Gaussian meteorological fields. - Proceedings of MGO, No. 525, p. 44-51 (in Russian).

Pedj, D. A., 1975: On the indicator of drought and moisture surplus. Proceedings of HMC of USSR, No. 156, p. 19-38 (in Russian).

Razuvaev, V. N., Apasova, E. G., Martuganov, R. A., Steurer, P. and Vose, R., 1993: Daily temperature and precipitation data for 223 USSR stations. ORNL/CDIAC, Numerical Data Package-40, Oak Ridge National Laboratory, Oak Ridge, Tennessee.

Trenberth, K. E., 1990: Recent observed interdecadal climate changes in the Northern Hemisphere. Bull. Amer. Meteor. Soc., 71, 988 - 993.

WMO, 1996: WMO Statement on the Status of the Global Climate in 1995, World Meteorological Organization, WMO, No. 838.

(Received 5 November 1997; in revised form 31 October 1998)

CHANGES IN THE PROBABILITY OF HEAVY PRECIPITATION: IMPORTANT INDICATORS OF CLIMATIC CHANGE

PAVEL YA. GROISMAN[1] , THOMAS R. KARL[2], DAVID R. EASTERLING[2],
RICHARD W. KNIGHT[2], PAUL F. JAMASON[2], KEVIN J. HENNESSY[3],
RAMASAMY SUPPIAH[3], CHER M. PAGE[3], JOANNA WIBIG[4], KRZYSZTOF
FORTUNIAK[4], VYACHESLAV N. RAZUVAEV[5], ARTHUR DOUGLAS[6],
EIRIK FØRLAND[7], and PAN-MAO ZHAI[8]

[1] *University Corporation for Atmospheric Research Visiting Scientist at the U.S. National Climatic Data Center, 151 Patton Ave., Asheville, NC 28801, U.S.A*
[2] *U.S. National Climatic Data Center, 151 Patton Ave., Asheville, NC 28801, U.S.A*
[3] *CSIRO Atmospheric Research, 107-121 Station Str., Aspendale, Victoria 3195, Australia*
[4] *Dept. of Meteorology and Climatology, University of Lodz, ul. Lipova 81, 90-568, Lodz, Poland*
[5] *Research Inst. for Hydrometeorological Information, 6 Koroleva Str., Obninsk, 601240, Russia*
[6] *Dept. of Atmospheric Science, Creighton University, Omaha, NE, 68178, U.S.A*
[7] *Norwegian Meteorological Institute, P.O. Box 43, Blindern, N-0313, Oslo, Norway*
[8] *National Climate Center, 46 Baishiqiao Rd., Beijing, 100081, China*

Abstract. A simple statistical model of daily precipitation based on the gamma distribution is applied to summer (JJA in Northern Hemisphere, DJF in Southern Hemisphere) data from eight countries: Canada, the United States, Mexico, the former Soviet Union, China, Australia, Norway, and Poland. These constitute more than 40% of the global land mass, and more than 80% of the extratropical land area. It is shown that the shape parameter of this distribution remains relatively stable, while the scale parameter is most variable spatially and temporally. This implies that the changes in mean monthly precipitation totals tend to have the most influence on the heavy precipitation rates in these countries. Observations show that in each country under consideration (except China), mean summer precipitation has increased by at least 5% in the past century. In the USA, Norway, and Australia the frequency of summer precipitation events has also increased, but there is little evidence of such increases in any of the countries considered during the past fifty years. A scenario is considered, whereby mean summer precipitation increases by 5% with no change in the number of days with precipitation or the shape parameter. When applied in the statistical model, the probability of daily precipitation exceeding 25.4 mm (1 inch) in northern countries (Canada, Norway, Russia, and Poland) or 50.8 mm (2 inches) in mid-latitude countries (the USA, Mexico, China, and Australia) increases by about 20% (nearly four times the increase in mean). The contribution of heavy rains (above these thresholds) to the total 5% increase of precipitation is disproportionally high (up to 50%), while heavy rain usually constitutes a significantly smaller fraction of the precipitation events and totals in extratropical regions (but up to 40% in the tropics, e.g., in southern Mexico). Scenarios with moderate changes in the number of days with precipitation coupled with changes in the scale parameter were also investigated and found to produce smaller increases in heavy rainfall but still support the above conclusions. These scenarios give changes in heavy rainfall which are comparable to those observed and are consistent with the greenhouse-gas-induced increases in heavy precipitation simulated by some climate models for the next century. In regions with adequate data coverage such as the eastern two-thirds of contiguous United States, Norway, eastern Australia, and the European part of the former USSR, the statistical model helps to explain the disproportionate high changes in heavy precipitation which have been observed.

Climatic Change **42**: 243–283, 1999.
©1999 *Kluwer Academic Publishers. Printed in the Netherlands.*

1. Introduction

Analyses of trends in mean precipitation during the past century reveal compelling evidence of the presence of trends over many regions of the world (Groisman and Legates, 1995; IPCC, 1996, 1998). In many countries (e.g., Russia, Norway, Sweden, Canada) the increase in precipitation was more pronounced in the cold season (about 10-15 %/100 yrs) than in the warm season (about 5%/100 yrs), but the absolute values of these changes are comparable because of the seasonal cycle of precipitation in most of the northern extratropics. In Poland and Australia, the century-long increase in precipitation was predominantly in the warm half of the year (Kozuchowski, 1985; Suppiah and Hennessy, 1998). In most parts of Norway the annual precipitation has increased by 8-14%/100yrs (Hanssen-Bauer et al., 1997), while the summer precipitation has increased less prominently (by 5-10%) and mostly in northern part of the country (Hanssen-Bauer, 1994).

Of particular interest, from both practical and theoretical considerations, are the analyses of precipitation change that reveal increases in extreme and very heavy precipitation from North America, Australia, and Japan. Karl et al. (1995) and Karl and Knight (1998) provide evidence for a statistically significant increase in extreme precipitation (greater than 50 mm per day) precipitation in the United States. Similarly for Australia, Suppiah and Hennessy (1996, 1998) show significant increases for the higher percentiles, e.g., the 90th and 95th percentiles. This was augmented by an increase in heavy-rain days in eastern Australia associated with East Coast cyclones reported by Hopkins and Holland (1997). Iwashima and Yamamoto (1993) analyzed daily precipitation data from 1890 to 1980 at 55 Japanese stations and found that more stations recorded their highest, 2nd highest or 3rd highest precipitation event in more recent decades. Thus, the frequency of years with extremely heavy daily precipitation is increasing at Japanese stations throughout the 20th century. Analysis of a small subset of 14 U.S. stations performed by Iwashima and Yamamoto (1993) suggests that this increase has occurred over the contiguous United States too. Tsonis (1996) shows that the variability of monthly precipitation totals over the United States, Europe, and Australia has also increased during the past 100 years. Beniston et al. (1994) concluded that "in a warmer global climate, precipitation in Alps would be generally reduced but the extreme precipitation events could be expected to increase significantly". This empirical conclusion was supported later by the modeling assessment of Schaer et al. (1996). Generally, climate model simulations consistently project increases in global precipitation due to global warming stemming from increases in greenhouse gases, particularly for the mid and high latitudes (IPCC, 1990, 1996). An increase in heavy precipitation is also simulated by climate models (IPCC, 1996; Schaer et al., 1996; Jones et al., 1997; Hennessy et al., 1997).

We are interested in heavy precipitation during the three warmest (and often wettest) summer months, which coincide with the period of the primary growing season. In this paper heavy precipitation changes during summer are assessed in eight countries: Canada, the United States, Mexico, the former Soviet Union, China, Australia, Poland, and Norway. We show that if the shape of the precipitation distribution (often well described by the gamma distribution) does not change as total precipitation increases, a disproportionate increase in heavy precipitation is expected.

2. Data Used

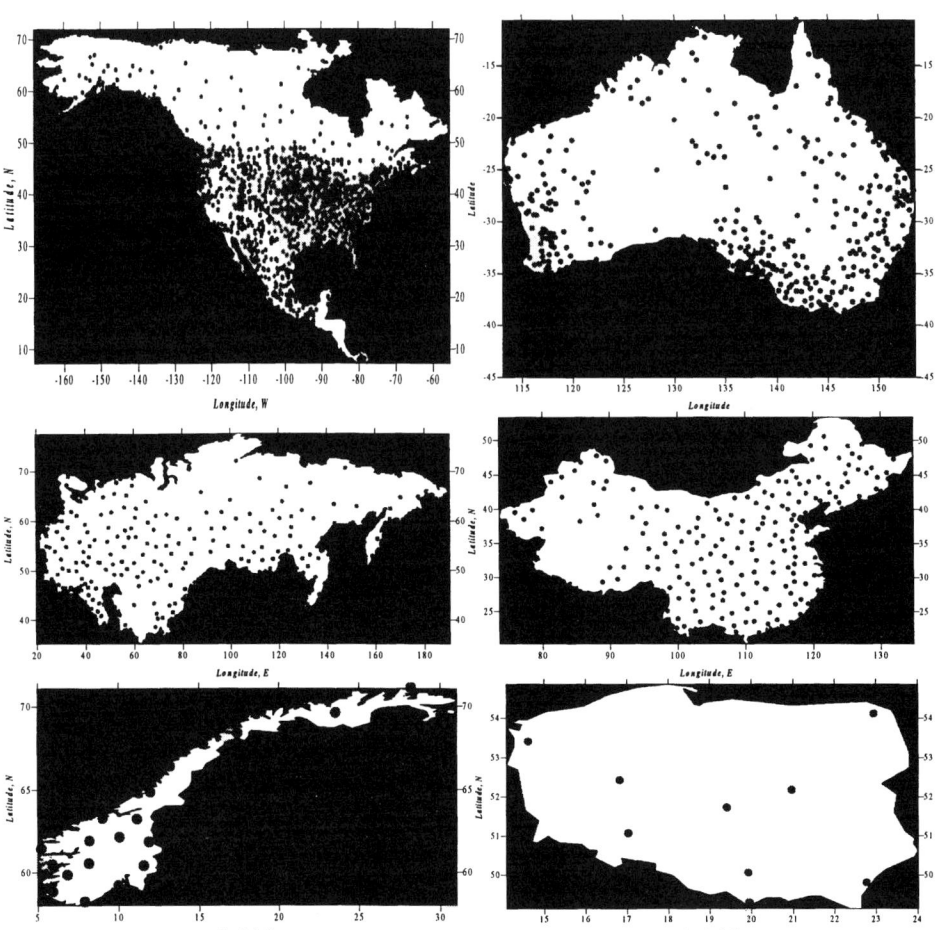

Figure 1. Maps of the stations with daily precipitation time series used in this study for North America (Canada, the United States, and Mexico), Australia, the former Soviet Union, People Republic of China (PRC), Norway, and Poland. Only the continental part of all these countries is shown. Several stations from adjacent islands were also used in the analyses. Note the different spatial scales in each map.

Daily precipitation data sets for eight countries were used in our analyses. For the former Soviet Union we used an archive of 223 stations of the international exchange available from the Carbon Dioxide Information Analysis Center (Razuvaev et al., 1993, updated) from the beginning of observations to 1994. For North America we employed a new daily precipitation data set accumulated at the National Climatic Data Center (Easterling, 1997; Easterling et al., 1998). Daily data from Canada (93 stations) and Mexico (202 stations) spanned the years 1900-1995 and 1950-1990 respectively. A subset of the highest quality stations from the U.S. Historical Climatology Network (HCN) of 134 stations with century-long daily precipitation time series (Hughes et al., 1992) comprised the data base for the contiguous United States, supplemented by an additional 53 stations to provide more representative spatial coverage. The U.S. time series, now updated through 1996, were previously used by Karl et al. (1995) and Karl and Knight (1998) in the analyses of extreme precipitation over the contiguous United States. Additionally, for mapping of precipitation distribution parameters only, we used 1060 HCN stations from the contiguous U.S. and 44 Alaskan stations spanning the years 1948-1995. Data for 198 Chinese stations of international exchange span the period from 1951 to 1994 (Baker et al., 1995). An extended high-quality historical precipitation data set for Australia comprises 379 stations from the beginning of observations (113 start as early as 1891) up to 1996 (Lavery et al., 1997). A subset of 13 century-long homogeneous daily precipitation time series was used for estimates of the precipitation extremes over Norway. Somewhat shorter homogeneous daily precipitation time series (40 to 60 years of data) from another 8 Norwegian stations were used mostly for mapping of precipitation distribution parameters. Data from ten first order stations well distributed over Poland (except the north-eastern) were available for the post-World War II period. The station networks are shown in Figure 1 and their pre-processing is described in Appendix 1.

3. Model of the Daily Precipitation Distribution

It is widely recognized that the distribution of daily precipitation totals, P, can be approximated by the gamma-distribution $\Gamma(\eta, \lambda)$ (Thom, 1951, 1958; Bagrov, 1965; Mooley, 1973; Crutcher et al., 1977, Buishand, 1978; Guttman et al., 1993) with the density function

$$p(\eta, \lambda, x) = \text{const}(\eta, \lambda) * x^{\eta-1} * \exp(-\lambda x), \qquad (1)$$

when x>0, and zero when x ≤0. For this family of distributions the η-parameter defines the shape of the distribution, while the λ-parameter characterizes the scale. The mean, μ, variance, σ^2, and the coefficient of variation C_v of this distribution are defined by these parameters:

$$\mu = \eta / \lambda; \qquad \sigma^2 = \eta / \lambda^2; \qquad\qquad C_v = \sigma/\mu, = 1/sqrt(\eta) \qquad (2)$$

Note that C_v is only a function of the shape parameter.

Since it does not rain every day, a mixed distribution model is considered for daily precipitation totals. Under this model, it is assumed that the occurrence of daily precipitation events has a binary distribution with the probability of a single event P_{pr} and the distribution function of precipitation totals $F(x)$ is expressed as:

$$F(x) = P(X \leq x) = (1-P_{pr}) + P_{pr} \int_0^x p(\eta, \lambda, t) \, dt \qquad (3).$$

The precipitation amount during this event is considered to have a gamma-distribution. For (3), we have three parameters: P_{pr}, η, and λ, where the density function (1) now characterizes a conditional distribution of daily precipitation. For this model, Eqs. 2 will be transformed into:

$$\mu = P_{pr}\eta/\lambda; \ \sigma^2 = (P_{pr})^2 \, \eta / \lambda^2; \ C_v = \sigma/\mu, = 1/sqrt(\eta) \qquad (2').$$

We shall use this model throughout and apply it to daily precipitation totals over Eurasia, Australia, and America with the following simplification: a precipitation event is defined as a non-zero 24-hour total. Analysis of the weather duration tables from the United States primary meteorological network shows that this is not exactly the case. In the summer time, rainy days are composed from two rain events on average divided by a short no-rain period. But we do not have weather duration tables for most of the network data we are using for most of the period under consideration, i.e., we have no better choice.

Another simplification is that we are assuming independent daily precipitation events. In fact, the probability of having a summer day with precipitation after a rainy day is higher than after a day without precipitation and, similarly, the probability of a day without precipitation after a dry day is higher than after a wet day (Katz, 1977, Richardson, 1981). No efforts were made throughout this paper to address temporal correlation of precipitation behavior: grouping of dry and wet days into spells. This could adversely affect our assessment of the probability of heavy rains in the framework of the model (3), although the theoretical analysis by Katz (1998) indicates that temporal correlation is not crucial for estimates of the probability of extreme precipitation events. Therefore, we specifically tested the goodness of fit of model (3) for estimates of the probability of heavy rains. In the regions with a dense network of long-term homogeneous precipitation observations with a sufficient amount of precipitation events (Eastern United States, Eastern Australia, European Russia, Southern Norway) we calculated empirical estimates of the probability of "heavy" precipitation (i.e., above a given threshold) and compared them with calculations based on model (3). Figure 2 presents an example of such a comparison for the contiguous United States. It shows that, for the threshold under consideration (here, 50.8 mm), model (3) reasonably well reproduces the pattern of probability but sometimes underestimates its absolute values. We cannot test empirically the goodness of fit of our model in the regions with short or few homogeneous precipitation time series, "dry" regions, and for the probability to exceed higher thresholds (e.g. 150 mm) due to the lack of sufficient heavy rain events.

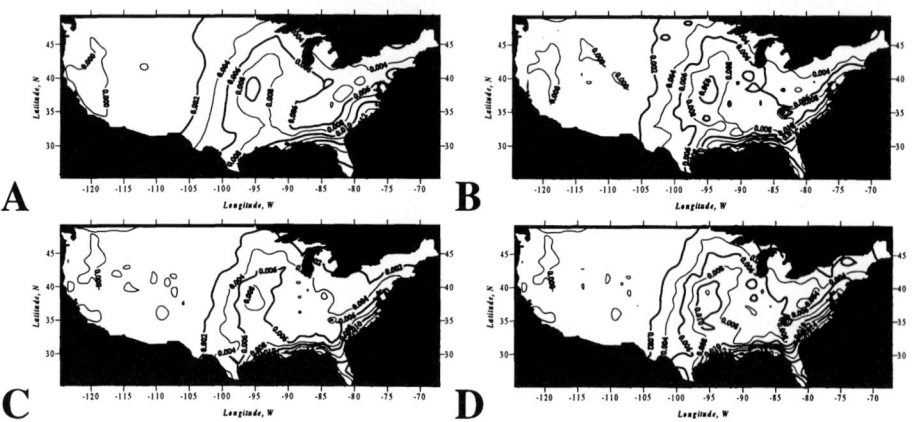

Figure 2. Probability of a summer day with precipitation exceeding 50.8 mm over the contiguous United States: (A) direct empirical estimates using the century-long homogeneous time series of 172 stations (Hughes et al., 1992, updated; Karl and Knight, 1998); (B) the same as (A) but using the time series of 1088 HCN stations for period 1950-1995 ; (C) calculations based on model (3) with parameters η, λ, and P_{pr} estimated from the data of 1088 HCN stations for period 1950-1995; and (D) the same as (C) but with the λ-parameter reduced by a factor of 1.05 to allow a 5% increase in mean daily precipitation.

The mean precipitation in model (3) is a product:

$$\mu = P_{pr}\eta/\lambda \tag{4}$$

and its change can be a result of the contribution of all three parameters. We are interested in the changes in the probability of heavy rains that can accompany changes in mean precipitation. Therefore, we tested the sensitivity of this probability to changes in μ that are introduced by the variation of each of these three parameters. In the regions with mean daily summer precipitation above 1 mm day^{-1} for typical combinations of η, λ, and P_{pr}, the change in the probability of exceeding heavy precipitation thresholds with a change in μ was analyzed. The strongest changes in heavy precipitation probability occur when the changes in μ are associated with variation of scale parameter, λ, and the smallest changes occur when the changes in μ are associated with variations of P_{pr}. For example, the probability of exceeding a 50.8 mm day^{-1} threshold over the eastern two-thirds of

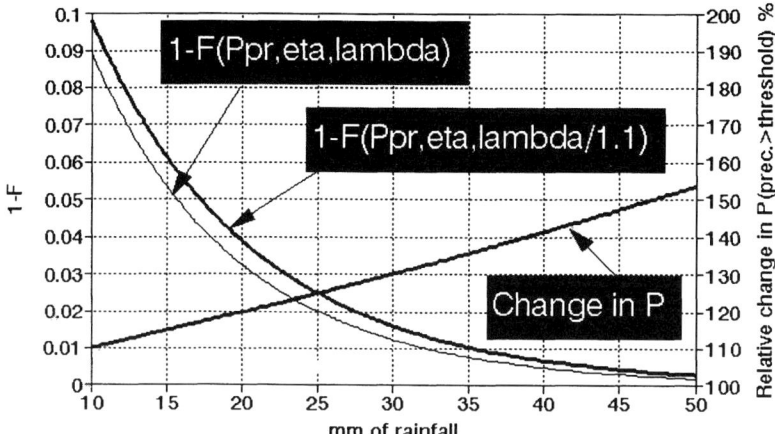

Figure 3 A. Tail of the distribution function [1-F(x, P_{pr}, η , λ)] that mimics July precipitation in Toronto, Ontario (P_{pr} =0.3; η = 0.76; λ= 0.09 mm⁻¹; μ = 2.5 mm day⁻¹) and the same function for a 10% increase in mean value, μ, of July precipitation assuming that P_{pr} and η do not change. The relative change in exceedance the x-threshold is also given as a function of the change in x.

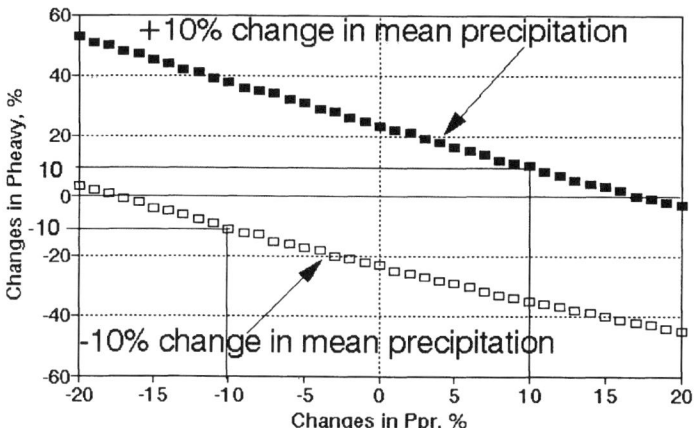

Figure 3 B. Changes in the probability of heavy rains (above 50.8 mm) in Guangzhou, PRC (P_{pr} =0.6; η = 0.56; λ= 0.04 mm⁻¹; μ = 8 mm day⁻¹) with a 10% increase/decrease in mean summer precipitation, μ, assuming that η does not change and the changes in μ are due to changes in P_{pr} and λ. Because μ =$P_{pr}\eta/\lambda$, the changes in λ in these scenarios are a function of $\Delta\mu$ (= ±10%) and ΔP_{pr} and are not shown.

the contiguous United States with a 10% increase in μ changes by approximately 40%, 20%, and 10%, if this increase in μ is produced in Eq. 4 by an appropriate change in λ, η, or P_{pr} respectively. Obviously, an increase/decrease in P_{pr} produces a linear 1:1 increase/decrease in probability to exceed any given threshold. Changes in the two other parameters produce disproportionally high changes in the probability of extreme precipitation compared to the corresponding changes in mean precipitation.

Figure 3A further illustrates how a 10% increase in μ due to a change in the scale parameter increases the probability of daily precipitation above 25.4 mm (1 inch) from 0.018 to 0.023, assuming $\mu = 2.5$ mm day^{-1}, $\eta = 0.75$, and $P_{pr} = 0.3$. This is a 25% increase in extreme precipitation occurrence compared to a 10% increase in the mean. This hypothetical example was selected to match the July daily precipitation distribution in Toronto, Ontario[*], and the precipitation increase documented over southern Canada by Groisman and Easterling (1994) (cf., also IPCC, 1996). Figure 3A also shows the effect of the threshold selection on the change in exceedance of this threshold with a 10% increase in mean precipitation due to a change in λ. In Figure 3B we sketch the changes in the probability, P_{heavy}, of summer daily precipitation above 50.8 mm in Guangzhou, PRC, when the mean precipitation, μ, changes by $\pm 10\%$ due to changes in P_{pr} and λ but without changes in η. It shows that depending upon the ratio of changes in these two parameters to the change in μ, P_{heavy}, can change

- with a higher than linear rate, when changes in P_{pr} are less than the changes in μ by absolute value;
- linearly, when changes in μ are solely due to changes in P_{pr} (fixed λ).
- with a lower than linear rate or inversely (in this example, when absolute values of ΔP_{pr} are above 17%), when changes in P_{pr} are higher than the changes in μ by absolute value.

Therefore, in constructing scenarios of a future climate change, we have to judge which of these three parameters will be responsible for the change in the mean precipitation. This will affect substantially the behavior of precipitation extremes in these scenarios and, in turn, will have important socio-economical and ecological consequences.

The above provides a rationale for our approach. We presume that for daily precipitation described by (3), the changes in λ, η, or P_{pr} which have occurred interannually and in the seasonal cycle during the past century as well as their spatial variability contain information about the stability of these parameters in moderate climate and weather variations. Then, using this information, we can apply a plausible scenario of the mean precipitation change and derive valuable

---

[*] Empirical estimates of the probability of the daily precipitation total in July to exceed a threshold of 25.4 mm are equal to 0.025 for the Toronto International Airport and 0.020 for the downtown Toronto meteorological stations.

information about the most probable change in precipitation extremes. We consider scenarios of the small/moderate increase of mean precipitation that match the precipitation changes during the past 100 years over the countries under consideration (IPCC, 1996). We estimate parameters of model (3) for the period of the mass data availability and then use them to test the present and future tendencies in extreme precipitation.

The next section describes the spatial distribution of parameters of summer daily precipitation over the countries under consideration. It is followed by analyses of temporal and spatial stability of one of these parameters, η, and possible trends in precipitation frequency, P_{pr}. The final section presents major results of this study: the effects of changes in mean precipitation on the extreme daily precipitation values under the assumption that the shape parameter of the precipitation distribution and the frequency of the precipitation events do not change. Other scenarios for changes in the parameters of the precipitation distribution model are also considered. The scenario results are compared with direct estimates of trends in heavy precipitation during the past 100 years over the United States, Australia, and Norway.

4. Summer Daily Precipitation and Its Parameters

The results of model (3) for the summer (JJA, for Australia DJF) daily precipitation distribution are shown in Figure 4 for all eight countries under consideration. We selected a schematic presentation of the mean seasonal precipitation, μ, in Figure 4A because this quantity is well documented in climatological literature. We selected regions with very different precipitation regimes: Arctic tundra, deserts of central Asia, southern Australia, north-west Mexico, and western USA receive less than 1 mm per day, while precipitation over tropical regions of southern Mexico, northern Australia, and southern China exceed 10 mm per day. Over vast agricultural areas of the northern and southern extratropics, summer precipitation of 2 to 3 mm per day is common; eastern China, southeastern USA, the Gulf coast of Mexico, western Norway, and coastal regions of east Australia receive on average 5 mm per summer day.

Figure 4B presents the probability distribution of summer daily precipitation, P_{pr}. It shows that the frequency of rainy days varies widely over the study area. It is close to 0.5 over the wet tropics of southern Mexico, southern China, and northern Australia, over the North Atlantic region (Norway, Labrador Peninsula, Northwest USSR and Poland), and over coastal and mountainous regions along the Pacific rim (southern Alaska, northwest China, Sichuan and Yunnan provinces of PRC, and the Russian Far East). It is less than 0.05 over the deserts of North America, Australia, and Eurasia.

Figures 4C and 4D show the distribution of the scale and shape parameters of the gamma distribution of daily summer precipitation on the days with measurable

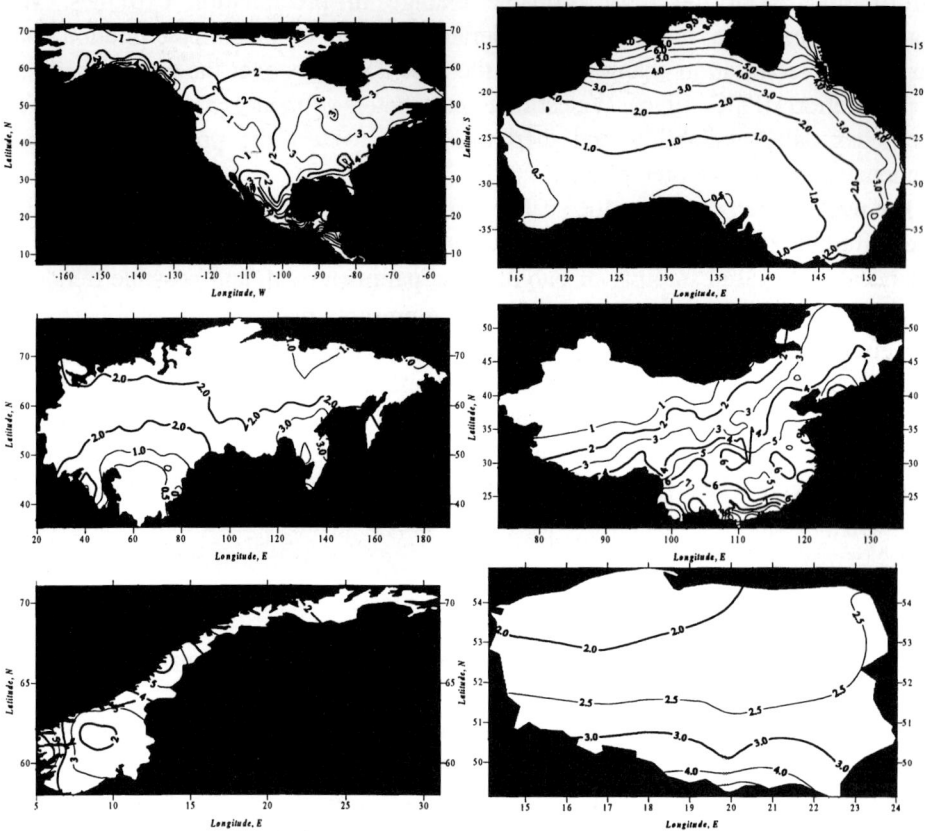

Figure 4 A. Mean daily summer (June-August) precipitation (mm day^{-1}) over North America (Canada, United States, and Mexico), Australia, the former Soviet Union, PRC, Norway, and Poland for the three parameter model of the precipitation distribution (3). In Australia, December through February are considered as summer months.

rain. The scale parameter, λ, has units of mm^{-1} for daily precipitation. Smaller values of λ indicate higher intensities of daily precipitation. It is a parameter that changes by an order of magnitude from subarctic regions and deserts (~ 0.30 mm^{-1}) to humid tropics (~ 0.03 mm^{-1}). However, the shape parameter, η, is dimensionless and has little spatial variation. The fact that the shape parameter is a spatially and temporally stable characteristic of regional precipitation has been shown at monthly and annual time scales (Shver, 1976; Groisman and Easterling, 1994). Figure 4D and the next section show that this is also true at the daily time scale. Over the eastern two thirds of the contiguous United States, Russia, and Canada with daily summer precipitation above 1 mm day^{-1}, this parameter varies by 10-15% around its mean value of approximately 0.8. Changes are very small over Poland, Norway, eastern Australia and eastern China. Over Australia, the η-values

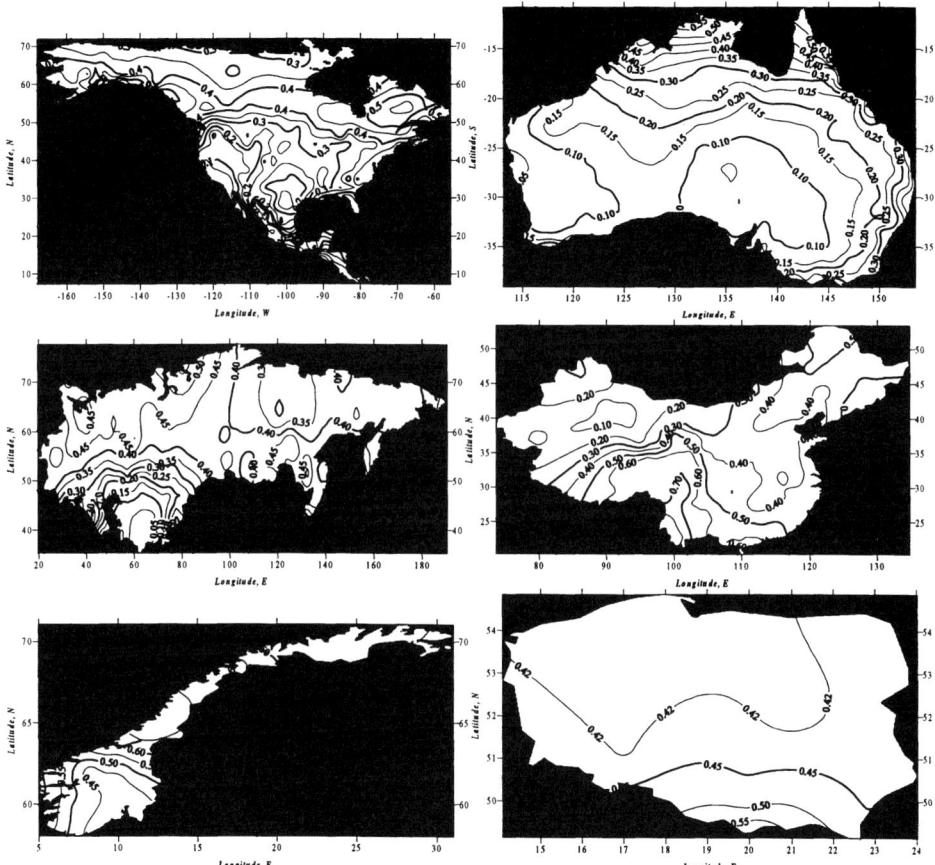

Figure 4 B. The probability of summer (June-August) daily precipitation over North America (Canada, United States, and Mexico), Australia, the former Soviet Union, PRC, Norway, and Poland for the three parameter model of the precipitation distribution (3). In Australia, December through February are considered as summer months.

for summer (DJF) are similar to those over North America and Eurasia and vary around a mean value of approximately 0.75. In monsoon regions of China and the Russian Far East the η-values for summer (JJA) are relatively low (varying from 0.5 to 0.6). Over regions with daily summer precipitation above 1 mm day^{-1}, the lowest values of η are estimated in eastern China (up to 0.45 on Shandong Peninsula) and the highest in the tropics of southern Mexico (up to 1.2 along the coast of the Gulf of Mexico). Generally, we found little spatial variation in our estimates of this parameter with the exception of mountainous and desert regions. This exception is further illustrated in Figure 5 where we single out the μ- and η-estimates over Mexico. Here, high gradients of mean precipitation (e.g., along the Gulf and Pacific coasts, it differs by more than an order of magnitude) are associated with a higher spatial variability of the η-parameter than over seven other countries: it varies from 0.7 to 1.2, i.e., by ±25%.

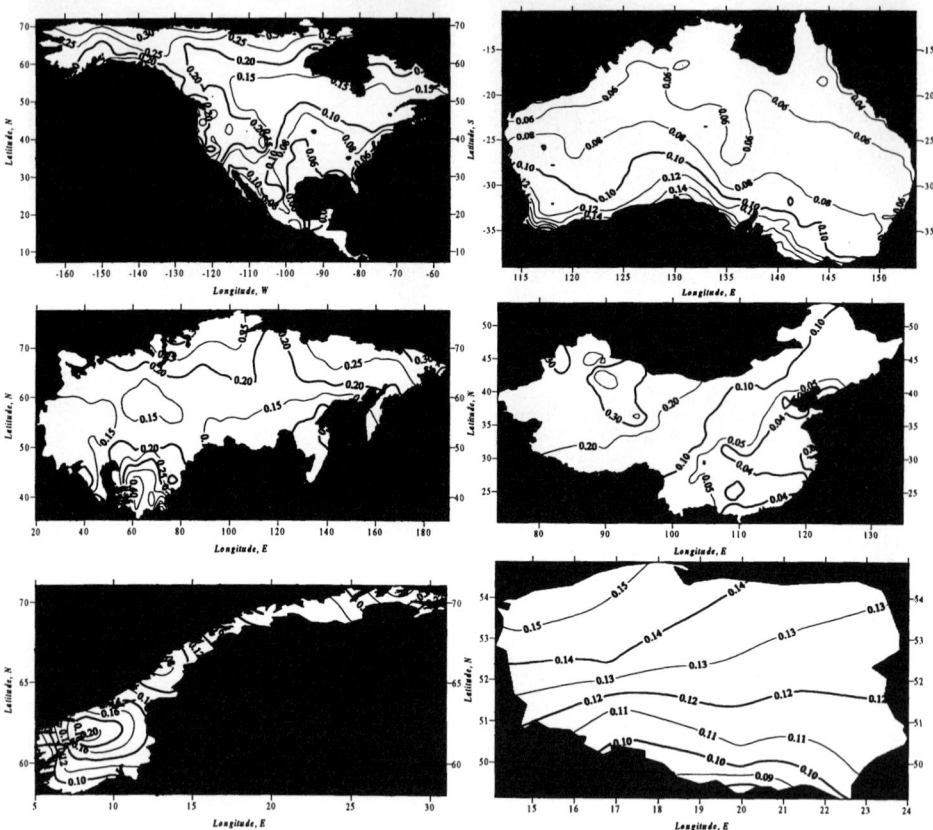

Figure 4 C. The scale parameter λ (mm⁻¹) over North America (Canada, United States, and Mexico), Australia, the former Soviet Union, PRC, Norway, and Poland for the three parameter model of the precipitation distribution (3). In Australia, December through February are considered as summer months.

5. Testing the Temporal Stability of the Shape Parameter

In the previous Section we have shown the spatial stability of the shape parameter of precipitation distributions. There are indications that an increase/decrease in mean precipitation at long-term stations is accompanied by an increase/decrease in precipitation variability (Bootsma, 1994). This leaves the coefficient of variation C_v of precipitation less affected by these changes and, according to equation (2'), the shape parameter, η, stays intact too. Now we test the temporal stability of the shape parameter in two ways:

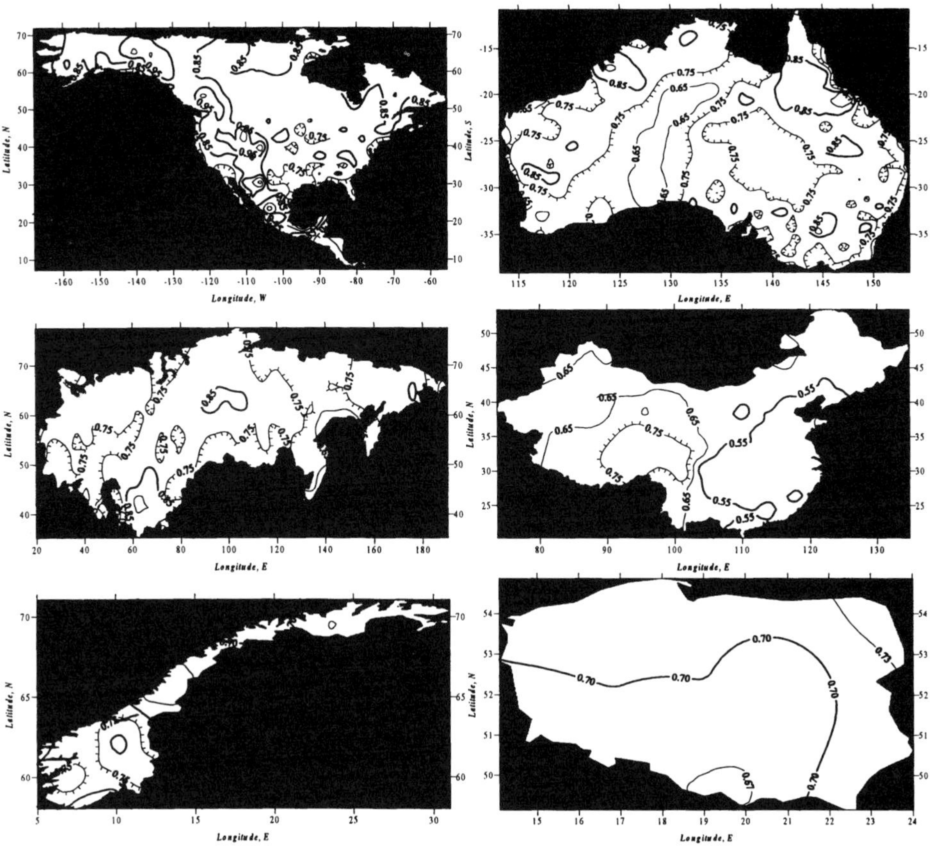

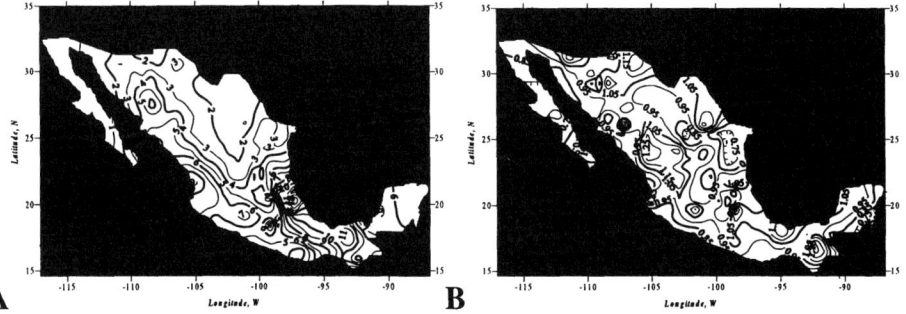

Figure 4 D. The shape parameter, η, over North America (Canada, United States, and Mexico), Australia, the former Soviet Union, PRC, Norway, and Poland for the three parameter model of the precipitation distribution (3). In Australia, December through February are considered as summer months.

Figure 5. (A) Mean daily summer precipitation (mm day⁻¹) and (B) the shape parameter, η, over Mexico for the three parameter model of the precipitation distribution (3).

- changes in the seasonal cycle and
- changes between "wet" and "dry" summers.

For the latter purpose the entire period of homogeneous summer observations at each station was divided into two groups of summers, those that have seasonal total precipitation below the long-term mean value and those that have seasonal total precipitation above this mean. This dichotomy essentially changed the mean precipitation values in each group. The difference between "wet" and "dry" summers was usually on the order of magnitude of the mean precipitation in "dry" summers, thus this partition imitated a large "climatic" change in precipitation. We then compared the parameters of the model (3) for each of these two periods to find out which of the three parameters changes the most.

Table I

The country-wide percentage differences, Δ, in parameters of daily precipitation between "wet" and "dry" summers for the United States, Australia, China, the former Soviet Union, Norway, and Poland over the regions with "dry" summer precipitation above 1 mm day^{-1}. Differences are presented in percent of the mean "wet" values [e.g., $\Delta\mu = 100\%$ $(\mu(\text{wet})- \mu(\text{dry}))/ \mu(\text{wet})$].

Country	$\Delta\mu$	ΔP_{α}	$\Delta\eta$	$\Delta\lambda$
USA	50	31	-10	-52
Australia	52	27	-10	-71
China	36	12	-3	-43
former USSR	39	17	-6	-45
Poland	36	15	-8	-44
Norway	38	17	1	-33

Analysis of the behavior of the shape parameter in the seasonal cycle shows that it is also relatively stable during the march of the seasons (Figure 6). In this figure we present the seasonal cycle of monthly precipitation and the shape parameter of the distribution of daily precipitation for a broad variety of regions spanning from Subarctic Canada to the Gulf of Mexico, from the North Atlantic to the South-China Sea, and over many climatic zones of the Australian continent. For comparison, the mean monthly precipitation is plotted on the same graphs to illustrate the fact that the long-term mean precipitation, μ, is much more variable than the η-parameter. However, a noticeable exception occurs in regions with a very strong seasonal cycle of precipitation during the "dry" season, when daily precipitation is much less than 1 mm/day. This exception is further illustrated by the spatial pattern of summer daily precipitation and the η-parameter of its distribution over Mexico (Figure 5). We are mostly interested in the probability of heavy rains and changes associated with moderate changes in mean precipitation. This goal allows us to omit areas with the mean summer precipitation below 1 mm per day from further consideration in this study and focus on other regions which essentially include major agricultural areas in each country.

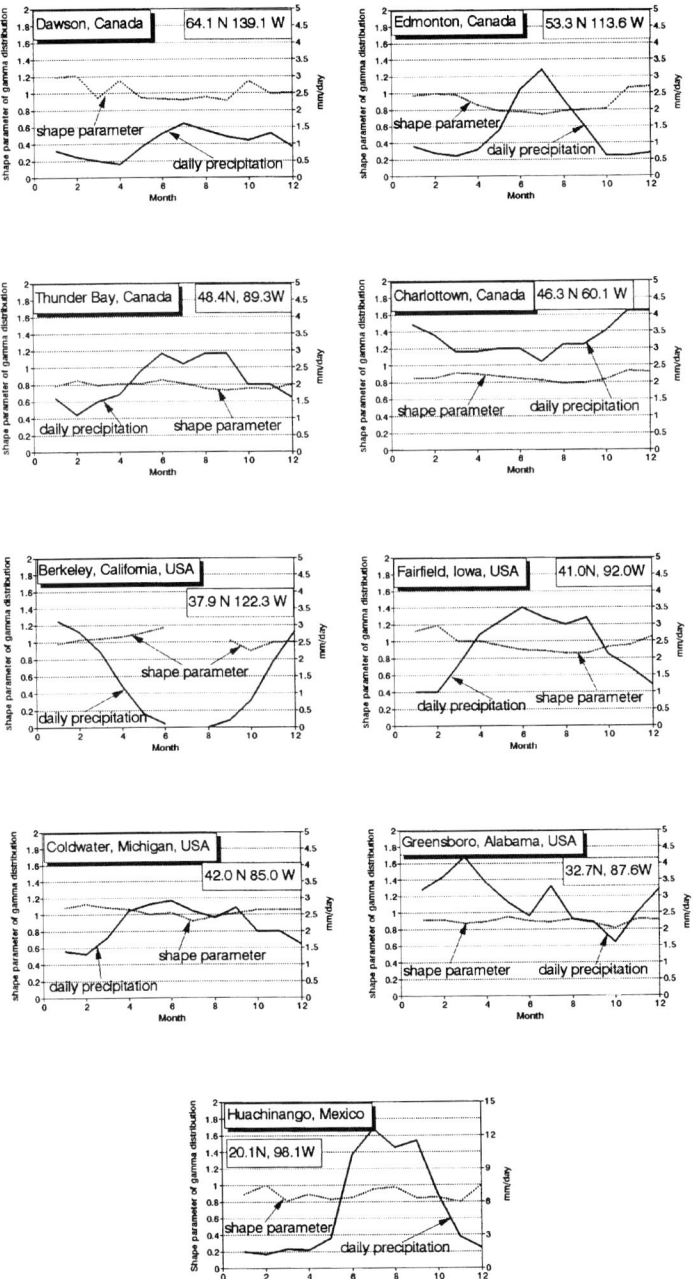

Figure 6 A. Seasonal cycle of the mean daily precipitation (mm day^{-1}), µ, and the shape parameter, η, for selected stations over North America for the three parameter model of the precipitation distribution (3). For Canada, only snowfall data were used instead of the gauge measurements to avoid the homogeneity problems with a cold season precipitation time series. The y-axis in the graph of Mexican daily precipitation was reduced threefold compared to other graphs in this Figure.

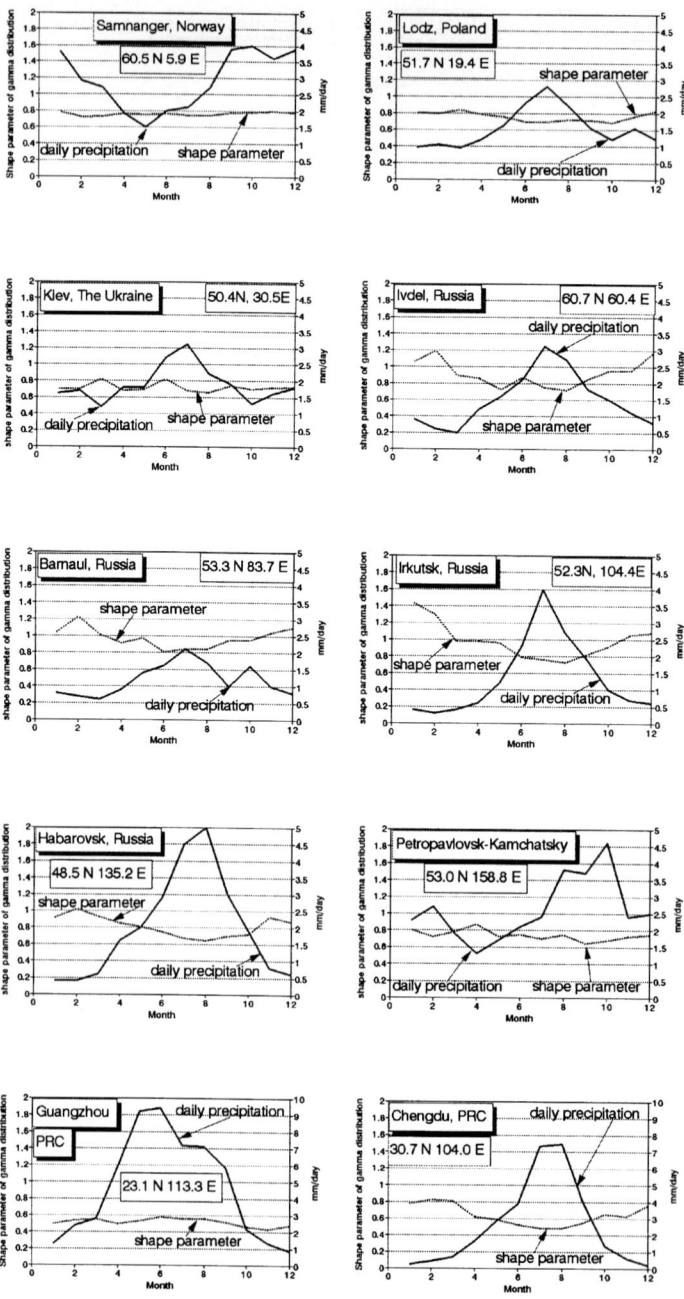

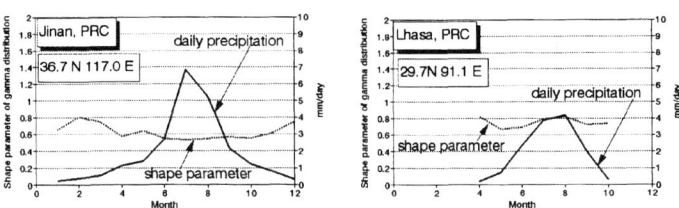

Figure 6 B. Seasonal cycle of the mean daily precipitation (mm day⁻¹), μ, and the shape parameter, η, for selected stations over Northern Eurasia for the three parameter model of the precipitation distribution (3).   To derive the graphs in this figure for the former Soviet Union, only the data after 1967 were used to avoid the homogeneity problems with a cold season precipitation time series. The y-axes in the graphs of Chinese daily precipitation were reduced twofold compared to other graphs in this Figure.

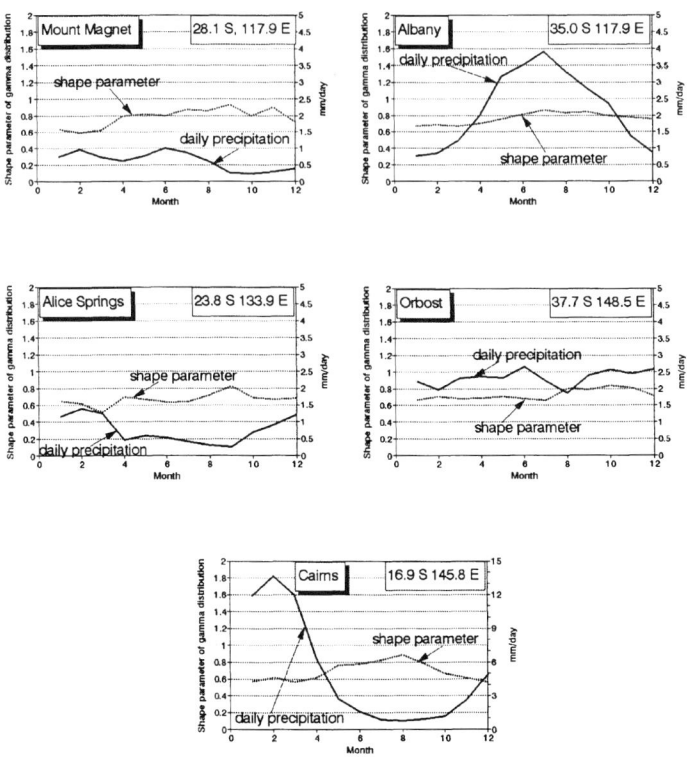

Figure 6 C. Seasonal cycle of the mean daily precipitation (mm day⁻¹), μ, and the shape parameter, η, for selected stations over Australia for the three parameter model of the precipitation distribution (3). The y-axis in the graph of daily precipitation in Cairns (northeastern Australia) was reduced threefold compared to other graphs in this Figure.

Figure 7 and Table I summarize our intercomparison of the parameters of daily precipitation distribution in "wet" and "dry" summers. This analysis was performed for six countries, excluding Canada and Mexico. The nature of our dichotomy forces the mean values, μ, to differ between "wet" and "dry" summers by 40 to 60% of the "wet" totals (Table I). The smallest variation between "wet" and "dry" summers was documented in the maritime climate of Norway (about 40% difference for a 100-year-long time series; in Poland, Russia, and China 40% differences were encountered for 30 to 60-year time series).   Over the continental

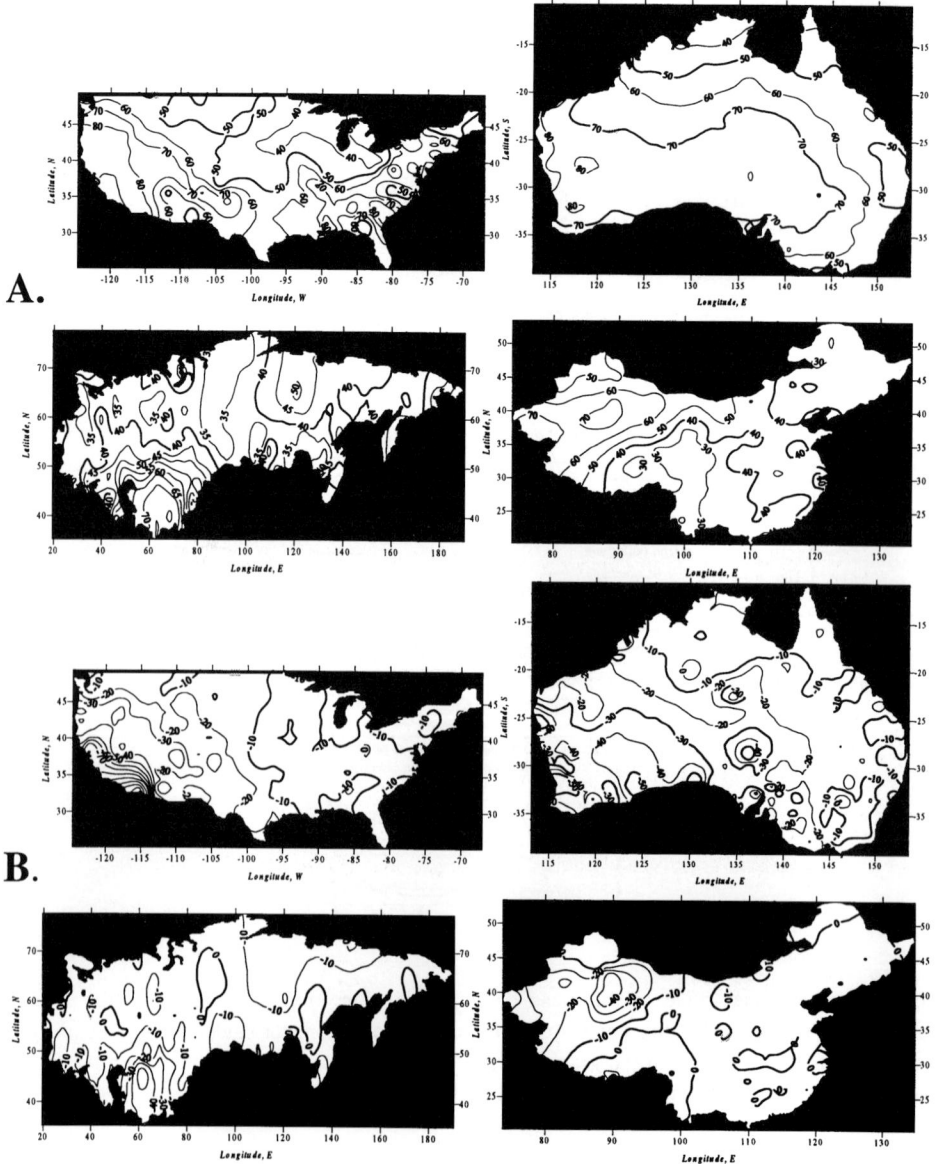

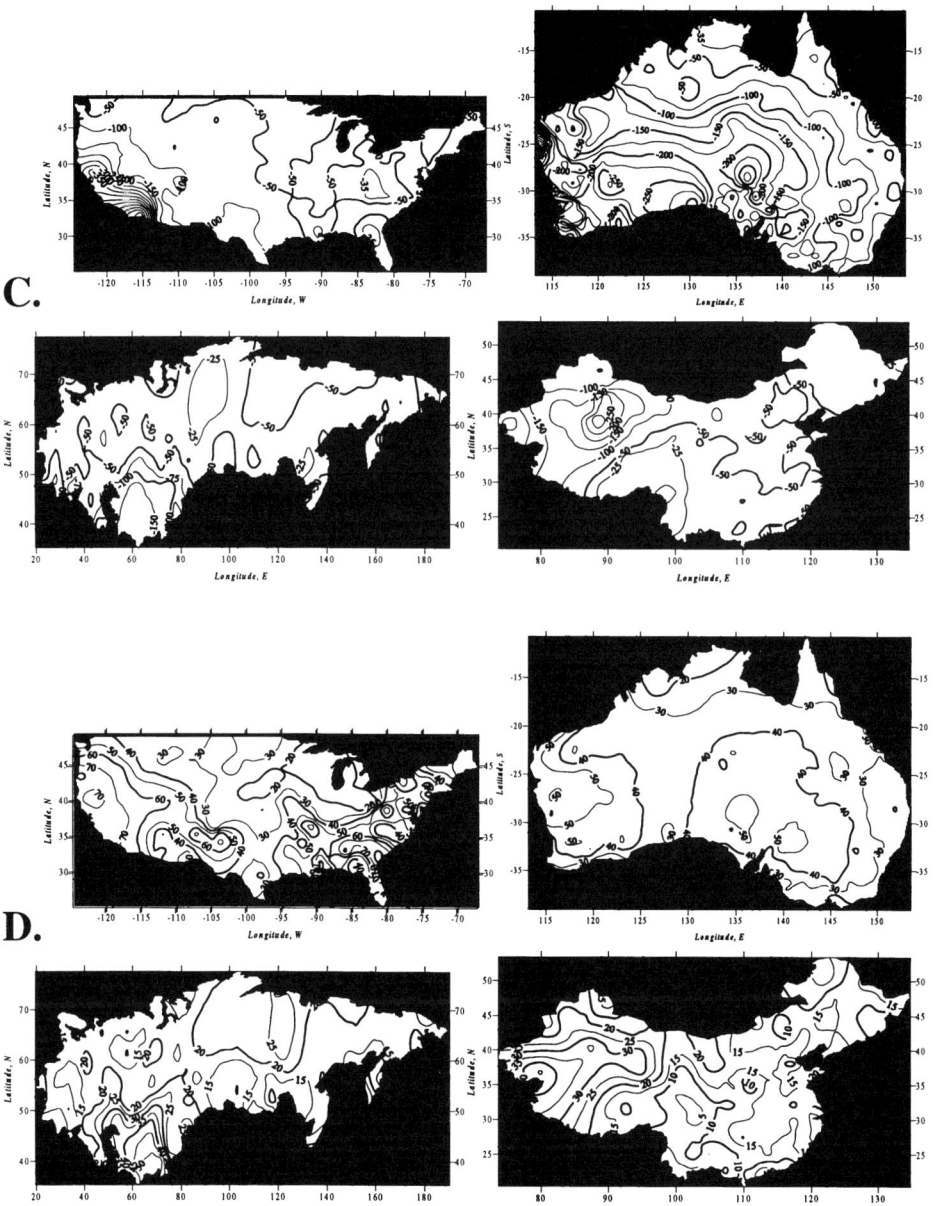

Figure 7. Differences (%) between "wet" and "dry" summers (JJA, DJF for Australia) in four countries (The United States, Australia, the former USSR, and China) as measured by (A) mean precipitation; (B) shape parameter of the distribution of precipitation totals; (C) scale parameter of the distribution; and (D) probability of a day with precipitation. "Wet" summers at a given station have the mean precipitation above the long-term mean value, while other summers are considered "dry".

areas with the mean daily precipitation above 1 mm, these differences are also
close to 40%. Over the contiguous United States and Australia the differences are
higher than 40% because we have longer (~90 years) time series (e.g., for Russia
only the last 27 years with homogeneous precipitation time series were used in this
intercomparison). The mean precipitation is a product: $\mu = P_{pr}\eta/\lambda$, and its change
is a result of changes in one or more of these three parameters. The average
differences in Table I and the patterns of these differences in Figure 7 show that
the most variable parameter, which contributes most to the difference between
"wet" and "dry "summers, is the scale parameter which may change by 100% or
more. The shape parameter is less variable and changes only slightly over eastern
China, former Soviet Union, Poland, and Norway. High precipitation variability in
Australia and the USA leads to twice as much precipitation in wet summers
relative to dry summers in the past 100 years, yet there is only a modest 10%
decrease in η which does not noticeably contribute to the change in μ over the
regions with daily precipitation above 1 mm day^{-1}.

6. Testing Changes in Precipitation Probability

The probability of daily precipitation, P_{pr}, can be estimated from the available data
sets even when we neglect the precipitation less than 1 mm (Bogdanova, 1987).
The number of days with precipitation in this category is closely related to the
lower threshold of the precipitation gauge measurements. This threshold was not
constant for many precipitation networks throughout the world including, e.g.,
Russia, Australia, and Canada and may introduce artificial trends in the number of
days *without* precipitation. To avoid the above mentioned inhomogeneities in
Russian, Australian and Canadian precipitation data, and keeping in mind that the
daily precipitation less than 1 mm usually contributes only a few percent to
monthly totals and is not of practical importance, we analyze the probability of

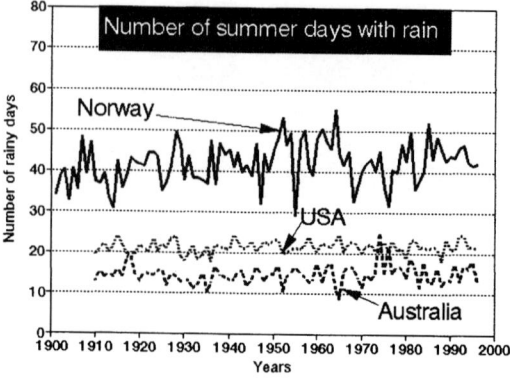

Figure 8. Average number of summer days with precipitation, area-averaged over the United
States, Australia, and Norway. Only the days with precipitation above 0.2 mm and 1 mm have
been counted at Norwegian and Australian stations respectively.

days with "measurable precipitation above 1 mm day^{-1}" for these three countries instead of P_{pr}.

For the United States, Australia, and Norway, we searched for century long trends in precipitation frequency (Figure 8, Tables II and III). For other countries our analyses are restricted to the post World War II period only. Before W.W.II, daily precipitation time series are unavailable for China and Mexico and there are inadequate data for the former Soviet Union, Poland, and Canada.

Table II

Linear trends in the number of precipitation days and mean precipitation for each season, area-weighted over the contiguous United States for period 1910-1996. Asterisk indicates a statistically significant difference from zero at the 0.05 significance level. The estimates are based on the century-long daily time series from the 187-station HCN data set.

Season	Mean number of days with precipitation	Linear trend (days/10yrs)	Correlation with seasonal precipitation totals
Winter	22	0.09	0.84
Spring	24	0.27*	0.91
Summer	22	0.12	0.90
Autumn	19	0.29*	0.93

Season	Long-term mean precipitation, mm	Linear trend, mm/10yrs	Linear trend, %/10yrs
Winter	170	0.0	0.0
Spring	200	2.0	1.0
Summer	205	1.5	0.7
Autumn	170	2.6*	1.5*

Table II shows the trends in area-weighted numbers of precipitation days per season and precipitation totals over the contiguous United States. Trends in summer and winter P_{pr} over the contiguous United States are not statistically significant at the 0.05 significance level. In spring and autumn upward trends in P_{pr} result in an annual increase of 5 to 6 precipitation days relative to the beginning of this century. The summer trend of P_{pr} over the U.S. is not statistically significant at the 0.05 level, but is significant at the 0.10 level[**]. In two regions of

---

[**] Here and throughout this paper, we employ a two-tail student t-test for testing linear trends for statistical significance. But, when we applied the non-parametric test for nonrandomness of the ranks in this time series based on Kendall's τ statistic (Kendall and Stuart, 1979), the upward trend in summer P_{pr} was found statistically significant at the 0.05 level.

the Mid-West (The Upper Mississippi and Missouri River Basins) the increase of 4 to 5 summer days per 100 yrs is statistically significant at the 0.05 level. Trends in precipitation totals have been discussed by Karl and Knight (1998). The numbers presented here differ from those shown in the first line of Table I of Karl and Knight (1998) for the special HCN network due to several improvements in this data set (infilling of missing values, inclusion of few additional stations to cover the data sparse areas, and additional quality control that allow us to reveal and fix some erroneous values).

There is no indication of statistically significant changes in winter (JJA) frequency of rainy days in Australia. However, during the summer (DJF) season in the southeast of the continent, a century-long statistically significant 20% increase of precipitation frequency has occurred (Table III). In a related study, Hennessy et al. (1998) found significant increases in the number of rain days in all seasons except winter in Australia from 1910-1995.

Table III

Linear trend in the number of precipitation days and mean precipitation for the summer season (DJF) area-weighted over the Australian continent for the periods 1910-1996 (continent) and 1900-1996 (eastern coastal regions). Asterisk indicates a statistically significant difference from zero at the 0.05 significance level.

Region	Mean precipitation, P_1	Linear trend in P_1 (%/10yrs)	Number of days with precipitation above 1 mm, N_1	Linear trend in N_1 (%/10yrs)
Entire continent	210	0.9	15	1.1
Coastal regions of Queensland and Northern Territory	590	0.6	33	0.8
Coastal regions of New South Wales and Victoria	180	3.1*	17	2.0*

Over Norway a century-long trend in annual precipitation was reported by Hanssen-Bauer and Førland (1994). P_{pr} at Norwegian stations exceeds 50% and the increase in summer total precipitation was accompanied by a further increase in precipitation frequency. Five more summer days with precipitation above 0.2 mm (three with precipitation above 1 mm day^{-1}) are registered now compared to the beginning of the century (Figure 8). Analyses of the number of summer days with precipitation over Poland, China, Russia, Canada, and Mexico in the post W.W.II period show no indication of trends.

Tables IV and V and Figure 9 summarize our analyses of trends in the number of days with "measurable precipitation above 1 mm day^{-1}" for Russia and Canada.

Table IV

The annual number of days with precipitation above 1 mm averaged over southern Canada (south of 55°N) separately for liquid and frozen precipitation. There are no changes that are statistically significant different from zero at the 0.05 significance level. Asterisk indicates a statistically significant difference from zero at the 0.10 significance level.

Precipitation Type	Period 1943-1975	Period 1976-1995	Difference, days
Rainfall	89	90	1
Liquid equivalent of snowfall	46	44	-2*
Summer Precipitation	31	31	0

Table V

The average number of summer days with precipitation above 1 mm averaged over several regions of the former USSR. Asterisks indicate statistically significant differences/trends at the 0.05 significance level.

	Number of stations	Period 1943-1975	Period 1976-1986	Difference (days)	Linear trend 1936-1994 (day/50 yrs)
European part of the former USSR	89	25	26	1	0.4
Asian part of Russia	99	28	27	-1*	-2.2*
Kazakhstan and Central Asian States	35	7	6	-1*	0.2
Former USSR	223	23	23	0	-0.5

In southern Canada (Table IV) we found no trends in the number of days with precipitation above 1 mm, but registered a small redistribution between precipitation in frozen and solid form. On average, after the mid-1970s the average number of days with rainfall has increased by a day compared to the three previous decades, while the average number of days with snowfall decreased by approximately two days. There is no trend in summer P_{pr} for southern Canada.

Analysis of the number of days with precipitation above 1 mm for Russia (Table V) shows that since 1936 there has been an absence of systematic changes in P_{pr} over the European part of Russia but we found a statistically significant trend (19%/50yrs) in frequency of summer daily precipitation events above 20 mm. Over the Asian part of Russia (Siberia) we found a statistically significant *decreasing* trend in precipitation frequency.

In summary, over the United States, Norway, and Australia we found an increase in summer precipitation frequency over the past century. The increase in precipitation frequency in Norway (5 days per 100 years) is larger than in the other seven countries. The highest *relative* change in precipitation frequency (about 20%) has occurred in southeastern Australia. When the same analyses are repeated only for the post- W.W.II period, they do not show statistically significant trends in P_{pr} over these countries. For all other countries we could not find systematic changes in summer values of P_{pr}. Therefore, we conclude that there is no evidence that P_{pr} during summer months has substantially changed during the past five decades over the large-scale regions considered with the exception of the Asian part of Russia. Regional time series of P_{pr}, however, deserve a more thorough analysis, which is beyond the scope of this paper.

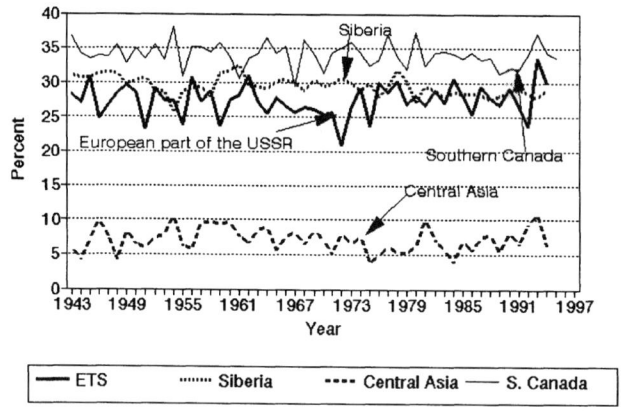

Figure 9. Percent of summer days with precipitation above 1 mm over the former Soviet Union and southern Canada. The numbers were arithmetically averaged over 58 first order stations in Southern Canada (south of 55°N), 89 first order stations in the European part of the former Soviet Union (ETS), 99 stations in Siberia, and 35 stations in Central Asian members of the Commonwealth of Independent States.

7. The Effect of Changes in Mean Precipitation on Heavy Daily Precipitation

Katz and Brown (1992) established that the probability of an extreme event (i.e., the probability of the meteorological variable exceeding an unusually high value threshold) becomes significantly larger for comparable increases in standard deviation compared to the mean. For daily summer precipitation during the post-W.W.II period, this general conclusion is superimposed over the spatial and temporal stability of the shape parameter (Figures 4 through 7 and Table I) and empirical evidence of the stability of the number of precipitation events (Tables IV and V and Figures 8 and 9). In moderately different climates (which can be associated with geographical shifts of climatic zones, the seasonal cycle of precipitation, cyclone tracks, etc.) we might expect that in each region the shape parameter will stay mostly intact. This, in turn, means that for summer precipitation over Australia, North Eurasia and America, changes in mean values will be approximately matched by changes in standard deviation (cf., Eqs. 2'), which will strongly affect the probability of extreme precipitation (cf., Figure 3).

In all eight countries considered (except China) at least a 5% increase in mean summer precipitation has been documented during the past 100 years (IPCC, 1996, 1998; Groisman and Easterling, 1994; Lettenmaier et al., 1994; Karl et al., 1993; Karl and Knight, 1998; Vinnikov et al., 1990; Groisman, 1991; Georgievsky et al., 1995, 1996; Lavery et al., 1992, 1997; Hanssen-Bauer and Førland, 1994; Hanssen-Bauer, 1994; Kozuchowski, 1985). In three countries (USA, Australia,

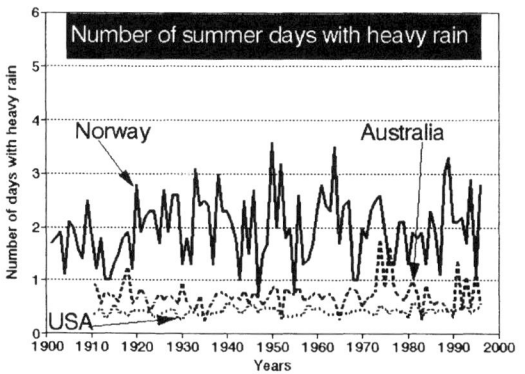

A.

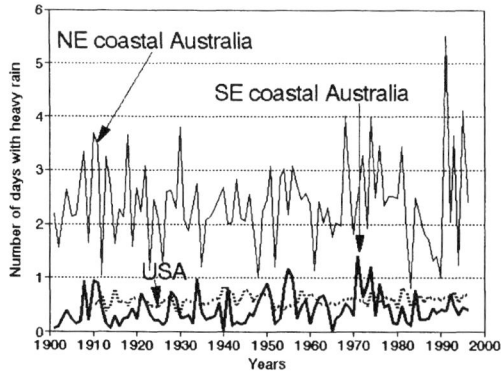

B.

Figure 10. (A) Average number of summer days with precipitation exceeding 50.8 mm (25.4 mm for Norway), area-averaged over Australia, the United States, and Norway. (B) The same, but area-averaging was conducted only over the eastern two-thirds of the contiguous United States and over the northeastern (north of 30°S, east of 130°E) and southeastern (south of 30°S, east of 140°E) coastal regions of Australia less than 350 km away from the coast line.

and Norway) we found a century-long increase in heavy precipitation frequency and in P_{pr} (Karl et al., 1996; Karl and Knight, 1998; Suppiah and Hennessy, 1996;

Tables II, III and VI; Figure 10). In other countries, where we have shorter and/or insufficient data, the direct detection of systematic changes in heavy precipitation using observational data is more difficult. Therefore, we exploit the findings of previous sections about the stability of η and P_{pr} (the temporal stability of P_{pr} is, however, dependent on the period of interest) and use various assumptions about how the mean precipitation may (or did) change to analyze the effect of these changes on extreme precipitation. As an example of this type of analysis, below we present the effect of a 5% increase in mean precipitation on the precipitation above selected thresholds assuming a scenario of no changes in η and P_{pr}. We apply this scenario to all eight countries although summer precipitation in some of them (Russia, Canada, Australia, Norway, Mexico) has increased at a higher rate during the past century, while over eastern China it decreased during the period from 1909 to 1993 (Ye et al., 1996).

Table VI

Country-wide linear trends of the number of summer days with heavy precipitation over the contiguous United States, Australia, and Norway. Asterisk indicates a statistically significant difference from zero at the 0.05 significance level.

Country	Period	Threshold used to define "heavy" rain	Average number of days with heavy rain	Linear trend, day/10years	Linear trend, %/10years
Contiguous USA	1910-1996	50.8 mm	0.4	0.007*	1.7*
Eastern two-thirds of the contiguous USA	1910-1996	50.8 mm	0.6	0.010*	1.7*
Australia	1910-1996	50.8 mm	0.7	0.018	1.1
Coastal regions of New S. Wales and Victoria	1900-1996	50.8 mm	0.4	0.019*	4.6*
Norway	1901-1996	25.4 mm	2.0	0.04	1.9

We define (somewhat arbitrarily) "heavy" precipitation, P_{heavy}, as a daily precipitation exceeding the 25.4 mm threshold in northern countries (Russia, Canada, Norway, and Poland) and exceeding the 50.8 mm threshold in mid-latitudes (the United States, Mexico, China, and Australia). Figures 11 and 12 provide the climatology of summer heavy precipitation estimated from model (3). Figures 13 through 15 summarize our estimates of the disproportionate increase in precipitation for heavy precipitation rates, compared to a 5% increase in mean precipitation, if the shape of the precipitation distribution and the probability of a precipitation event do not change.

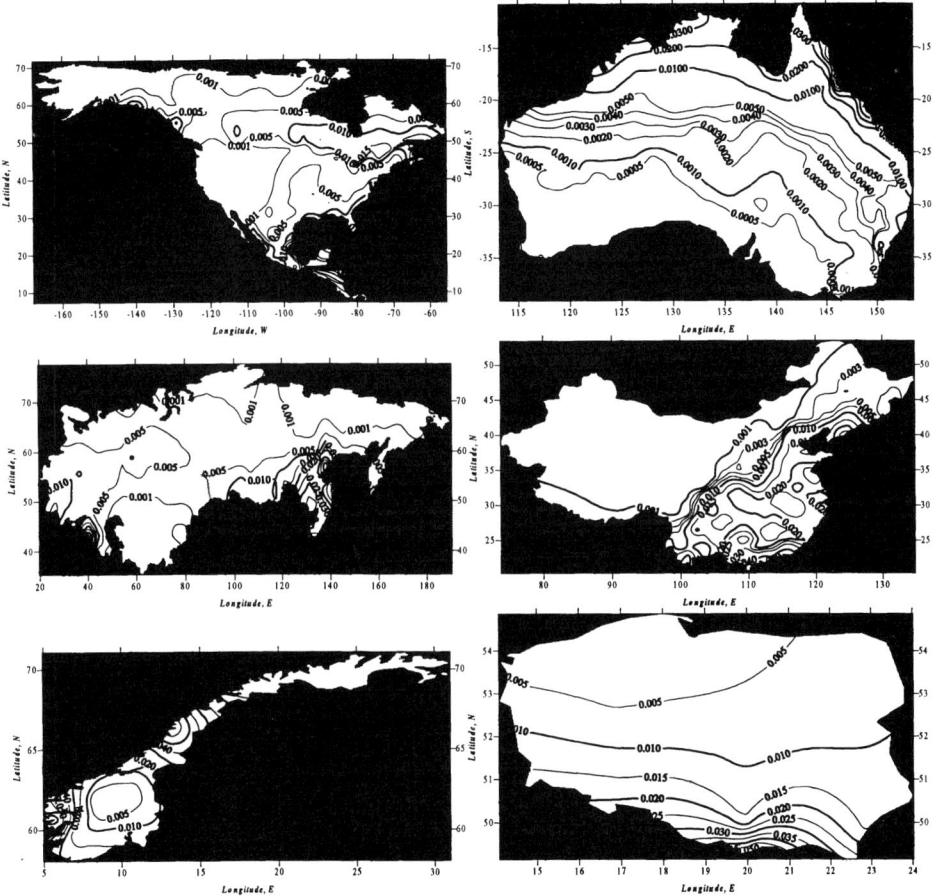

Figure 11. Probability of a day with precipitation exceeding 25.4 mm (Canada, the former Soviet Union, Poland, and Norway) and 50.8 mm (the United States, Mexico, PRC, and Australia). Estimates are based on model (3). Direct estimates of these probabilities based on century-long precipitation time series (e.g., Figure 2) resemble the pattern shown in this figure but can be produced only for a small part of the area under consideration (in regions with daily precipitation above 1 mm day^{-1} in contiguous United States, Australia, Norway, European Russia, and southeastern Canada).

Over all of southern Canada, the former Soviet Union, Poland, and Norway a 5% increase in mean summer precipitation manifests itself in a 20% increase of the probability of days with precipitation above 25.4 mm (Figure 13). These heavy precipitation events (which on Figure 12 contribute less than 5% of summer precipitation totals of Norway, Russia, Canada, and Kazakhstan and less than 10% of summer precipitation of Belarus, Poland, and The Ukraine) contribute up to 30% (Russia, Canada, northern Norway) and more than 40% (southern Norway, Belarus, Poland, and The Ukraine) of the increase of mean daily precipitation (Figure 14).

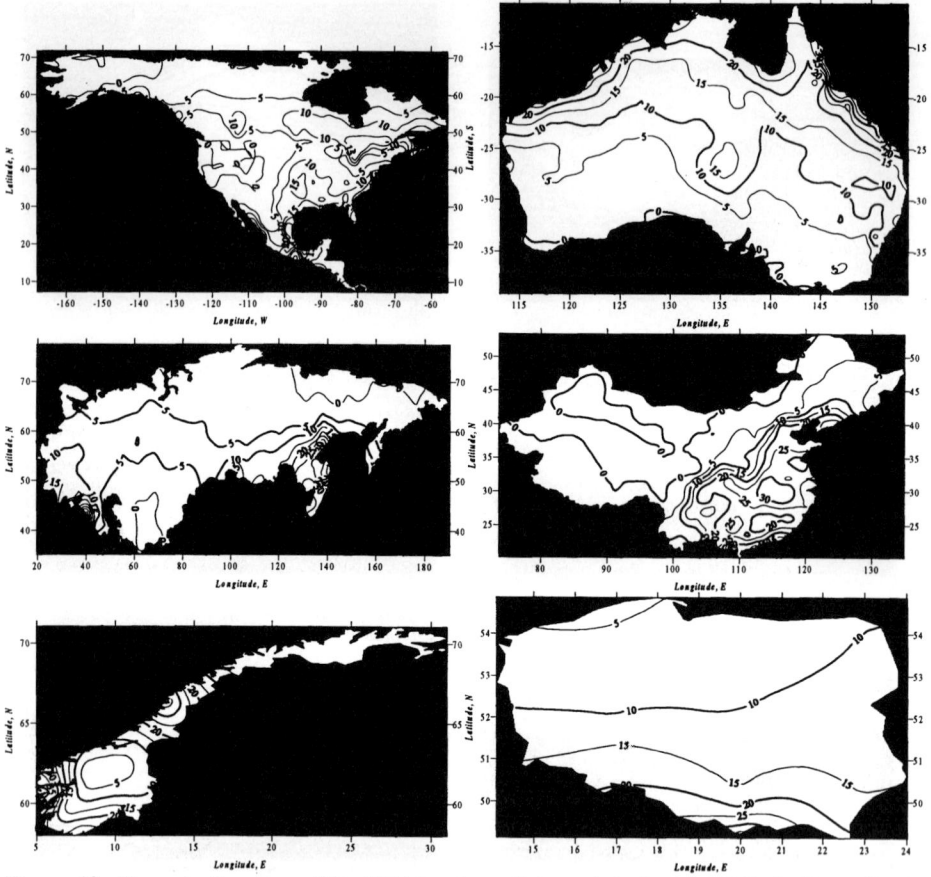

Figure 12. Percent of summer (JJA, DJF for Australia) precipitation that falls in "heavy" rains, i.e., with daily rates above 25.4 mm (Canada, the former Soviet Union, Poland, and Norway) and 50.8 mm (the United States, Mexico, PRC, and Australia). Estimates are based on model (3).

In the eastern United States, in regions with mean summer precipitation above 2 mm per day, an increase in mean daily precipitation by 5% yields an increase in the probability of daily precipitation above 50.8 mm (2 inches) by approximately 20%[***]. In the Mississippi River Basin up to half of the increase in mean summer precipitation is contributed by heavy rains (Figure 14). This helps explain why recent studies by Karl et al. (1995) and Karl and Knight (1998) were able to detect significant increases in extreme precipitation over the contiguous United States, while the century-long increases in summer precipitation totals over the same region were non-significant (Karl et al., 1993, Groisman and Easterling, 1994).

IPCC (1998) shows an increase in mean annual (summer) precipitation of 10 to 20% during the 20th century over most of Mexico. The scenario of a 5% increase

---

[***] By 15% to 20% in the Southeast and 20% to 30% in the Northern part of the country.

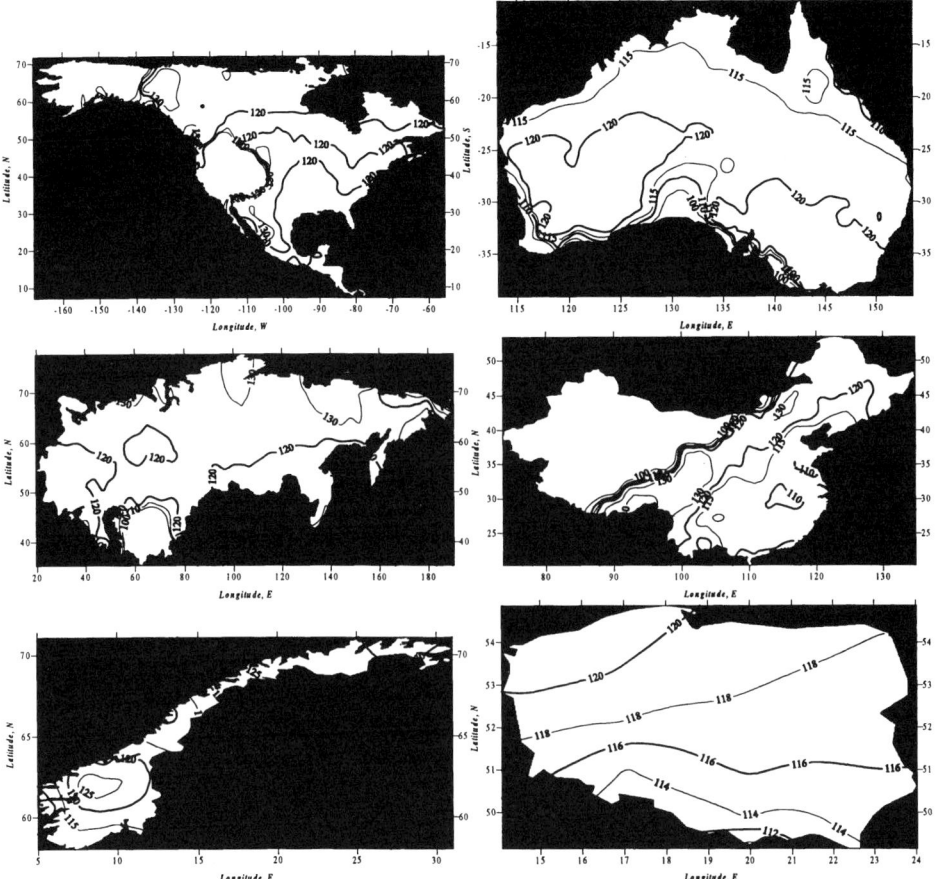

Figure 13. Percentage change of the probability of summer daily precipitation exceeding the heavy rainfall thresholds, P_{heavy}, (defined in text) when the mean daily precipitation increases by 5% assuming that P_{pr} and η do not change. The change is expressed as a ratio P_{heavy}(scenario)/P_{heavy}(climate).

in summer mean precipitation over Mexico yields a 20-30% increase in the probability of daily summer precipitation above 50.8 mm (2 inches) except in the desert regions of the country (Figure 13). In this scenario, more than 70% of the increase in mean summer precipitation over the tropical regions of Mexico is contributed by heavy rains (Figure 14). This is not a surprise because heavy precipitation is typical in the tropics (Figure 12).

For China, we considered a scenario of a 5% increase in mean summer precipitation although this scenario is not supported by real trends in mean precipitation during the past 50 years. For eastern China this scenario yields a 10 to 30% increase in the probability of summer precipitation exceeding 50.8 mm (Figure 13). This increase is less pronounced in the coastal areas with higher precipitation rates and more pronounced inland. Over the Tibetan Plateau and

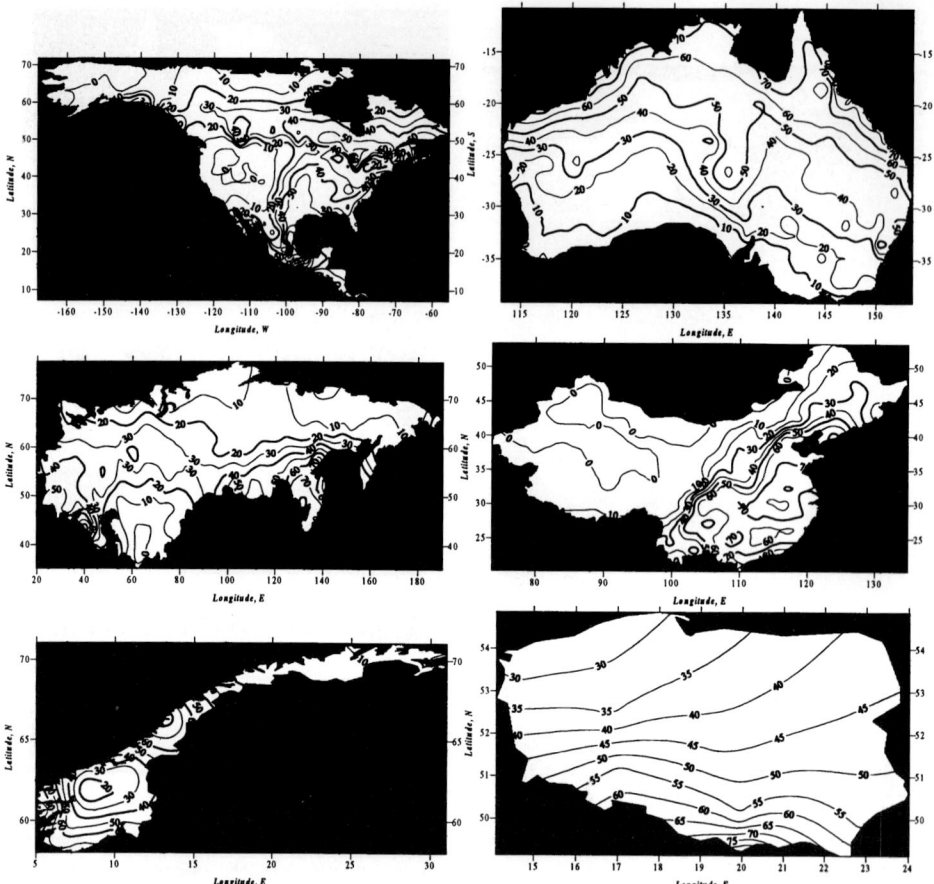

Figure 14. The contribution (percent of total increase) of summer daily precipitation exceeding the heavy rainfall thresholds (defined in text) when the mean daily precipitation increases by 5% assuming that P_{pr} and η do not change (same scenario and thresholds as in Figure 13).

Sinkiang Province it is too arid and there is no heavy precipitation. More than 50% of the scenario-increase in mean summer precipitation over eastern China is contributed by heavy rains (Figure 14). This contribution increases to 70% in tropical parts of southern China.

Hennessy et al. (1998) found a 9% increase in mean summer precipitation from 1910-1995 over Australia. Our scenario of a 5% increase in summer mean precipitation over Australia yields similar results to those for China. We found a 10 to 20% increase in the probability of summer precipitation exceeding 50.8 mm, more pronounced over relatively dry interior regions of the country and less pronounced over tropical coasts (north and northeastern Australia). Over the regions with low summer precipitation (south and southwestern Australia) 50.8 mm is not exceeded so no changes were found. The contribution of heavy rains to the 5% increase in mean precipitation gradually decreases from 70% in the north to 10% in the south of the continent (Figure 14).

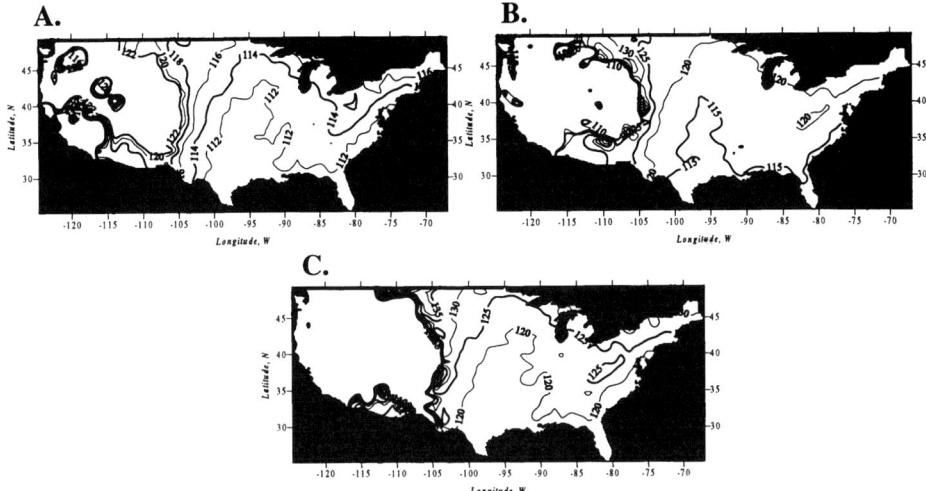

Figure 15. Percentage change of the probability of summer daily precipitation over the United States exceeding (a) 50.8, (b) 76.2, and (c) 101.6 mm thresholds when the mean daily precipitation increases by 7% assuming that P_{pr} is also increased by 5% (to match the estimate in Table II) and η does not change. The change is expressed as a ratio P_{heavy}(scenario)/P_{heavy}(climate).

For all countries, we calculated the change in the contribution of heavy rain events to total precipitation when mean precipitation increases by 5%. This involved calculating the ratio of the percentage of heavy rains associated with a 5% increase to the current percentage shown in Figure 12. The ratio indicates an increase in the contribution of heavy rains to the summer precipitation totals by a factor of 1.1 to 1.2 (i.e., a 10-20% increase). Karl et al. (1995) showed that changes in the proportion of precipitation contributed by heavy extreme precipitation events were increasing relative to the total annual precipitation received in the United States during the past century. Here we show that such a result is also consistent with an increase in mean daily precipitation with no change in the probability of precipitation and the shape parameter. Such a result is not intuitive and implies that not only does more precipitation occur in heavier rain events as precipitation increases, but that the increase is disproportionately larger than the change in the mean.

In order to better match the observed precipitation changes for the United States, Norway, and Australia we should split the observed increase in mean precipitation between increases in intensity and frequency. Below we present our best guess scenarios for extreme precipitation changes in these three countries with changing P_{pr}. In these scenarios we continue to fix η. The following scenarios are considered:

- the contiguous United States: a 7% increase in mean summer precipitation and a 5% increase in P_{pr} (scenario to match "century-long" trends shown in Table II).

- Norway: a 7% increase in mean summer precipitation in the "windward" part of the country (all northern Norway and coastal ocean-front part of southern Norway) and a 5% increase in the number of rainy days; a 7% decrease in mean summer precipitation in a "leeward" part of the country (interior and southeastern part of southern Norway, south of 62.5°N and east of 6°E) without any changes in precipitation frequency; this scenario is based on the analysis of the summer precipitation changes over the past 60 years (Førland et al., 1996) and the trends in P_{pr}.

- Australia: a 9% increase in mean summer precipitation and a similar increase in P_{pr} (see Table III and Section 2 in Appendix).

Figure 15 shows that the scenario increase in the number of days with summer precipitation, P_{pr}, found in the United States (Table II) in conjunction with a higher increase of the mean summer precipitation does not significantly affect the results reported in Figure 13. We use this scenario to further investigate the changes in probability of extremely heavy rains (above 76.2 mm and 101.6 mm). This extrapolation can be used for practical purposes in hydrological calculations of the consequences of the contemporary climatic change.

Table VII

Scenario of the most-probable changes in the probability of summer heavy rainfall, P_{heavy} above 25.4 mm over Norway derived from the recent (past 60 years) variations in mean precipitation, μ, and frequency, P_{pr}. Average numbers of days with heavy rainfall and their linear trends estimated from the century-long homogeneous time series for period 1901-1996 are also shown. The partition of the country into two regions (windward and leeward) was accomplished by separation into a leeward part of the country the region south of 62.5°N and east of 6°E.

Region	$\Delta\mu$ (%)	ΔP_{pr} (%)	ΔP_{heavy} (%)	Average number of days with heavy rains	Average linear trend, %/10years
Windward part of the country	+7	+5	12	2.9	1.4
Leeward part of the country	-7	0	-26	1.3	-2.0

Table VII shows that when the scenario mean precipitation increase/decrease is accompanied by the change in frequency of precipitation events (as found in Norway), changes in P_{heavy} are less prominent but, nevertheless, still higher than the changes in mean precipitation. Empirical estimates of century-long trends in the number of days with heavy rainfall (the last column in Table VII) support this analysis and the order of magnitude of these trends is consistent with theoretical estimates of changes in P_{heavy}.

When a change in mean summer precipitation is solely due to P_{pr}, a proportional change in the probability of heavy precipitation is expected. Thus, a 9% increase in mean summer precipitation over Australia should give a 9% increase in the probability of heavy rains. Table VI and Figure 10 show that the increase in heavy rainfall over Australia observed during the past century was about 10% for the entire continent and about 45% in the southeast. However, unusually high precipitation variability during the past three decades[****] strongly affects all statistics computed from these time series and makes a linear trend of the all-Australian heavy rainfall statistically insignificant at the 0.05 level. For the coastal region of southeastern Australia (New South Wales and Victoria; Tables III and VI) we found statistically significant increases in μ (3.1%/10yrs), P_{pr} (2.0%/10yrs), and P_{heavy} (4.6%/10yrs). Each of these changes is statistically significant and the 45% increase in heavy rainfall in the southeast is particularly impressive and matches our model estimates of the P_{heavy} increase for the stations in this region.

In Figure 3B we sketch the changes in the probability of heavy rains, P_{heavy}, when the mean precipitation, μ, is changing without changes in the shape parameter, η, but with changes in P_{pr} and λ. It shows a variety of P_{heavy} changes depending upon changes of the ratio of these two parameters. But in our analyses, only *one* combination/realization has been observed in each region where we have sufficient homogeneous precipitation data on a century time-scale (the eastern two-thirds of the United States, coastal regions of southeast Australia, European part of the former USSR, and southern Norway): the changes in P_{pr} are of the same sign and less than the changes in μ by absolute value. This implies (according to our model) that the changes in P_{heavy} will be in the same direction as changes of μ with a higher than linear rate. This is exactly what we have observed in our empirical estimates of P_{heavy}.

The scenarios discussed above have been calculated using gamma distribution parameters of daily precipitation and observed trends in summer country-wide precipitation totals and frequency. There is no guarantee that these trends will continue in the future. However, the purpose of such exercises is twofold:

(1) We show how important parameters of daily precipitation events have contributed to the historical changes in mean seasonal precipitation without actual monitoring the changes in these parameters (which is otherwise arduous and often impossible due to data paucity and inhomogeneity problems); and

(2) If changes in mean precipitation can be predicted (e.g., by climate models), the revealed relationships between mean and extreme precipitation will assist us in the assessment of the hydrologic, ecological, and socio-economic consequences of these changes.

For the United States and eastern Australia, we possess sufficient century-long homogeneous time series of daily precipitation to evaluate the trends in heavy precipitation directly, i.e., without the help of model (3) (Figure 10, Table VI). These data support our conclusions about the century-long disproportionate

---

[****] The most humid northern part of the continent received a record high intensity and frequency of summer (DJF) heavy rains in 1991 and 1995 years and a record low intensity and frequency of heavy rains in 1983 and 1990 years.

increases in heavy precipitation over these two countries (cf., Karl et al., 1996; Karl and Knight, 1998; and Suppiah and Hennessy, 1996, 1998). To be conclusive, analyses of country-wide trends in extreme precipitation have to be based on a more dense network than similar analyses for mean precipitation, because of the low ratio of signal to noise in the data. For example, high variability due to the small number of stations used in the analysis of Norway and Australian continent heavy precipitation (cf., Figure 10) makes the trend estimates shown in Table VI statistically insignificant. However, the approach used in scenarios shown in Figures 13 through 15 can handle the data paucity problem.

8. Conclusions

A simple statistical model of daily precipitation applied to the data of eight countries shows that the shape parameter of the precipitation distribution remains regionally and temporally stable, the number of days with precipitation remains more or less stable, while the scale parameter is highly variable in time and space. This implies a likelihood that changes in mean monthly precipitation in these countries will be associated with disproportionately large changes in the extremes.

When mean summer precipitation increases by 5%, similar to what has occurred in several regions during the past century, with no change in the number of precipitation days and no change in the shape parameter of the precipitation distribution, there is a 20% increase in the probability of summer daily precipitation over a 25.4 mm threshold in northern countries (Russia, Canada, Norway, and Poland) or a 50.8 mm threshold in mid-latitudes, tropics, and subtropics (the United States, Mexico, China, and Australia). That is, the increase in the probability of "heavy" precipitation is four times the increase in mean precipitation.

Increases in atmospheric water vapor have been documented in North America, China, and a large portion of the tropical Oceans (IPCC, 1996; Ross and Elliot, 1996). Such increases in water vapor suggest an enhanced hydrological cycle. These changes have accompanied a widespread increase in temperature over the last century. Our results complement the above findings and indicate that in a warmer and wetter world, as projected by climate models driven by increasing greenhouse gases, increases in extreme precipitation are likely to be disproportionately large compared to any change in the total precipitation[******].This is likely to have important socio-economic and ecological impacts. This feature of

---

[******] Various lines of evidence (climate models, observations and theory) indicate that future greenhouse warming will lead to increases in mean rainfall (IPCC, 1996), and our statistical model suggests that this will be accompanied by large increases in heavy rainfall. Moreover, future increases in heavy rainfall derived from our study are consistent with coarse resolution results from global climate models (Schaer et al., 1996; Jones et al., 1997; Hennessy et al., 1997).

summer precipitation may already be manifested   in recent increases in precipitation extremes over some regions, e.g., the United States and Australia.

Acknowledgments

National Weather Services of the United States, Russian Federation, People Republic of China, Norway, Mexico, and Australia provided their data for this research. Canadian data were furnished to the U.S. National Climatic Data Center by the Atmospheric Environment Service of Canada. The data from Polish stations were provided to the research team by the Polish Institute of Meteorology and Water Management. The authors express their special thanks to two anonymous reviewers for thoughtful comments and suggestions.

Appendix 1. Some Details Related to Quality Control, Pre-processing, and Parameter Estimation of the Gamma Distribution Parameters of Daily Precipitation Data.

1. Several quality control procedures had been performed with daily precipitation data before they were used in our analyses. The U.S. procedure is described in Hughes et al. (1992), the North American data set pre-processing is described by Easterling et al. (1998). The Australian procedure is described by Lavery et al. (1992, 1997) and Plummer et al. (1997, 1998). The Norwegian procedure is described by Hanssen-Bauer and Førland (1994), the former USSR method is described in Razuvaev et al. (1993), and the Polish procedure is described in Fortuniak (1996). We are not aware of inhomogeneity problems in precipitation time series for Mexico. The original data, however, have passed quality control procedures (Easterling and Peterson, 1995) and numerous small scale (i.e., on a station level) adjustments to the data have been made. Specifically, Mexican data were checked for extreme outliers by comparison with neighboring stations and existing climatologies.

There are no instrumental inhomogeneity problems with precipitation observations in Chinese and Polish stations of international exchange during the period of data availability. Careful analysis revealed some inhomogeneity problems at several Norwegian stations around 1900 which related to the introduction of wind shields on rain gauges or station relocations. Therefore, we use the precipitation time series from these stations only after the wind shields have been installed or after the date of the last relocation.

2. There is a very specific problem with century-long Australian precipitation time series. Sometimes rain gauges were not read for several days, so rainfall records appear as an accumulated total followed by flags indicating the accumulation period and the number of rain-days in that period. A significant number of

Australian stations often reports accumulated precipitation values most of which coincide with Mondays and/or days after holidays. On days when accumulations are not reported, all-Australian average summer rainfall is 10% to 15% lower on Sundays than on any other day. This suggests that Sunday rainfall has not been consistently reported especially in the first half of the century. Our estimates show that this effect can affect (increase) the trends in P_{pr} and P_{heavy} by 5 to 10%. Since the 1970s, there has been an increase in the number of rainfall accumulations recorded at many stations. The National Climate Centre (Neil Plummer, Personal communication) believes this reflects a decline in recording of weekend rainfall. Plummer et al. (1997) show a marked decrease in frequently reporting rainfall stations around the early to mid 1970s. At many of the post office stations, this is because Australia Post went from a 5.5 day a week operation to 5 days from 23 Feb. 1974.

Keeping in mind the problems with accumulated rainfall and the need to analyze daily rainfall trends, we selected the following approaches: (a) we treated the accumulated totals as 24-hourly totals and (b) we split the accumulations into daily amounts determined by the reported number of rain-days in each accumulation period. Method (a) was tested by comparison with the trends in three- and seven-day precipitation totals (that should not be affected by accumulation). In this test, we found that the effect of our treatment of accumulation periods on daily trends in P_{pr} and P_{heavy} is minimal. Moreover, the comparison of P_{pr} and P_{heavy} for each day of the week shows that our estimates of P_{pr} and P_{heavy} in Tables III and VI can be inflated only by a few percent due to our treatment of accumulation reports in trend analysis. In method (b) each rain-day in the accumulation period was assigned the value of the accumulated total divided by the number of rain-days. For example, a total of 20 mm accumulated over three days including two rain-days would be replaced by two days of 10 mm and one day of 0 mm. Suppiah and Hennessy (1996) found that calculated rainfall trends were insensitive to the method of replacement of accumulated values. While there are problems and potential biases in both methods, the results presented in Tables III and VI are based on method (a) which is more likely to underestimate mean values of P_{heavy} and P_{pr} but keeps their trends intact.

3. Data for Canada and the former Soviet Union required special attention due to inhomogeneities (Goodison and Louie, 1986; Groisman et al., 1991; Groisman and Easterling, 1994; Metcalfe et al., 1997). In this paper we focus on summer precipitation. Thus, no efforts were made to homogenize cold season daily precipitation time series. An inhomogeneity problem that affects precipitation measurements in all seasons in Russia is the absence of a wetting correction in the data prior to 1966. This was mitigated by introducing this correction into the data prior to 1966 so they would be in agreement with the current observational practice (Struzer, 1975).

The daily Canadian precipitation data were quality controlled by the Canadian Atmospheric Environment Service. These data required an additional adjustment to homogenize the precipitation time series. We did not use the reported precipitation records because they contain an inhomogeneity related to a change in the technique of measuring frozen precipitation in the Canadian primary network during the early 1960s. Instead, to obtain mean daily precipitation, P, we used separate records of rainfall, R, and snowfall, S. The measurements were combined using the formula

$$P=R'+0.1*S, \text{ where}$$
$$R' = 1.02 * (R+0.2 \text{ mm}) \text{ for } R>0 \text{ before 1975 and } R'=R \text{ otherwise.}$$

The conversion R' was used to accommodate the results by Struzer (1975), Metcalfe et al. (1997), Mekis and Hogg (1997), Bogdanova and Mestcherskaya (1998), and our understanding of wetting rainfall losses from the old Canadian rain gauge before the mid-1970s. For further discussion of this issue see (Sevruk, 1982; Gray and Male, 1981; Groisman and Easterling, 1994, Metcalfe et al., 1997; and Groisman and Legates, 1995).

4.   Difficulties related to estimation of the shape parameter of the gamma-distribution with limited data are reduced by use of the maximum likelihood estimators. These estimators are considered the best, but for small sample sizes they are biased (Crutcher and Joiner, 1980). Therefore, we considered our estimates valid only for the sample size, N, of more than one hundred rain events. This leaves us with a bias of less than 5% (Crutcher and Joiner, 1980). Usually in humid regions, however, we were able to accumulate samples ten times larger to prevent this problem from affecting our results and conclusions. The following expression for asymptotic variance of the maximum likelihood estimate of the shape parameter, $^\wedge\eta$, was used:

$$\text{Var}(^\wedge\eta) = \eta \ N^{-1}(\eta d^2\Gamma(\eta)/d\eta^2 - 1)^{-1},$$

where N is the sample size and Γ is the gamma function. It was derived from the general formulae for this variance (cf., Kendall and Stuart, 1979). For typical values of the shape parameters that we calculated, an asymptotic standard deviation of our estimates was much less than 4%.

   The only difficulties with parameter estimation in our analyses resulted from problems in the low intensity precipitation data. The maximum likelihood estimators of the two parameter gamma distribution (in our case, a conditional distribution of daily precipitation) are functions of two sufficient statistics:

$$S_1 = \Sigma x_i \qquad \text{and} \qquad S_2 = \Pi x_i,$$

where the first is a sum and the second is a product of all sample values. While S_1 is relatively robust to the high and low precipitation values, S_2 is extremely sensitive to them. We are focused on extreme precipitation and the information carried by S_2 is of interest. However, changes in light precipitation also strongly affect this statistic, so if the number of light precipitation events changes due to factors unrelated to weather, this statistic will be severely contaminated. For example, the change in the threshold of measured precipitation with transition from imperial to metric measurements (Australia, Canada), introduction of wetting corrections to the data in an attempt to measure precipitation "up to the last drop" (Russia), and modification of the gauge to "make it more precise" (Canada, Russia) make the daily precipitation time series inhomogeneous.

By adjusting the Russian and Canadian data, we were able to create homogeneous time series of precipitation totals and some other statistics (e.g., number of days with precipitation above 1 mm). However, the time evolution of the shape parameter of the gamma distribution is strongly affected by the S_2-statistic and was not preserved by these adjustments. Therefore, in our estimates of the shape parameter for the former Soviet Union, we use the precipitation data only after 1967, when the last significant change in the instrumentation was introduced (Groisman et al. 1991). For Canada, we avoid assessment of the changes in the shape parameter after 1975, when the new rain gauge was introduced to the Canadian primary network (Metcalfe et al., 1997; Mekis and Hogg, 1997). In our assessment of the trends in the number of days with precipitation over Australia, Russia, and Canada, we considered only the days with precipitation above 1 mm to eliminate/reduce the contribution of the above mentioned inhomogeneities to our conclusions. For Norway, where precipitation has been consistently measured with a 0.1 mm accuracy, we nevertheless consider only the trends in the number of days with precipitation above 0.2 mm. This was done after we had discovered there a 100% increase in the number of days with precipitation equal to 0.1 mm in the 1930s and then found evidence (Bruun, 1949) that this was not a climate-related change but a result of improved observational diligence.

References

Bagrov, N.A.: 1965, 'On the distribution of monthly precipitation', *Trudy* TsIP (Trans. of the Central Institute for Weather Forecasting), **139**, 3-21.

Baker, C.B., Quayle, R.G., and Wanlin, W.: 1995, 'The influence of night time cloud cover on the observed minimum temperature in China', *Atmos. Res.*, **37** (1-3), 27-35.

Beniston, M., Rebetez, M., Giorgi, F., and Marinucci, M.R.: 1994, 'An analysis of regional climatic change in Switzerland', *Theor. Appl. Clim.*, **49**, 135-159.

Bogdanova, E. G.: 1987, 'On the possibility of quality control and restoring the information about the number of days with precipitation of different intensity', *Transactions of the Main Geophysical Observatory*, Leningrad, **512**, 16-19 (in Russian).

Bogdanova, E.G. and Mestcherskaya, A.V.: 1998, 'Estimation of the effects of wetting losses on the homogeneity of annual precipitation', *Russian Meteorology and Hydrology* (in press).

Bootsma, A.: 1994, 'Long-term (100 yr) climatic trends for agriculture at selected locations in Canada', *Climatic Change*, **26**, 65-88.

Bruun, I.: 1949, 'The precipitation in Norway 1895-1943, Vol. I', The Norwegian Meteorological Institute, Oslo, 114 pp. (In Norwegian with English summary).

Buishand, T.A.: 1978, 'Some remarks on the use of daily rainfall models', J. Hydrol., 36, 295-308.

Crutcher, H.L., McKay, G.F., and Fulbright, D.C.: 1977, 'A note on a gamma distribution computer program and computer produced graphs', NOAA Tech. Rep. EDS 24, Environmental Data Service, NOAA, Dept. Of Commerce, Washington, D.C. 55 pp. + Appendices.

Crutcher, H.L. and Joiner, R.L.: 1980, 'Gamma distribution shape parameter bias', NOAA Tech. Memorandum EDIS 29, U.S. Dept. of Commerce, Washington, D.C., 38 pp.

Easterling, D.R. and Peterson, T.C.: 1995, 'A new method for detection undocumented discontinuities in climatological time series', Internat. J. Climatol., 15, 369-377.

Easterling, D.R.: 1997, 'Data for studying changes in climate extremes over the USA and Mexico', Proc. of GCOS Workshop "Indicators and indices for monitoring trends in climate extremes", Asheville, NC, June 3-6, 1997.

Easterling, D.R., Diaz, H.F., Douglas, A.V., Hogg, W.D., Kunkel, K.E., Rogers, J.C., and Wilkinson, J.F.: 1998, 'Long-term observations for monitoring climate extremes in the Americas', Climatic Change (this issue).

Førland, E. J. and 14 others: 1996, 'Changes in "normal" precipitation in the North Atlantic Region', DNMI Report 31/96 KLIMA. 27 pp.

Fortuniak, K.:1996, 'Stochastic and deterministic aspects of variability of the selected climate elements' (in Polish), PhD Thesis. 166 pp. (Available from The Library of the University of Lodz, ul. Jana Matejki 34/38, 90-237, Lodz, Poland).

Georgievsky, V. Yu., Zhuravin, S.A., and Ezhov, A. V.: 1995, 'Assessment of trends in hydrometeorological situation on the Great Russian Plain under the effect of climate variations', AGU Proc., Fifteen Annual Hydrology Days, April 3-7, 1995, Fort Collins, Colorado, USA, 47-58.

Georgievsky, V.Yu., Ezhov, A.V., Shalygin, A.L., Shiklomanov, I.A., and Shiklomanov, A.I.: 1996, 'Evaluation of possible climate change impact on hydrological regime and water resources of the former USSR rivers', Russian Meteorol. and Hydrol., 1996 (11), 89-99.

Goodison, B.E. and Louie, P.Y.T.: 1986, 'Canadian methods for precipitation measurement and correction', WMO/TD-No. 104. Instruments and observing methods Report No. 25. Papers presented at the Workshop on the Correction of Precipitation Measurements, Zürich, Switzerland, 141-145.

Gray, D.M. and Male, D.H. (Eds.): 1981, 'Handbook of Snow', Pergamon Press, 776 pp.

Groisman P.Ya.: 1991, 'Data on present-day precipitation changes in the extratropical part of the Northern hemisphere', p.297-310 in: Schlesinger M.E. (editor) "Greenhouse - Gas - Induced Climatic Change: A Critical Appraisal of Simulations and Observations". Elsevier, Amsterdam, 615 p.

Groisman, P.Ya., Koknaeva, V.V., Belokrylova, T.A., Karl, T.R.: 1991, 'Overcoming biases of precipitation measurement: a history of the USSR experience', Bull. Amer. Meteorol. Soc., 72, 1725-1733.

Groisman, P.Ya. and Easterling, D.R.: 1994, 'Variability and trends of precipitation and snowfall over the United States and Canada', J. Climate , 7, 184-205.

Groisman, P.Ya. and Legates, D.R.: 1995, 'Documenting and detecting long-term precipitation trends: where we are and what should be done', Climatic Change , 31, 601-622.

Groisman, P.Ya., Karl, T.R., Easterling, D.R., Knight, R.W., and Jamason, P.B.: 1997, 'Changes in the probability of extreme precipitation: Why they can be so prominent in ongoing climatic changes', Paper presented at the GCOS Workshop on Indices and Indicators for Monitoring Trends in Climate Extremes (Asheville, NC, June 3-6, 1997).

Guttman, N.B., Hosking, J.R.M., and Wallis, J. R.: 1993, 'Regional precipitation quantile values for the Continental United States computed from L-moments', J. Climate, 6, 2326-2340.

Hanssen-Bauer, I.: 1994, 'Regional trends in Norwegian precipitation series', In: *"Climate Variations in Europe"*, Proceedings of the European Workshop on Climate Variations held in Kirkkonummi (Majvik), Finland, 15-18 May 1994. Painatuskeskus, Helsinki, 170-182.

Hanssen-Bauer, I. and Førland, E. J.: 1994, 'Homogenizing long Norwegian precipitation series', *J. Climate,* **7,** 1001-1013.

Hanssen-Bauer, I., Førland, E. J., Tveito, O.E., and Nordli, P. Ø.: 1997, 'Estimating regional precipitation trends. Comparison of two methods', *Nordic Hydrology,* **28,** 21-36.

Hennessy, K.J., Gregory, J.M., and Mitchell, J.F.B.: 1997, 'Changes in daily precipitation under enhanced greenhouse conditions', *Clim. Dyn.,* **13**(9), 667-680.

Hennessy, K.J., Suppiah, R., and Page, C.M.: 1998, 'Australian rainfall changes, 1910-1995', *Aust. Meteorol. Mag.,* (in press).

Hopkins, L.C. and Holland, G.J.: 1997, 'Australian heavy-rain days and associated East Coast cyclones: 1958-92', *J. Climate,* **10,** 621-635.

Hughes, P.Y., Mason, E.H., Karl, T.R., and Brower, W.A.:   1992, 'United States Historical Climatology Network Daily Temperature and Precipitation Data', ORNL/CDIAC-50, NDP-042, ESD Publ. No. 3778, Carbon Dioxide Information Data Center, Oak Ridge National Lab., Oak Ridge, TN. 54 pp. + Appendices.

Intergovernmental Panel on Climate Change (IPCC): 1990, *'Climate Change. The IPCC Scientific Assessment'*, Ed. J.T. Houghton, G.J. Jenkins, and J.J. Ephraums, Cambridge University Press N.Y., 362 pp.

Intergovernmental Panel on Climate Change (IPCC): 1996, *'Climate Change 1995: The Science of Climate Change. The Second IPCC Scientific Assessment'*, J.T. Houghton, L.G. Meira Filho, B.A. Callendar, N. Harris, A. Kattenberg, and K. Maskell, Eds., Cambridge University Press N.Y., 572 pp.

Intergovernmental Panel on Climate Change (IPCC): 1998, *'The Regional Impacts of Climate Change'*, R Watson, M. Zinyowera, R. Moss and R. Dokkien, Eds., Cambridge University Press N.Y., 517 pp.

Iwashima, T. and Yamamoto, R.: 1993, 'A statistical analysis of the extreme events: Long-term trend of heavy daily precipitation', *J. Meteorol. Soc. Japan,* **71,** 637-640.

Jones, R.G., Murphy, J.M., Noguer, M., and Keen, A.B.,: 1997, 'Simulation of climate change over Europe using a nested regional-climate model. II. Comparison of driving and regional model responses to a doubling of carbon dioxide', *Quart. J. Roy. Meteorol. Soc.,* **123,** 265-292.

Karl, T.R., Groisman, P.Ya., Knight, R.W., and Heim, R.R. Jr.: 1993, 'Recent variations of snow cover and snowfall in North America and their relation to precipitation and temperature variations', *J. Climate,* **6,** 1327-1344.

Karl, T.R., Knight, R.W., and Plummer, N.: 1995, 'Trends in high-frequency climate variability in the twentieth century', *Nature,* **377,** 217-220.

Karl, T.R., Knight, R.W., Easterling, D.R., and Quayle, R.G.:   1996, 'Indices of climatic change for the United States', *Bull. Amer. Meteorol. Soc.,* **77,** 279-292.

Karl, T.R. and Knight, R.W. 1998: Secular trends of precipitation amount, frequency, and   intensity in the USA. *Bull. Amer. Meteorol. Soc.,* **79,** 231-241.

Katz, R.W.: 1977, 'Precipitation as a chain-dependent process', *J. Appl. Meteorol.,* **16,** 671-676.

Katz, R.W.: 1998, 'Extreme value theory for precipitation: Implications for climate change', AMS Proceedings of the Ninth Symposium on Global Change Studies, Phoenix, Arizona, 11-16 Jan. 1998, Amer. Meteorol. Soc., Boston, Mass., j11-j15.

Katz. R.W. and Brown, B.G.:1992, 'Extreme events in a changing climate: variability is more important than averages', *Climatic Change,* **21,** 289-302.

Kendall, M.G. and Stuart, A.: 1979, 'The Advance Theory of Statistics. Vol. 2, Inference and Relationship', Ch. Griffin and Co., London. 748 pp.

Kozuchowski, K.:1985, 'Variation in precipitation in the years 1881-1980 in Poland' (in Polish with English abstract), *Acta Geographica Lodziensia,* **48,** 158 pp.

Lavery, B. M., Kariko, A. P., and Nicholls, N.: 1992, 'A high-quality historical rainfall data set for Australia', *Aust. Meteorol. Mag.*, **40**, 33-39.

Lavery, B.M., Joung, G., and Nicholls, N.: 1997, 'An extended high-quality historical rainfall dataset for Australia', *Aust. Meteorol. Mag.*, **46**, 27-38.

Lettenmaier, D.P., Wood, E.F., and Wallis, J.R.: 1994, 'Hydro-climatological trends in the continental United States, 1948-88', *J. Climate*, **7**, 586-607.

Mekis, E. and Hogg, W.D.: 1997, 'Rehabilitation and analysis of Canadian daily precipitation time series', Proceedings of the 10th AMS Conference on Applied Climatology, October 20-23, 1997, Reno, Nevada, 300-304.

Metcalfe, J.R., Routledge, B., and Devine, K.: 1997, 'Rainfall measurements in Canada: Changing observational methods and archive adjustment procedures', *J. Climate*, **10**, 92-101.

Mooley, D.,A.: 1973, 'Gamma distribution probability model for Asian summer monsoon monthly rainfall', *Mon. Wea. Rev.*, **101** (2), 160-176.

Plummer, N., Salinger, M.J., Nicholls, N., Suppiah, R., Hennessy, K.J., Leighton, R.M., Trewin, B.C., and Lough, J.M., : 1998, 'Twentieth century trends in climate extreme over the Australian region and New Zealand', *Climatic Change* (this issue).

Plummer, N., Trewin, B.C., Hicks, R., Nicholls, N., Torok, S.J., Lavery, B.M., and Leighton, R.M.: 1997, 'Australian data for documenting changes in climate extremes', Paper presented at the GCOS Workshop on Indices and Indicators for Monitoring Trends in Climate Extremes (Asheville, NC, June 3-6, 1997).

Razuvaev, V.N., Apasova, E.G., and Martuganov, R.A.: 1993, 'Daily Temperature and Precipitation Data for 223 USSR Stations', ORNL/CDIAC-56, NDP-040, ESD Publ. No. 4194, Carbon Dioxide Information Data Center, Oak Ridge National Lab., Oak Ridge, TN. 47 pp. + Appendices.

Richardson, C.W.: 1981, 'Stochastic generation of daily precipitation, temperature, and solar radiation', *Water Resour. Res.*, **17**(1), 182-190.

Ross, R.J. and Elliot, W.P.: 1996, 'Tropospheric water vapor climatology and trends over North America: 1973-93', *J. Climate*, **9**, 3561-3574.

Schaer, C., Frei, C., Lüthi, C., and Davies, H.C.: 1996, 'Surrogate climate change scenarios for regional climate models', *Geophys. Res. Lett.*, **23**, 669-672.

Sevruk, B.: 1982, 'Methods of correction for systematic error in point precipitation measurement for operational use', Operational Hydrology Report No. 21, Publ. 589, World Meteorological Organization, Geneva, Switzerland, 91 pp.

Shver, Ts. A.: 1976, '*Precipitation Over the USSR Territory*,' (in Russian), Gidrometeoizdat, Leningrad, 302 pp.

Suppiah, R. and Hennessy, K.J.: 1996, 'Trends in the intensity and frequency of heavy rainfall in tropical Australia and links with the Southern Oscillation', *Aust. Meteorol. Mag.*, **45**, 1-17.

Suppiah, R. and Hennessy, K.J.: 1998, 'Trends in total rainfall, heavy rainfall events, and number of dry events in Australia, 1910-1990', *Int. J. Climatol.*, (in press).

Struzer, L.R.: 1975, 'On the changes of the long-term precipitation values with the transfer to 6-hourly observations', *Transactions of the Main Geophysical Observatory*, **341**, 65-72.

Thom, H.C.S.: 1951, 'A frequency distribution for precipitation (abstract)', *Bull. Amer. Meteorol. Soc.*, **32** (10), 397.

Thom, H.C.S.: 1958, 'A note on the gamma distribution', *Mon. Wea. Rev.*, **86** (4), 117-122.

Tsonis, A.A.: 1996, 'Widespread increases in multi-decadal variability of precipitation over the last century', *Nature*, **382**, 700-702.

Vinnikov, K.Ya., Groisman, P.Ya., and Lugina, K.M.: 1990, 'The empirical data on modern global climate changes (temperature and precipitation)', *J. Climate*, **3**, 662-677.

Ye, P.Z. et al.: 1996, 'Study on patterns and causes of draught and flood in the Yangtze and Yellow River Valley', Shandong Sci. and Tech. Press, 387 pp.

(Received 5 November 1997; in revised form 28 August 1998)

LONG-TERM OBSERVATIONS FOR MONITORING EXTREMES
IN THE AMERICAS

DAVID R. EASTERLING[1], HENRY F. DIAZ[2], ARTHUR V. DOUGLAS[3],
WILLIAM D. HOGG[4], KENNETH E. KUNKEL[5],
JEFFRY C. ROGERS[6], and JAIME F. WILKINSON[7]

[1] *NOAA/NESDIS/National Climatic Data Center, Asheville, NC, U.S.A*
[2] *NOAA/ERL/Climate Diagnostics Center, Boulder, CO, USA*
[3] *Department of Atmospheric Science, Creighton University, Omaha, NE, U.S.A*
[4] *Climate Research Branch, Atmospheric Environment Service, Downsview, Ontario, Canada*
[5] *Midwestern Climate Center, Illinois State Water Survey, Champaign, IL, U.S.A*
[6] *Department of Geography, Ohio State University, Columbus, OH, U.S.A*
[7] *Servicio Meteorologico Nacional, Buenos Aires, Argentina*

Abstract. In this paper we discuss data available from Argentina, Mexico, the United States and Canada that are suitable for the analysis of extreme temperature and precipitation events. We also discuss some of the problems of homogeneity and quality control that can potentially affect the observation of extreme values. It is clear that even in countries like the United States and Canada that have a potentially rich source of climate data there are still problems in obtaining homogeneous data necessary to perform thorough studies of time varying changes in extreme events. These types of problems may be compounded if data are needed from countries that do not have a tradition of maintaining large climate archives or observing networks.

We also provide two examples of analyses that can be performed with these types of data: (1) the development of climate extremes indices for Canada, and (2) analyses of freezing events for Florida and the effect on the citrus industry. The Canadian example provides a good basis for countries to take indices developed for one country or region and through minor modifications make the index relevant to their own needs. The analysis of freezing events in Florida is a timely example of how extreme events have both ecological and societal impact.

1. Introduction

Observational studies into possible greenhouse induced climate change have traditionally focused on changes in mean and variance statistics (e.g. Easterling et al., 1997). However, recent events in many parts of the world have underscored the need to examine fluctuations and changes in extreme events. Although potential changes in long-term means are important from a number of standpoints, extreme events usually have the greatest and most immediate societal impact (Berz, 1997). Because of the high human and monetary costs often associated with extreme weather events many parts of society have become increasingly

Climatic Change **42**: 285–308, 1999.
©1999 *Kluwer Academic Publishers. Printed in the Netherlands.*

concerned about extreme events and their possible consequences. Some climate modeling studies involving enhanced greenhouse gases have suggested that if the climate changes over the next century, these changes will result in increases in extreme events, particularly increases in extreme temperature and precipitation events (Nicholls, 1995; Karl and Knight, 1998).

Monitoring the climate for changes is a difficult task. Often, the long-term observations available in computer-compatible form are taken from meteorological networks that usually were originally designed to observe the weather for the development of forecasts. Furthermore, the time-scale of the observations is usually monthly, seasonal, or annual averages. These types of data are useful for examining extremes such as prolonged, wide-spread droughts, or large-scale temperature anomalies (Karl et al., 1993) and developing climate indices. Short-term extreme events, such as heat or cold waves require high temporal resolution observations, such as daily maximum and minimum temperature or precipitation. In this paper we discuss a number of types of data suitable for examining climate extremes in both North and South America, and provide some examples of analyses. Nicholls (1995) provides a discussion of various types of extreme events, such as tropical and extratropical cyclones, and some data useful for their analysis. Here we focus solely on temperature and precipitation with a high temporal resolution. In the second section some data issues that can affect the results of analyses are discussed, in the next section we consider climate networks and data for the Mexico, the U.S., Canada, and Argentina. The fourth section provides two analyses: the development of climate extremes indices for Canada, and extremes in precipitation and freeze events in the southern U.S., and the last section is a brief summary and conclusions.

2. Data Problems Affecting the Analysis of Extremes

Often the analysis of extremes involves examining the tails of a statistical distribution. In other words, we are examining very high or very low temperatures, or very heavy precipitation totals or, as in the case of drought, zero values. There are a number of data problems and processing procedures that can potentially affect values in the tails of a distribution, or can affect the means and variances. In this section we discuss problems of homogeneity of the data, and quality control procedures and how these problems can affect observed extremes.

Although many problems of homogeneity may not have a large effect on the analysis of some types of extreme events there can be certain problems and certain types of extremes analyses that may have a noticeable effect. Instrument changes is one example. Record temperature observations have been known to be affected by a change in instrumentation. For example, in the U.S. First-order network the primary temperature sensor starting in the mid-1980s through to the installation of the Automated Surface Observing System (ASOS) has been the HO-83 Hygrothermometer. This is an aspirated thermistor-based system that includes both temperature and dewpoint temperature sensors. One instance where the installation of this system caused suspiciously high temperature readings was at the NWS First-order station at Tucson International Airport. A number of record high temperatures observed in the early 1990s lead to a number of studies that concluded that many of these records were likely due to inaccurate readings from the HO-83 (Gall et al., 1992). During the 1980s nearly half of the stations in the U.S. Cooperative network had their liquid-in-glass thermometers replaced with the Maximum/Minimum Temperature System (MMTS), an unaspirated electronic temperature system. Quayle et al. (1991) showed that, on average, the introduction of the MMTS reduced monthly maximum temperatures and increased monthly minimum temperatures, which would likely also have an affect on observed daily extremes.

Other inhomogeneities may also affect the temporal distribution of observed extremes. Station moves may cause an artificial change in observed extremes, particularly in areas of heavy terrain. Changing the siting of an instrument shelter, say from the side of a hill to the bottom, can easily cause more cold extremes to be recorded in minimum temperatures. The same can be said for extreme warm observations, if a shelter is moved from a shaded area into an area of constant sunshine.

Precipitation measurements are particularly susceptible to inhomogeneities that may affect extremes. Two main problems with precipitation measures are: (1) gauge undercatch in windy conditions, and (2) the use of gauges with different measurement mechanisms, particularly the use of tipping-bucket raingauges. Gauge undercatch is a problem with both solid and liquid precipitation (Sevruk, 1982; Groisman and Easterling, 1994). Wind-induced turbulence over the gauge orifice can result in much of the precipitation blowing over the gauge rather than settling into it. This is particularly a problem with snow and light rain where undercatch can be as much as 50% or more. However it can also be a problem with heavier precipitation if there are exceptionally windy conditions. One way of dealing with gauge undercatch is to install a wind shield on the gauge, which can dramatically increase the amount of precipitation caught by the gauge. However, this introduces a large discontinuity in the record that may have a large effect on the incidence of extreme precipitation events. The use of tipping-bucket raingauges also can affect the observation of extreme precipitation events. Tipping-bucket gauges are well-known to undercatch precipitation during heavy rainfall events when the rain rate exceeds the capability of the tipping mechanism to keep up with the water flowing through the gauge (Sevruk, 1982).

There are a number of methods available for adjusting data for inhomogeneities (Alexandersson, 1986; Karl and Williams, 1987; Easterling and Peterson, 1995; Vincent, 1997). These methods are appropriate for use with lower temporal resolution data (e.g. monthly, seasonal, annual) when there is a high-enough spatial resolution in the station network to allow statistically robust comparisons between a station and its neighbors. These methods are designed to adjust for random station changes, such as station moves, and work well with variables that are spatially conservative, such as temperature. However, for parameters such as precipitation, that are much more variable over space, these methods are often not appropriate and another approach is necessary. In both Canada and the United States adjustments for known measurement program-induced biases (e.g. network-wide installation of wind-shields) have been applied. This eliminates network-wide or regional biases, discontinuities, artificial trends and other inhomogeneities but leaves inhomogeneities in individual station time series as induced by local conditions such as location and exposure changes. The assumption is that these local changes will be uncorrelated in space. Thus, individual station time series of precipitation are unsuitable for trend analysis, and may in extreme cases induce artificial trends on the regional time series. But in general regionalized, mapped or gridded time series can be analyzed confidently, provided conditions of spatial coherence are met. Inherent in this approach is the conservative assumption that the time series parameter of interest (trend, change in variability or extremes, etc.) is more spatially consistent than the parameter itself.

Quality control procedures, in particular outlier checks, can also affect the incidence of extreme values in a dataset. For example, quality-control of the data for Las Cruces, NM in the National Climatic Data Center's TD-3200 dataset of daily observations resulted in the record one-day precipitation amount being excluded from the final dataset. This value, of greater than 150 mm was due to a localized, slow moving storm, and has been verified using newspaper accounts, and a New Mexico State University report (K. Kunkel, personal communication). Quality control of extreme precipitation values is particularly a problem in arid and semi-arid areas, such as New Mexico, where localized convective storms can often give large amounts of precipitation at one observing station and a nearby station receives nothing. Since many QC procedures are based on comparisons with nearby stations (e.g. Reek et al., 1992), legitimate extremes may be excluded, as in the Las Cruces case.

3. Climate Networks and Data

3.1. MEXICO

Mexico has had a long history of climate data collection. Near the turn of the Century the Mexican Meteorological Service (Servicio Meteorologico Nacional or SMN) established observing stations in primary towns and state capitals.

From this original group of stations, the SMN developed a nationwide climate observing system of more than 2,000 stations during the 1920s and 1930s. In the early 1940s the newly created secretariat of Hydrology (Secretaria de Recursos Hidraulicos or SRH) established a rural network of stations, bringing the nationwide number of observing sites (SMN and SRH) to over 5,000 by the early 1950s. This system remained in place until the early 1980s when government cutbacks resulted in suspension of payment to observers. Additionally, Mexican agencies became decentralization under the austerity programs of the 1980s and this lead to data collection efforts being shifted from the central archives to state agencies. Unfortunately, a considerable amount of data is not being received from these state agencies and the effective Mexican observing network appears to have been cut by 40% of the possible 5,000 stations in operation in the early 1980s.

The climate observing system in Mexico was standardized by the SMN in the early 1900s. The same procedures are followed by the SRH. Sites are selected by the government agency and observers are trained to take observations at the site. Compared to the actual site locations, there are apparently very few moves as the government finds new observers to maintain the exact site that has been established. Each station includes maximum and minimum thermometers in a standard shelter. Standard 10 inch rain gages, pan evaporation equipment and wind vanes are installed at the sites. The government agencies require that all observations be taken at 8 A.M. Local Standard Time.

The data set currently available through the National Climatic Data Center (NCDC) was collected through the assistance of Ing Enrique Ortega and Ing. Carlos Espinosa, past directors of the SMN. Long term data sets contain both SMN and SRH stations. The SMN has been responsible for the archiving of both the SMN and SRH data sets. Ing. Othon Cervantes, head of the digital climate archives provided technical assistance in the copying of the data sets. The daily data sets were key entered by a contractor in Mexico City. Quality control of the data was applied using WMO provided CLICOM software. The key entered data came from original paper copy and often the original data were hand written. In the southern two-thirds of the country, the key entered data sets contain data from the date when the station was first established. In some cases, original paper copy was not found for the early periods (especially the 1920s and early 1930s). Apparently funds on the project began to run out as the project began to work on state data from peripheral sections of the country (far east and north). Thus, key entry for far eastern and northern states was restricted to the period 1961-1990. While 1990 was the cutoff date for all key entry, as noted above, data receipt had become a problem beginning in the mid-1980s. Consequently, the data sets frequently end earlier than 1990, with most stations having end dates between 1984 and 1988. At present the SMN is updating the files, but this project will not be completed for a number of years. Additionally, data reception from the state capitals continues to be erratic, so data sets after the mid-1980s are likely to have gaps. Due to these problems, the SRH has developed a daily set of rainfall data that are collected each day. This data set of about 700 stations is primarily composed of SRH stations and some SMN observatories and is available for the period 1991 to June 1997.

3.2. THE UNITED STATES

The U.S. Cooperative Observer Network has been in operation since the late 19th Century. Although most of the daily observations have been published in Climatological Data since then, very little of the data obtained prior to 1948 has been put into a digital electronic format. Thus, it is usually not practical to include these pre-1948 data into climate trend analyses requiring daily data. This often severely limits the value of analyses of trends in extremes since there were important fluctuations in global mean temperature during the 50 years prior to 1948 that likely were accompanied by changes in extremes (Nichols et al., 1995). Two recent efforts were undertaken to expand the availability of pre-1948 data in digital electronic format.

In the early 1990s, the NCDC initiated and coordinated a project to key daily climate data for selected stations with very long records approaching 100 years. This project, intended to celebrate the 100th anniversary of the cooperative network, included the participation of many state climatologists and resulted in the keying of period of record daily climate data for several hundred long term stations. Unfortunately, these data were keyed in a variety of formats. Because of lack of resources, they have not been translated into a common standard format and, for this reason, they remain unavailable to the general scientific community for use in studies of climate extremes.

A second project was initiated by the Midwestern Climate Center in 1994. The purpose of this project was to key all daily climate data that were published in Climatological Data for the period 1896-1947 for nine Midwestern states: Illinois, Indiana, Iowa, Kentucky, Michigan, Minnesota, Missouri, Ohio, and Wisconsin. All nine state climatologists collaborated closely and this project was successfully completed in 1997.

The data were keyed directly from the publication Climatological Data. The keying of each state's data was performed at the state climatologist's home institution under the supervision of the state climatologist. Data files were sent to the Midwestern Climate Center where a set of quality control checks were applied uniformly to the entire dataset. These quality control procedures identified questionable values. These were inspected by the state climatologist who made a judgement about the validity of the values.

The key characteristics of the resulting data set are as follows:

- The climatic elements were daily precipitation, maximum temperature, and minimum temperature.
- Altogether, more than 2×10^7 values were keyed.
- The number of stations with 90 or more years of data in digital form increased from 105 to 299 for precipitation and from 103 to 221 for temperature.
- The total amount of pre-1948 data increased by about a factor of 3.
- Perhaps more important than the absolute increase in the amount of pre-1948 data, the spatial uniformity improved substantially. Prior to this project, most of the pre-1948 digital data was concentrated in a few states, notably Illinois, Indiana, and Iowa. With the addition of these data, the spatial distribution of stations is quite uniform throughout the period.

With respect to short-duration extremes, it is now possible to perform spatially detailed studies for the Midwest for periods of 100+ years. For example, we have analyzed secular trends in the frequency of heavy multi-day precipitation events in the Midwest for 1896-1995. Specifically, we looked at 7-day precipitation events exceeding the station-specific threshold for a 1-year recurrence. A Midwest composite was constructed by arithmetically averaging the values for all long-term stations in a climate division. The climate division averages were then combined through areal-weighting. This was possible because all climate divisions in these nine states have at least one long-term station; prior to this project many climate divisions did not have a long-term station. This analysis, shown in Figure 1, indicates that the last 15 years have been characterized by a very high frequency of heavy precipitation events, higher on average than other decadal or longer period of the last 100 years. An examination of only the last 50 years of record indicates a pronounced upward trend. However, there is also evidence of variability on longer time scales. Although most of the century is characterized by a lower frequency than the last 10-15 years, the frequency of heavy events was almost as high during the period of 1896-1906. The addition of the pre-1948 data provides an informative and revealing perspective on the recent trends.

3.3. CANADA

The Canadian Meteorological Service was established in 1871, and at that time there were 126 observing stations. Currently, there are three main types of stations: daily climatological stations, principal synoptic and hourly stations, and supplementary network stations. The data with the longest period-of-record are the daily climatological stations, which mainly take observations of temperature and precipitation. These stations are operated mainly by volunteer observers, and

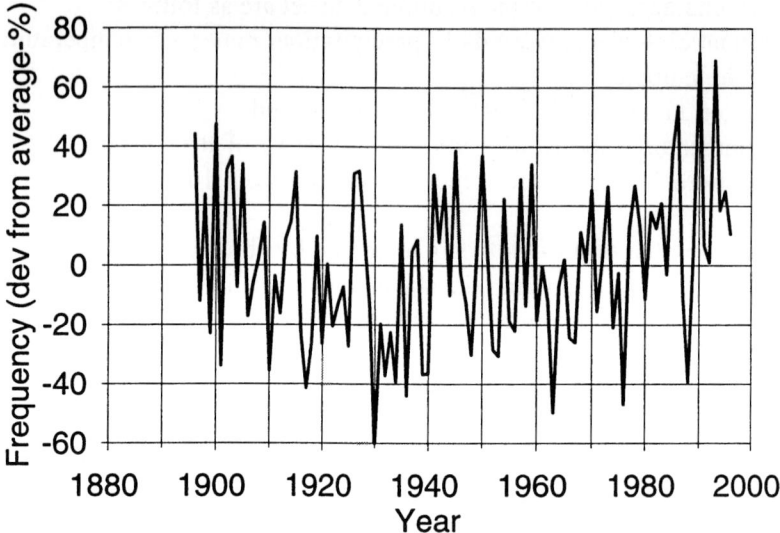

Figure 1: The frequency of heavy precipitation events (expressed as a deviation from the 1896-1996 period) for the Midwest U.S. states of Illinois, Indiana, Iowa, Kentucky, Michigan, Minnesota, Missouri, Ohio, and Wisconsin. At individual stations, events are defined as 7-day periods with precipitation totals exceeding the threshold for a 1-yr recurrence.

closely follow the spatial distribution of population with most of the stations in the southern part of Canada. Currently there are 2100 active observing sites in the network. Digital data for this network is generally available back to the early part of the twentieth century, and the station with the longest period of operation began in 1840 in downtown Toronto. Observations of temperature are taken once a day using liquid-in-glass max/min thermometers sheltered in a Stevenson screen (Stirling, 1994). Precipitation measurements are taken using two gauges: the Canadian Type-B 11.28 cm raingauge for liquid precipitation and the Nipher Snow Gauge system (at the 124 synoptic stations) or snow-ruler measurements with a 10:1 assumed ratio for solid precipitation (Metcalfe, et al., 1997).

There are a total of 738 principal and hourly stations, most of which also include daily observations. Nearly half of these stations, however, are automated observing sites. The spatial distribution of the network shows that most of the stations are located in southern Canada.

3.4. ARGENTINA

The Argentine meteorological network covers the southern tip of South America from 22°S to 78°S and from 34°W to 73°W, and is surrounded by the South Atlantic, Chile, the South Pacific and the Antarctic Oceans and the Wedell Sea. The Andes Mountain range runs along the western border of Argentina and acts

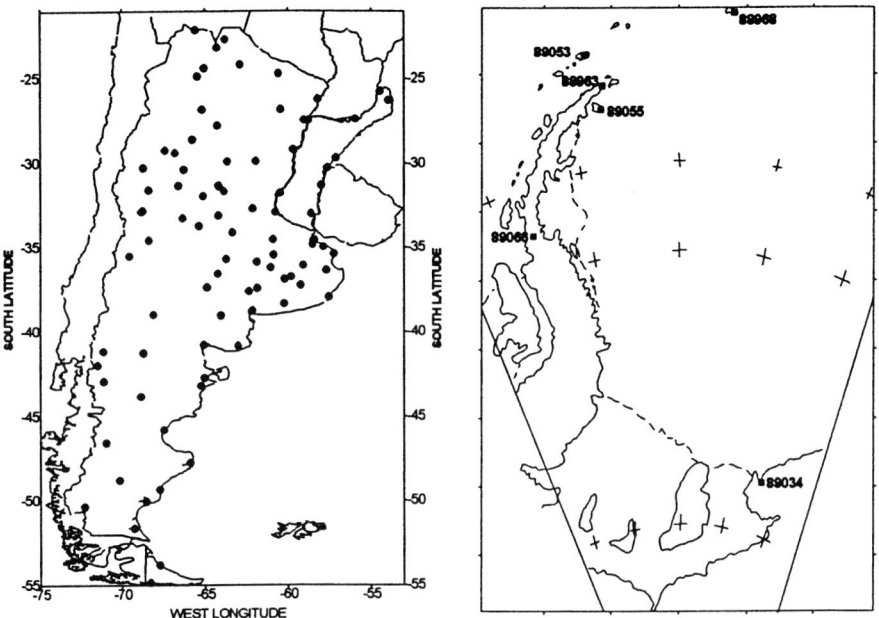

Figure 2: Argentinean CLIMAT network for (a) continental Argentina, and (b) the adjacent sub-antarctic region and in the Antarctic.

as a barrier to the mixture of air masses from the Pacific Ocean with those of Atlantic or Amazonian origin. As a consequence, there exists quite a wide variety of climates in Argentina: from tropical to polar ones and from dry to wet climates. One can observe strong and dry westerlies south of 40°S and wet northerlies in the central and northern regions. So, an appropriate network is necessary to study climates.

The Argentine CLIMAT network is composed of 88 stations in the continent and 6 additional stations in the adjacent sub-antarctic and antarctic regions (Figures 2a and 2b). This network can be complemented by additional synoptic stations that enlarge it to about 150 surface stations. However, in order to study climate changes or variability, the following three criteria have to be applied for the selection of stations:

(1)   continuity of the meteorological record in the same station location;
(2)   no significant changes in the environment of the station since its establishment;
(3)   temperature records not affected by the change of location of the station.

Using the resulting network, Hoffmann et al. (1997) have found a significant temperature increase over the southern tip of Argentina, south of 50°S, and in the adjacent sub-antarctic region. This increase is as much as +1.2°C in Rio Gallegos (southern tip of Argentina) and +1.5°C in Base Orcadas (adjacent sub-Antarctic region) when comparing the decades of 1981-1990 with the decade of 1941-1950 (the warmest decade of the first half of the present century in the Argentine Pampa and Chaco Plains). In the case of Base Orcadas station, the warming started in the 1930s resulting in a mean temperature increase of +2.1°C between the decades of 1931-1940 and 1981-1990.

Lastly, Table I summarizes the meteorological data available that could be used for climatic studies. The Argentine Servico Meteoroligica National, Argentina is conscious of the impacts of climate and its variability on societies and on the local and regional economy. As a consequence, it is making great efforts in order to improve the availability of climatological data and products for the international scientific community.

4. Example Analyses for Canada and the United States

In the following sections the analysis of extremes is illustrated in two ways: first by the development of climate extremes indices focusing on Canada, and secondly, through examination of frequencies of freezing events affecting fruit growing regions in the southern United States.

4.1. CLIMATE EXTREMES INDICES IN NORTHERN CLIMATES: THE CANADIAN EXPERIENCE

In this section we discuss the development of climate extremes indices, particularly for high latitude areas such as Canada. These indices are based on work performed by Karl et al. (1996) who developed two indices for the United States: a Greenhouse Climate Response Index (GCRI) and a Climate Extremes Index (CEI). The CEI has the most relevance to the current paper, and for the U.S. is comprised of the following five factors based on the percentage of area of the conterminous U.S.:

(1)   with maximum temperatures much below or much above normal,
(2)   with minimum temperatures much below or much above normal,
(3)   in severe drought or severe moisture surplus,
(4)   twice the percent area of the U.S. with a much greater than normal propor
       tion of precipitation derived from extreme (more than 50.8 mm) 24-hour
       precipitation events.
(5)   with a much greater than normal number of days with precipitation and
       days without precipitation.

Data Type	Period	Number of Stations	Storage Media
surface meterological	1950-1994: 4 obs/day, some incomplete, 1995-1997: 24 obs/day	250 synoptic and climatological surface stations; 150 synoptic stations	magnetic tape
daily precipitation	1933-1994, validated	1000 stations	magnetic tape
radiosonde	1958-1990, standard levels	10 stations	magnetic tape
surface climatological, monthly means	1961-1990	150 stations	magnetic tape
surface climatological, monthly means	1916-1958	80 stations	magnetic tape
surface climatological	beginning POR to 1990.	250 stations	microfiche
synoptic description and extremes	from Sept. 1995		magnetic tape
monthly climatological bulletin	from 1991		paper
climate forecast	from 1995		paper
surface synoptic maps	1980-1991		CD-ROM

TABLE I: Summary of data available from the Servicio Meteorologico Nacional, Argentina.

For the U.S. since 1910 this index (see Karl et al., 1996, Figure 13) has large decadal fluctuations, but since the mid-1970s the CEI has been consistently higher than the average, indicating that the area of the U.S. under some form of climate extreme has increased about 1.5% .

In the development of these types of indices it is highly desirable to use data that have been adjusted for inhomogeneities. Homogeneous data are being used in the production of a gridded time series for Canada of monthly temperature and precipitation for this century, using an approach described by Hogg et al. (1996). Isotropic and homogeneous monthly anomalies are interpolated to a 50 km grid and added to a statistical model of normal temperature and precipitation for the grid. The grid values will then be used to estimate other water balance parameters (evapotranspiration, runoff, Palmer Drought Severity Index, etc.) for the grid. Provided the parameter fields are isotropic and homogeneous, other statistics for the 500-600 stations could also be mapped to the same grid. Computation of time series of the area of Canada affected by various extreme conditions will be performed using this gridded dataset.

4.1.1. *Desirable Characteristics*

The ideal extremes index will have a number of desirable characteristics. It will improve the signal to noise ratio for trend. It will be associated with significant economic or social impacts of climate. It will be sensitive to changes associated with greenhouse gas warming. It must be easily mapped. It should be logically definable and meaningful everywhere. That is, in most cases thresholds etc. should be defined by the statistics of the climate parameter (e.g. percentiles) and not by some arbitrary constant which will have varying significance as an extreme event, as the climate varies. This is illustrated by extreme rainfall in Canada, where 25 mm rainfall in a day occurs regularly every year in the south but only once every 10 years in the north. There are, however, instances where absolute thresholds are useful, such as freezing and agriculture. Finally, the ideal extremes index should be compatible with both historical and GCM data.

There are at least two ways to characterize the degree of exposure to extreme events: either the frequency that a threshold is exceeded in a fixed time period, or the magnitude of a low probability occurrence as calculated using data from a fixed time period. These statistics can be calculated from the original hourly or daily data or by applying extreme value sampling theory and using only the extremes themselves for some limited time moving window.

4.1.2. *Important Extremes in Northern Climates*

Extremes of climate parameters which are socially and economically important in high latitudes consist of most of the ones important in temperate latitudes plus a few additions. High winds, intense rainfall and persistent drought conditions are universally important. In addition, intense snow storms, extreme seasonal snow accumulations and rapid melt events (rain plus high temperatures) are critical in northern climates. Extremes indices for all of these climate parameters can be developed from simple temperature and precipitation data and some examples have been generated by Karl et al. (1996). Severe thunderstorm conditions and associated lightning, hail, intense short duration rainfall, damaging winds and tornadoes are equally important to society but cannot be characterized by the basic datasets of daily data available most readily on a global basis. The same is true for economically and socially debilitating yet complex winter phenomena like freezing rain, severe wind chill and blizzards.

By now, the consensus GCM global warming predictions of increased temperatures and intensified hydrologic cycle in high latitudes are familiar (e.g. IPCC, 1995). Using output from the Canadian Climate Center coupled ocean/atmosphere GCM, Zwiers (1997) has shown that 100-year return period temperatures and daily precipitation rates are projected to increase, while findings for extreme winds were inconclusive. Combining this information with knowledge of important climate impacts in northern regions means that the most significant changes in historical climate extremes should be found in parameters like higher springtime temperatures, faster or earlier snowmelt, more intense rain and snow storms, stronger synoptic scale winds, and more severe droughts. These parameters will be used to examine the behavior of some potentially useful indices for climate extremes in northern latitudes.

4.1.3. *Rainfall Indices*

Indices for extremes of seasonal or longer accumulations and based upon area affected by accumulations in some top or bottom percentile (Karl et al., 1996) appear to make good sense for quantifying the impact of extreme conditions. In northern climates, it is logical to separately track rainfall and snowfall because of the implications for stored water, transportation, etc., and to isolate data by season because of expected differences in seasonal responses to greenhouse forcing. Trends in the final index may or may not be sensitive to the threshold percentile selected, but a threshold which identifies only a few events per season is desirable.

For tracking intense daily events, Karl et al. (1996) postulated that the fraction of annual rainfall occurring in extreme events is increasing. He selected a threshold of 2 inches (50.8 mm) and examined the percentage of the United States with greater than normal proportion of precipitation derived from events exceeding this threshold. Mekis and Hogg (1997) attempted to examine the same ratio for 65 Canadian precipitation stations that have been adjusted for inhomogeneities, but quickly identified difficulty in using the same threshold in higher latitude areas. The mean intensity of extreme events decreases rapidly with latitude above 50 degrees and for large portions of the country a 50 mm event is extremely rare. Since a different threshold was required, 25 mm was selected as a compromise for the whole country but a threshold based upon the statistics of extreme rain events at each station (e.g. the normal 95 percentile) is a better alternative. Contrary to Karl et al. (1996) for the United States, the proportion of rain falling in events exceeding 25 mm has decreased between 1910 and 1995 at the majority of the 65 Canadian stations analyzed.

Some work has been done to identify the most useful single station statistic to use as an index of extreme rainfall. Time series of the annual mean, standard deviation, coefficient of variation (standard deviation/mean) and Karl's ratio (the ratio of rainfall in events exceeding 25 mm/day over annual precipitation) for annual extreme daily rain at Medicine Hat, Alberta (-0.00003/yr), an average negative trend, are shown in Figure 3. Figure 4 shows the annual extreme 1-day rain event for Medicine Hat along with an estimate of the 100-year

Medicine Hat Daily Rain

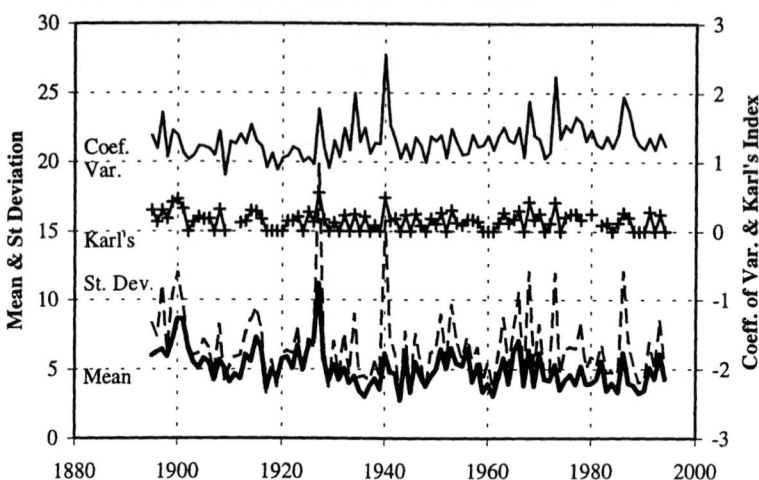

Figure 3: Mean, standard deviation (dashed line), coefficient of variation and Karl's index for Medicine Hat, Alberta daily rain, computed annually. Karl's index is the ratio of rainfall in events exceeding 25 mm/day over annual rainfall.

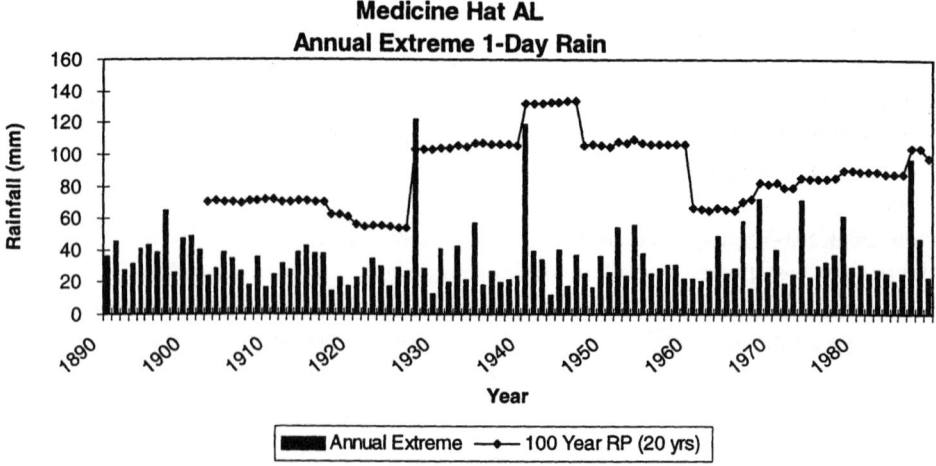

Figure 4: Annual extremes of daily rainfall for Medicine Hat, Alberta, along with Gumbel estimates of 100-year return period amount based upon a 20-year moving window.

Harrow Daily Snow

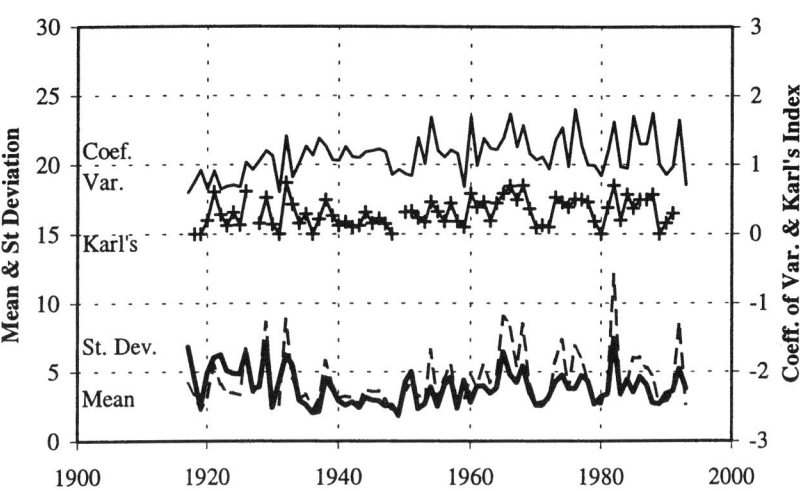

Figure 5: Same as Figure 3 except statistics are for Harrow, Ontario snowfall. Karl's index is the ratio of snowfall in events >10 mm/day over annual snowfall.

return period value based upon a moving 20-year extreme value analysis (Gumbel, 1958). Extreme thresholds can be envisioned for all curves in Figure 3 except the mean, which would identify virtually the same years with extreme events in Figure 4. There seems little reason to show preference for any of these statistics as an indicator of climate. The coefficient of variation is preferred at this stage, in the untested hope that it will be more easily mapped in data sparse regions.

The number of stations required to reasonably determine the area affected by significantly above normal values of these statistics, may make use of this rainfall index difficult, or even impractical, in Canada. The time series signal is very noisy, which suggests it may be too noisy in space to permit accurate inter-polation and area determination in data sparse regions. The work has not been done to verify this.

As Karl et al. (1996) implies, some sort of threshold exceedence may be more revealing than a linear trend computation. The WMO-CLIVAR-GCOS Asheville Workshop on Indices and Indicators for Extremes concluded that a time series of the simplest conceivable extremes index, a count of events exceeding a percentile defined threshold, would ensure the largest possible availability of data globally and still provide valuable information on variability and trends in extremes. Such an index is easy to implement for individual stations but methods to sensibly integrate them areally or regionally require further development.

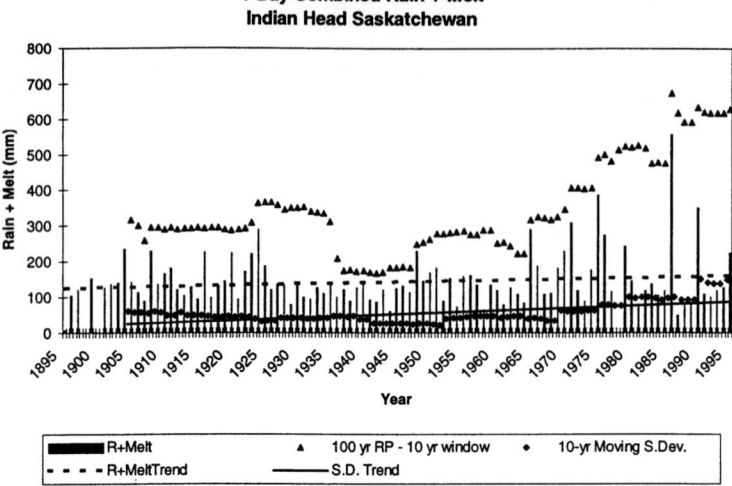

Figure 6: Annual maximum of daily combined rain + snowmelt for Indian Head, Saskatchewan. The 100-year return period values are calculated from a 10 year moving window, plotted at the end of the period and are sensitive to the standard deviation of data from those 10 years.

4.1.4. *Snow and Snowmelt*

Figure 5 shows the same statistics for daily snowfall at Harrow, Ontario. The threshold for Karl's ratio for snowfall was selected at 10 mm per day for this exercise but a percentile for each station would be more appropriate. The chosen index for extreme snow events should be similar to the one for rain events. As previously mentioned, an index for extremes of accumulated snowfall could take the same form as the annual rain index.

For many northern rivers, the critical water input to floods comes from snowmelt. Extreme snowmelt floods are almost invariably combined events involving a large snow pack, high temperatures for rapid melting, and significant rain to increase melt and add water to the runoff. The spring 1997 floods across the northern Great Plains are an example of such conditions. There are ways to historically track such conditions using only daily temperature and precipitation data. Louie and Hogg (1980) describe a technique employed widely in Canada over the last 20 years to assess the probability of combined rain and snowmelt events. For each year of data, a synthetic snowpack is built using snowfall information and depleted using temperature data and a degree day melt model. Rainfall is added to melt to create a database of rain+melt. Figure 6 shows the annual maximum of this value for 110 years of data at Indian Head, Saskatchewan. The linear trend is slightly downward but the 10-year moving standard deviation increases from the 1950s onward. The standard deviation dominates return period calculations. Unfortunately, this is a very noisy parameter in time and space because of the combined ingredients, and mapping and areal calculation of the index is very difficult. Further effort is required to determine its usefulness in an extremes index.

4.1.5. *Temperature*
In a northern country, two types of temperature extremes are most significant: extreme warmth in winter and spring to melt snow and ice and extreme cold in fall, winter and spring to damage crops and even endanger life. Extreme warmth in summer can occasionally cause difficulties in southern portions of the country but is not a nationwide problem and observed trends in summer temperature are small. However, GCMs do predict an increase in annual maximum temperature extremes (Zwiers, 1994). Therefore, in Canada, it is proposed that the gridded time series of monthly temperatures will be used to determine areas exceeding the 90th percentiles for above and below normal, as suggested by Karl et al. (1996), on a seasonal basis. Extreme daily temperatures will be considered for an index in the cold seasons only.

4.1.6. *Drought and Excess Moisture*
Drought will also be assessed following the basic approach suggested by Karl et al. (1996) using the Canadian gridded time series of monthly temperature and precipitation. Values of the Palmer Drought Severity Index (PDSI) will be estimated at each grid square and areas in the 90th percentile of drought, tabulated. Karl et al (1996). recommend the PDSI for moisture excess as well, which could also be calculated on the Canadian gridded data. Other simple water balance schemes are being evaluated and will be considered for use in a Canadian extremes index.

4.1.7. *A Combined Extremes Index*
Two extremes indices are proposed for land areas in high latitudes: a general index, sensitive to all-season and annual impacts and a cold season index emphasizing winter impacts. In each, the goal is to evenly weight critical extremes based on both global change expectations and important impacts. The two proposed combined indices are composed of the following:

Annual
- (1)   % area much above normal precipitation (highest 10%)
- (2)   % area much above and much below normal average temperature (highest & lowest 10%)
- (3)   % area in extreme drought (highest 10% PDSI)
- (4)   % area with yearly number of extreme temperature or precipitation events in highest 10%
- (5)   Wind.

Cold Season
    (1)   % area much above and much below normal snowfall (highest &
               lowest 10%)
    (2)   % area much above and much below normal average tempera-
               ture in spring (highest & lowest 10%)
    (3)   % area much above and much below normal cold season water
               budget (highest & lowest 10%)
    (4)   % area with yearly number of extreme temperature or precipita-
               tion events in highest 10%
    (5)   Wind.

In the Annual Combined Extremes Index (ACEI), only above normal precipitation is included because low precipitation is dealt with more effectively in the
PDSI computation. The coefficient of variation of daily rainfall is included to
reflect areas affected by intense short rain events in the yet to be proven hope
that it will be more spatially conservative than proportion of rainfall occurring in
extreme events.

    In the Cold Season Combined Extremes Index (CSCEI), both above and
below normal snowfall are included because of the importance of both too much
and too little snow while the cold season water budget includes the combined
effects of rain and snow. It is anticipated that the yearly coefficient of variation
of daily snowfall will reflect exposure to extreme snow storms. The temperature
index contribution reflects risk of severe melt events.

    Canadian experience has shown that high latitude countries both share sen-
sitivities to universally important climate extremes and have unique sensitivities.
Recommendations for both an Annual Combined Extremes Index and a Cold
Season Combined Extremes Index are made to reflect these sensitivities while
still being compatible with readily available data.

4.2. CHANGES IN COLD WAVE FREQUENCIES AND INTENSITIES IN THE UNITED STATES

Cold waves have long attracted the attention of weather forecasters and climatologists (Garriott, 1906; Wendland, 1987) due to their numerous economic and
societal impacts and hazards. Cold waves can be defined in terms of (1) severity
or rapidity of the drop in air temperature (Wendland, 1987), (2) the numerical
values of the low temperatures or temperature departures that ultimately occur,
and (3) the duration of the cold wave (varying time spans can be used). To a
lesser extent, cold waves can be characterized by the intensity of the polar anticyclone that accompanies it (Cox, 1916; Rogers and Rohli, 1991). Important
questions are whether cold waves tend to cluster in time, whether their frequen-

cies exhibit interdecadal variability, and how their frequencies may change in the future.

One societal impact associated with severe cold waves in the southeastern United States is the destruction caused in the citrus industry (Miller, 1991; Rogers and Rohli, 1991). The Florida citrus industry was crippled in the period 1977-1989 by a series of six severe winter freezes leading to widespread citrus-tree destruction across northern Florida. Cold waves are synoptic-scale weather events, occurring across a sizeable area (e.g., North America) on time scales of 5 - 10 days. Table 2 is an annotated list of the worst citrus freeze events since 1766. The most severe freezes primarily occur in clusters, including those from 1977-1989 and 1886-1899. For the citrus industry, the most lethal events occurred in pairs during these clusters, including the back-to-back freezes in the winter of 1894-1895 and those of December 1983 and January 1985. The late nineteenth century freezes kept Florida citrus production below 1894 levels until 1909-10 and, as with the recent cluster, contributed to a net southward migration in the citrus production belt. The milder decades between the severe citrus freeze clusters appear associated with a gradual northward migration in citrus production.

Unusually harsh freezes are rather sporadic (e.g., those of 1835, 1917, 1962) outside of the two late-century clusters, and others freezes on Table 2 are relatively benign throughout these milder periods. The 1962 freezes are, for example, described as the "worst since 1899." Similarly in the nineteenth century, the 1880 freeze is the coldest since 1857 but the far more severe freeze of 1886 at the start of the first cluster is compared to the event of 1835, which is among the worst on the list. The data suggest that at least a half-century of comparatively milder winters precedes each severe cold-wave cluster.

Tropical fruits had also been uninjured in more than half a century at St. Augustine at the time of the freeze of 1835 (Blodget, 1857). One historical account indicates that many of the trees destroyed in 1835 around St. Augustine were nearly 100 years old. A cluster of severe winters prior to 1835 over the eastern and southeastern United States occurred during the last quarter of the 18th century, including the winters of 1776-77, 1779-80, 1783-84, during the Revolutionary War, and 1786-87, 1796-97 and 1798-99 (Blodget, 1857; Ludlam, 1966). If the St. Augustine Florida reports are characteristic however, any cold waves during these winters either may not have greatly affected Florida citrus production, or the citrus trees may have been able to withstand them due to their age or other factors such as abundant soil moisture.

The occurrence of severe cold waves may be governed partly by low-frequency atmospheric circulation variability and the tendency toward trends and interdecadal variations in the frequency of the two extreme modes comprising individual low-frequency atmospheric circulation patterns. The work of Rogers and Rohli (1991), Rohli and Rogers (1993), and Downton and Miller

TABLE II

Historical hard and severe citrus freezes in Florida since 1880 including two earlier freezes of historical note. All are abbreviated from those appearing in Rogers and Rohli (1991).

Dates	Description
1766 Jan. 3	Earliest recorded freeze, frost destroyed limes, citron and banana trees around St. Augustine (Fairbanks, 1895).
1835 Feb. 7-8	Many kinds of fruit and trees killed, many of them never started again. Destroyed all oranges/lemons to 28°N.
1880 Dec. 30	Coldest since 1857 at Jacksonville, where oranges were frozen on trees.
1886 Jan.10-12	Over $3 million citrus damage, many young trees killed. The coldest period since 1835 (Garriott, 1906).
1894 Dec.28-29	Most fruit frozen, unharvested orange crop lost, estimated at 2-3 million boxes.
1895 Feb. 8-9	Majority of orange trees killed (which were unprotected/defoliated in December, 1894 and had started re-growing.
1899 Feb.12-13	Severe damage in many millions of dollars throughout the southeastern U.S., most young citrus damaged in northern Florida.
1905 Jan. 26-27	Considerable loss of unprotected fruit, but generally oranges only suffer defoliation.
1909 Dec. 30	Many oranges frozen in central and northern Florida, young growth cut back but no orange trees killed.
1917 Feb. 3-5	Citrus frozen in north, central, some damage in south. Trees defoliated, much grapefruit & pineapple damage.
1928 Jan. 3-4	Defoliation, many young trees and new growth killed but not older groves.
1934 Dec. 12-13	Heavy losses citrus, oranges frozen; young trees killed, older ones frozen back, shedding fruit and foliage.
1940 Jan. 28-29	Most severe FL freeze since 1917. Defoliation and near-total loss of tangerines/grapefruit, over half oranges lost.
1947 Feb. 10-11	$50 million damage to citrus and truck crops, central and northern FL.
1957 Dec. 12-13	The worst of several freezes this winter. Trees defoliated with heavy losses.
1962 Jan. 9-13	Damage to low-lying, young/new growth citrus, especially in the northern Florida damage minor compared to severe damage in Texas, Louisiana.
1962 Dec. 13-14	Most damaging Florida freeze this century; 50 million boxes lost, 7-10 million trees killed.
1977 Jan. 19-21	Extensive fruit injury and leaf kill, but limited wood damage. 98% of all oranges frozen.
1981 Jan. 13-14	Widespread citrus and foliage damage; some bark splitting and wood damage, overall a production loss of 30 million boxes.
1982 Jan. 11-12	Hard freeze to most of FL where citrus and vegetable damage was severe; defoliation, and wood damage.
1983 Dec. 25-26	Trees were not cold-hardened. Severe freeze all areas, fruit frozen, wood damage and tree mortality in the north. Total fruit loss of 51 million boxes.
1985 Jan. 21-22	Severe freeze; fruit leaf and wood damage occur even after a distinct post-1983 southward shift in FL citrus production.
1989 Dec. 24-25	Hard freeze all citrus areas. 20% reduction in FL citrus production.

(1993) indicate that the positive mode of the Pacific-North American (PNA) circulation teleconnection pattern occurs frequently during winters characterized by synoptic-scale cold waves that cause severe citrus freezes. The PNA positive mode is characterized by strong meridional circulation over central and eastern North America producing substantially below normal mean air temperatures over the southeastern United States. Although data for the PNA are limited to the recent 50 years, the occurrence of positive and negative modes of this circulation pattern appear to exhibit interdecadal variability. The PNA was persistently in an enhanced positive mode from 1977 to 1989 while conversely it was in a climatically benign negative mode from 1948 to 1957 when citrus freezes did not occur (Table II). Correlations between monthly mean circulation indices and Florida air temperatures records (Downton and Miller, 1993) show that the majority of tree-damaging freezes occur during cold spells of several weeks duration associated with a strong positive PNA pattern. The three studies agree that El Niño-Southern Oscillation, although having some physical linkages to the PNA, is not tied to the incidence of severe Florida freezes. The duration of each PNA mode (positive or negative) typically exceeds that of any single synoptic event and often many anticyclones and cyclones traverse the continent during a spell in which one PNA mode prevails in the atmospheric circulation. It would seem reasonable to suggest that the PNA positive mode favors the development and movement of anticyclones linked to cold waves in the United States that lead to citrus damage. This would support the notion that there appear to be periods of time when the positive mode persists among winters, seemingly those in the last quarter of recent centuries.

One key problem confronting climatologists is the comparative lack of long-term daily air temperature records as discussed previously. Other data of value include daily pressures, useful for reconstructing anticyclone and storm tracks associated with cold waves. These tracks represent important information in learning which geographical regions experience the brunt of a cold wave and, in the case of Florida citrus freezes, even determine the type of freeze that has occurred (radiative versus advective; Rogers and Rohli, 1991). Any improvement in the temporal coverage of surface synoptic pressure maps from the current once-daily coverage over much of this century, will be very useful for future analyses.

The impact of cold waves can be quite extensive and regionally dependent, determined by the intensity and track of the accompanying synoptic weather system. Impact studies are needed to learn more about the critical levels at which cold waves affect human mortality, produce damage in cities, to homes and buildings, and how they affect natural ecosystems. Climatological work should also focus on long-term changes in cold wave frequencies and intensities and on the regional importance of low-frequency circulation (teleconnection) patterns and the ways in which they serve as a backdrop for the production of cold wave events and how they affect the trajectories of weather systems. Improved statistical studies of cold waves are needed, including analyses of temporal and regional variability of the statistical interpretations, and especially those incorporating data from the first half of this century.

5. Summary and Conclusions

It is becoming increasingly clear that changes in extreme events will accompany any climate change that may occur (Karl et al., 1996). We have described data available for examining extreme climatic events in North and South America as well as some problems that can affect analyses. In particular, problems of homogeneity of the data, and methods of quality control should be considered when performing these types of analyses.

Data currently available for extremes analysis that we discuss include observations of daily, monthly, seasonal, and annual values of temperature and precipitation for Argentina, Mexico, the United States and Canada. Our discussion not only highlights the data available in digital form, but also discusses potential sources of long-term, high temporal resolution data that should be digitized in the immediate future. This mainly includes pre-1948 observations of daily temperature and precipitation for the U.S. that are currently available only on paper. This discussion also highlights the need to make available to researchers more and longer-term observations from all parts of the world. If we experience a human-induced change of climate over the next 100 years, there will likely be regions that experience increases in some extreme events but also, in the same regions other types of extreme events may decrease or remain unchanged. Therefore analyses of extreme events are critical to identifying which regions, and which types of events are changing and which remain unchanged.

Acknowledgements

Partial support for this work was provided to the National Climatic Data Center by a U.S. Department of Energy Interagency Agreement and the NOAA Office of Global Programs.

References

Alexandersson, H.: 1986, 'A homogeneity test applied to precipitation data', *J. Clim.* **6**, 661-675.
Berz, G.: 1997, 'Changing weather extremes: The viewpoint of an international reinsurer', *Proceedings, Workshop on Indices and Indicators for Climate Extremes*, Asheville, NC, June 3-6,

1997, sponsored by CLIVAR, GCOS, WMO, and the National Climatic Data Center, Asheville, NC.

Blodget, L.: 1857, *Climatology of the United States. Philadelphia*, J.B. Lippincott, 145-154.

Cox, H.J.: 1916, 'Cold Waves.' In *Weather Forecasting in the United States*, A.J. Henry, E.H. Bowie, H.J. Cox and H.C. Frankenfield (eds.), Washington: Government Printing Office, 143-167.

Downton, M.W. and Miller, K.A.: 1993, 'The freeze risk to Florida citrus. Part II: Temperature variability and circulation patterns', *J. Clim.* **6**, 364-372.

Easterling, D.R. and Peterson, T.C.: 1995, 'A new method for detecting undocumented disconti-nuities in climatological time series', *Int. J. Clim.* **15**, 369-377.

Easterling, D.R., Horton, B., Jones, P.D., Peterson, T.C., Karl, T.R., Parker, D.E., Salinger, M.J., Razuvayev, V., Plummer, N., Jamason, P., and Folland, C.K.: 1997, 'Maximum and minimum temperature trends for the globe', *Science* **277**, 364-367.

Fairbanks, G.R.: 1895, 'Florida freezes for a century and a half', *Proc. Florida State Hort. Soc.*, **8**, 16-20.

Gall, R., Young, K., Schotland, R.,and Schmitz, J.: 1992, 'The recent maximum temperature anomalies in Tucson: Are they real or an instrumental problem?', *J. Clim.* **5**, 657-665.

Garriott, E.B.: 1906, 'Cold waves and frosts in the United States.' Bulletin P, Washington, D.C.: United States Department of Agriculture, Weather Bureau, 22 pp., plus 328 charts.

Groisman, P.Ya., and Easterling, D.R.: 1994, 'Variability and trends of total precipitation and snowfall over the United States and Canada', *J. Clim.* **7**, 184-205.

Gumbel, E.J.: 1958, *Statistics of Extremes*, Columbia University Press, New York.

Haddock, D.: 1981, 'Florida's severe freeze affects citrus: January 12-14, 1981', *Weekly Wea. Crop Bull.* **68**, 11-14.

Hoffmann, J.A.J. et al.: 1997, 'Temperature, humidity and precipitation variation in Argentina and the adjacent sub-antarctic region during the present century', *Meteorol. Zeitschrift, N.F.* **6**, 3-11.

Hogg, W.D., Louie, P.Y.T., Niitsoo, A. and Milewska, E.: 1996, 'Time series of water balance parameters for the Canadian Mackenzie Basin GEWEX Study Area', *Preprints, 2nd Int'l Sci-entific Conf. on GEWEX*, Washington DC, pp. 188, WCRP.

IPCC: 1995, *Climate Change 1995 - The Science of Climate Change. The Second IPCC Scientific Assessment.* J.T. Houghton, L.G. Meira Filho, B.A. Callendar, N. Harris, A. Kattenberg and K. Maskell (eds.), Cambridge University Press, N.Y., 572 pp.

Karl, T.R., and Williams, C.N.: 1987, 'An approach to adjusting climatological time series for discontinuous inhomogeneities', *J. Clim. Appl. Meteor.* **26**, 1744-1763.

Karl, T.R., Quayle, R.G., and Groisman, P.Ya: 1993, 'Detecting climate variations and change: New challenges for observing and data management systems', *J. Clim.* **6**, 1481-1494.

Karl, T.R., Knight, R.W., Easterling, D.R., and Quayle, R.G.: 1996, 'Indices of Climate Change for the United States', *Bull. Amer. Meteor. Soc.* **77**, 279-292.

Karl, T.R., and Knight, R.W.: 1998, 'Secular trends of precipitation amount, frequency, and inten-sity in the United States', *Bull. Amer. Meteor. Soc.* **79**, 231-242.

Louie, P.Y.T. and Hogg, W.D.: 1980, 'Extreme value estimates of snowmelt', *Proc. Canadian Hydrology Symposium*: 80. NRCC, Toronto, pp. 64-78.

Ludlam, D.M.: 1966, *Early American Winters 1604-1820.* Boston: American Meteorological Soci-ety, 285 pp.

Mekis, E. and Hogg, W.D.: 1997, 'Rehabilitation and analysis of Canadian daily precipitation time series', *Preprints, 10th Conf. Applied Climatology.* Oct. 20-23, 1997, Reno, NV, AMS Bos-ton.

Metcalfe, J.R., Routledge, B., and Devine, K.: 1997, 'Rainfall measurements in Canada: Changing observational methods and archive adjustment procedures', *J. Clim.* **10**, 92-101.

Miller, K.A.: 1991, 'Response of Florida citrus growers to the freezes of the 1980s', *Clim. Res.* **1**, 133-144.

Nicholls, N.: 1995, 'Long-term climate monitoring and extreme events', *Clim. Change* **31**, 231-245.

Quayle, R.G., Easterling, D.R., Karl, T.R., and Hughes, P.Y.: 1991, 'Effects of recent thermometer changes in the cooperative network', *Bull. Amer. Meteor. Soc.* **72**, 1718-1723.

Reek, T., Doty, S.R., and Owen, T.W.: 1992, 'A deterministic approach to the validation of historical daily temperature and precipitation data from the cooperative network', *Bull. Amer. Meteor. Soc.* **71**, 826-831.

Rogers, J.C. and Rohli, R.V.: 1991, 'Florida citrus freezes and polar anticyclones in the Great Plains', *J. Clim.* **4**, 1103-1113.

Rohli, R.V. and Rogers, J.C.: 1993, 'Atmospheric teleconnections and citrus freezes in the southern United States', *Phys. Geog.* **14**, 1-15.

Sevruk, B.: 1982, 'Methods of correction for systematic error in point precipitation measurement for operational use', Operational Hydrology Report No. 21, World Meteorological Organization, Geneva, 91 pp.

Stirling, L.M.: 1994, 'Atmospheric observing systems and their data management support structure', *Proceedings of the Workshop on Canadian Climate System Data*, Quebec, May 1994, Environment Canada (NWRI), Burlington, Ontario.

Vincent, L.A.: 1998, 'Technique for the identification of inhomogeneities in Canadian temperature series', *J. Clim.*, **11**, 1094-1104.

Wendland, W.M.: 1987, 'Prominent November coldwaves in the north central United States since 1901', *Bull. Amer. Meteor. Soc.* **68**, 616-619.

Zwiers, F.W., 1994: Changes in screen temperature extremes under a doubling of CO_2, In: *Research Activities in Atmospheric and Oceanic Modelling*, G.J. Boer (ed.), pp. 7.44-7.46, CAS/JSC Working Group Numerical Experimentation, Report **19**, February 1994, WMO/TD-No. 592.

Zwiers F.W. and Kharin, V.V.: 1998, 'Changes in the extremes of the climate simulated by CCC GCM2 under CO2 doubling', *J. Clim.*, **11**, 2200-2222.

(Received 5 November 1997; in revised form 18 November 1998)

CLIMATE EXTREMES: SELECTED REVIEW AND FUTURE RESEARCH DIRECTIONS

THOMAS R. KARL and DAVID R. EASTERLING

NOAA/NESDIS/National Climatic Data Center, Asheville, NC, USA

Abstract. Trends and multi-decadal variations of weather and climate extremes have only recently received attention from the climate community. Interest has stemmed from exponentially increasing economic losses related to climate and weather extremes, and apparent increases in deaths attributed to these events, suggesting that key decision makers need a better understanding of the potential uses of climate information. The need for data on climate extremes in disaster mitigation activities such as the International Decade for Natural Disaster Reduction also has provided another motivation for focus in this area.

The losses cited above raise questions as to whether extreme weather events are actually increasing in frequency, whether society as a whole is becoming more vulnerable to extreme weather events, whether public perception has been unduly influenced by enhanced media attention, or some combination. Given these questions, of particular interest here is the extent to which we can document changes in climate and weather extremes. Attribution of ongoing trends to specific climate forcings, such as anthropogenic effects or other factors related to natural climate variability are still equivocal.

For some areas and variables increases in the frequency of extreme events are apparent, while in other areas there are suggestions of declines in these events. A review of this information suggests that further understanding of the cause(s) of the apparent changes in climate and weather extremes is strongly dependent upon progress in our ability to monitor and detect these multi-decadal trends. Based on these analyses we show that this will likely require increased attention in the following areas: 1) The development of more effective international data exchange for high resolution historical climate and weather records, 2) Increased emphasis on rescuing data with appropriate resolution from deteriorating manuscripts and other non-electronic media, 3) A greater emphasis on removing inhomogeneities[1] in the instrumental record and ongoing weather monitoring programs (that provide much of our information about changes and variations of weather and climate extremes), 4) More effective use of space-based measurements and reanalysis products derived from models, 5) More robust monitoring of local extreme weather events such as tornadoes, hail, lightning, and wind, and 6) More effective means to integrate and communicate information about what we know and do not know about changes in climate extremes. Progress in each of these areas is reviewed in context with outstanding remaining challenges, and the benefits that can be expected if we meet these requirements.

---

[1]Inhomogeneities are defined as changes and variations in the record that are non-climatic or are not representative of the time and space scales of interest, e.g., urban heat island effects are climate-related, but are not the scales of interest for global temperature change analyses.

Climatic Change 42: 309–325, 1999.
© 1999 *Kluwer Academic Publishers. Printed in the Netherlands.*

1. Introduction

Each year extreme climate and weather events take tens of thousands lives, cause untold human hardship, and result in enormous economic losses. Since the late 1980's the insurance and re-insurance industry has pointed out an exponential increase in economic losses due to these events (Munich Re, 1996). In many countries the general public has also become concerned as press reports, first-hand experience, and anecdotal information all appear to suggest an increase in the frequency and severity of these events. Over the globe, economic losses have continued to increase during the 1990s, and although the number of disaster- related deaths has increased over the past 25 years (IFRCRCS 1997), the relative increase in weather-related deaths has not been as dramatic as the rapid increase in economic losses. For example, the enormous loss of life from the 1991 floods in Bangladesh, where an estimated 140,000 deaths occurred, was still considerably fewer than the 300,000 during the 1970 catastrophic floods. It is quite likely that improvements in communications and warning systems have played an important role in moving people out of harms way. Human infrastructure however, is not so mobile. The losses cited above raise questions as to whether extreme weather events are actually increasing in frequency, whether society as a whole is becoming more vulnerable to extreme weather events, whether public perception has been unduly influenced by enhanced media attention, or some combination. Given these questions, of particular interest here is the extent to which we can document changes in climate and weather extremes. There is little doubt that our vulnerability to extreme events is increasing as society continues to inhabit and develop vulnerable areas such as coastal margins and floodplains, and existing populations in these types of areas increase. Although there is some controversy as to whether urban areas are overly vulnerable to natural disasters (e.g. see Mitchell 1990 and Knovitz 1990), by nature of the fact that cities are areas of high population concentration with complicated interdependent infrastructure, there is little doubt that they are vulnerable to exceptional extreme weather events such as hurricanes, heavy flooding, or extreme temperature events. Moreover, changes in the natural landscape associated with mans infrastructure, (roads, parking lots, buildings, reservoirs, dams, alterations in streamflow, sewage and storm water routing runoff, etc.) can often contribute to major catastrophes during extreme rainfall events and flooding situations. The United Nations estimates that nearly half of the world's population lives in cities, up from 30% in 1950, and that by 2025, 60% of the world's population is expected to reside in urban areas.

In light of these socio-economic trends there are four fundamental issues related to variations and changes of climate extremes that are reviewed in this paper. This includes:

Can we detect any change in climate and weather extremes?
•   Are these changes unusual in light of natural climate variability?

- Is there any evidence to link observed changes in extremes to anthropogenic effects?
- What priorities are needed to reduce uncertainties?

2. Observed Trends

The concern over potential impacts of climate change by various parts of society has been heightened by increases in weather related impacts that have occurred in recent years. Understanding potential climate change both in terms of trends, and changes in extreme events is critically important for a wide range of policy decisions (Pielke and Landsea 1998). Therefore, it is useful here to first examine observed trends in various parts of the climate that may have an impact, either directly or indirectly, on society. Furthermore, it should be made clear that a trend in one individual variable, such as the annual global temperature, does not necessarily confirm that climate change is occurring. However, it is the continued documentation of trends and changes in a number of key variables that adds to the body of evidence that there is a discerable anthropogenic impact on the climate. In the following sections we examine trends in various aspects of temperature, precipitation and storms, particularly as they relate to climate extremes. Due to a shortage of available data and subsequent analyses there currently is not strong evidence that on a global basis extreme weather events are increasing in severity or frequency. However, in some regions where data are available to examine these types of events there is clear evidence of changes in some extremes and overall climate variability (IPCC 1996). Lastly, it is becoming increasingly evident that human society is going to have to learn to live with whatever climate is produced by a substantial increase in CO_2. Although there is still debate, it is possible that the earth may see a doubling or even a tripling of atmospheric CO_2 sometime around the turn of the next century (Schneider 1998). If this is the case, then continued documentation of climate trends, particularly in terms of climate extremes, will be critical for decision makers in the future as they deal with environmental changes and their impacts.

2.1. TEMPERATURE

There is now clear evidence for an observed increase in global average temperatures of about 0.5 °C since the start of the 20th century (IPCC, 1996). It is not as well appreciated however, that on regional scales, especially over land, the observed rates of temperature change are often several times larger. Clearly, if there are large changes in the mean, changes in the extremes of temperature are also likely (IPCC 1996). A recent analysis of 50% of the global landmass by Easterling et al. (1997) shows that indeed, the mean daily maximum and minimum temperatures are both increasing, but the rate of increase of the minimum temperature for the 1950-1993 period is more than twice the increase of the maximum (1.8 °C/100 years versus 0.8 °C/100 years). The increase in the mean minimum temperature has been demonstrated to have affected the length of the frost free period, which has potential impacts for a number sectors such as agriculture (growing season length and pest control) or power generation and consumption. For example, Cooter and LeDuc (1995) report that in the northeastern USA over the 1950-1994 period the frost-free period begins about 11 days earlier in the 1990's compared to the 1950's, and Easterling (1998) has shown that the northern Great Plains, western and northwestern USA have experienced a significant decrease in the number of days below freezing for the 1910-1997 period. Evidence for a significant reduction in the number of Twentieth Century frost-days in many portions of Australia has been documented in several reports, e.g., Plummer et al, (1999); Karl et al., (1997). Salinger (1997) also reports a decrease in the number of frost days over much of New Zealand during the Twentieth Century.

In Australia, the increase in the annual mean minimum temperature is quite consistent with reduced frost days, but in the Northeast USA the change in the mean minimum temperature is quite small relative to the change in the spring frost date. This is not an unusual circumstance, and points to the danger of broad generalizations based on changes in the mean. For example, the work of Rogers and Rohli (1991) and Downton and Miller (1993) document an increase in the frequency of major freezes affecting Florida during the late 1970s and 1980s, yet the mean minimum winter temperature during that time was comparable to values during the 1950s and early 1960s. Nonetheless, in New Zealand, Salinger (1997) reports a good relationship between the mean annual temperature and the number of days below freezing or above 30 °C.

Extreme high temperature events are also responsible for highly publicized weather impacts such as heat-wave mortality. Karl and Knight (1997) found that elevated nighttime apparent temperatures (an index of both temperature and humidity) coupled with a variety of societal factors (Changnon, et al. 1996) were responsible for the impacts of the unusual and deadly heat wave that gripped Chicago, Illinois during the summer of 1995. Prior to 1995 the trends of elevated nighttime apparent temperatures had only slightly increased in the Central USA, making the 1995 event even more unusual.

Aside from the few examples cited above, there is a surprising dearth of analyses addressing changes in extreme temperatures. Perhaps partly because temperature is generally regarded as following a normal distribution, relatively few analyses of changes in growing seasons and temperature extremes have been undertaken. Many who have addressed this issue (Karl and Knight, 1997; Katz and Brown, 1992, Mearns et al., 1984) have generally applied normal distribution functions to their analyses, and made inferences about changes in the extremes of temperature, based on changes in the mean.

So, perhaps the greatest uncertainty about changes in temperature extremes for any specific location relates to the properties of extremes themselves. Although it is true that temperatures approximate a normal distribution, where the behavior of extremes ought to be well-approximated from changes in the mean, the data suggest that such inferences cannot consistently be relied upon. For example, despite an increase of mean temperatures in the USA of about 0.4 °C over the past Century, annual extreme maximum has decreased by 0.2 °C, while the annual extreme minimum increased by the same amount. In the former USSR, the contrast is also most apparent during the spring. For example, Karl et al. (1991) found an increase in the spring **mean** minimum temperature, averaged across the country, of 1.4 °C from 1951-1986, but the **1-day extreme** minimum temperature increased by 2.2 °C, and both increases were statistically significant. This suggests that even a relatively minor increase in mean temperature may result in more frequent extremes, for example more heat waves or more extreme cold events.

For extremes, Katz and Brown (1992) point out that changes in the variability are more important than changes in the mean. IPCC (1996) comprehensively discusses what is known about changes in temperature variability. Since variability of temperature can be defined in several ways it is important to understand what aspects of variability are being analyzed. For example, the variance of the two series of 5,5,5,0,0,0,-5,-5-,5 and 5,-5,0,5-5,0,5,-5,0 representing annual anomalies are identical, but the absolute value of the interannual annual differences are quite dissimilar indicating a difference in persistence. In a global study, Parker et al. (1994) compared spatially averaged variances of annual temperature anomalies between the two periods 1974 to 1993 with 1954 to 1973 and found evidence for an increase in temperature variability in the 1974-1993 period of between 4 and 11% depending on the season. In some areas the increase was considerably larger, especially over North America. Karl, et al. (1995) analyzed changes in variability on a variety of times-scales from 1-day to 1-year for much of the Northern Hemisphere (USA, China, and the former USSR) during the Twentieth Century. Their analysis was based on the absolute value of time-averaged differences from one period to the next. Using this statistic they found evidence for a decrease in temperature variability on short time-scales (e.g., up to a few days), but no broad scale increases in interannual variability. Therefore, although the year-to-year variability may not be affected, this could have implications for the length of certain types of multi-day events, such as heat waves, or cold snaps.

2.2. PRECIPITATION

2.2.1. *Intense Precipitation*
There have been a number of large flooding events in the 1990's in Europe, Asia, and the United States that have highlighted a renewed emphasis on changes in precipitation extremes. Work resulted from this renewed emphasis is beginning to suggest that there have been some important changes and variations related to a

variety of precipitation extreme statistics. Recent work by Groisman et al. (1999) provides a framework for understanding a number of recent analyses that have pointed toward an increase in precipitation extremes in North America, portions of Europe, Japan, Australia, South Africa, the former Soviet Union, and elsewhere (IPCC, 1996). Groisman et al. (1999) demonstrate, using daily precipitation data from North America, a large portion of Asia, portions of Europe, and Austalia that any change in the mean monthly total precipitation will influence the extremes more than any other precipitation rate. With an increase in total precipitation, a disproportionate increase in precipitation for higher daily precipitation rate is expected, compared to more moderate precipitation rates. However, changes in the total number of raindays remains somewhat inconclusive..

IPCC (1996) has demonstrated that precipitation has generally increased across much of the mid-to-high latitude land areas during the past Century. Karl and Knight (1998) find a very significant increase in extreme precipitation events during the Twentieth Century in the USA. The increase has occurred due to both an increase in the frequency of very heavy and extreme precipitation events as well as an increase in their intensity. Similar analyses for the USA have now been run for Canada (since 1941), the former Soviet Union (since 1967), and Australia (since 1910) with less striking changes, but clear evidence for an increase in heavy and extreme precipitation events. Extensions (Suppiah et al., 1997) of earlier work by Suppiah and Hennessy (1996), show that in Australia the 90th, 95th, and 99th percentiles of daily precipitation totals has increased by 20, 6, and 4% respectively, when averaged across the country. In a recent analysis of six long-term stations in Germany an increase in daily extreme precipitation amounts has also been detected in all but one of the stations, and the trends are statistically significant at three of the stations with increases during the Twentieth Century of over 25mm/day (Rösner et al., 1997). Over South Africa, Mason et al. (1998) indicate that significant increases in extreme rainfall events have taken place between the two 30-year periods, 1931-60 to 1961-90. The intensity of the 10-year high rainfall event has increased by over 10% over large areas of South Africa. Mason et al. (1998) find that percentage increases are largest for the heaviest rainfall events. Iwashima and Yamomota (1993) also found an increase in the likelihood of extreme precipitation events in recent decades in Japan.

There are regions however, were little or no change in the intensity or frequency of extreme precipitation events has been identified. For example, an analysis of 1-day, 2-day, and 3-day precipitation total in India do not reveal any general trend toward more intense events (Kumar, et al. 1997), as increases in the west are balanced by decreases in the east. Similarly, for China Zhai et al. (1999) do not find evidence for an increase in precipitation extremes for 1 and 3-day events, and there is little change in total annual precipitation in China. Analyses of changes in short-term precipitation extremes for the former Soviet Union have been limited to date due to data inhomogeneity problems (Karl and Knight, 1995), but IPCC

(1996) suggests a net overall increase in precipitation for this region of the world, which suggests that extreme precipitation amounts may have increased prior to 1967 (the beginning date of the Karl and Knight analysis).

2.2.2. *Droughts and Floods*

The question was posed by IPCC (1990) whether there has been a tendency for an increase in the area of the globe affected by droughts and floods. Analyses at that time showed little suggestion of an increase. More recent work however, e.g., (Karl et al., 1995) indicates that in the USA the increase in precipitation during the past few decades leading to more wet spells and floods has not been accompanied by commensurate decrease in the frequency or intensity of droughts. In other words, at least in the USA there has been an increase in the percent area of the country experiencing a climate extreme. In a broader analysis Dai et al. (1998) find that there has been an increase in the frequency and intensity of droughts or wet spells in areas that are influenced by ENSO. This is especially notable in the tropics and subtropics. Furthermore, Mantua, et al. (1998) have recently identified a long-period oscillation centered over the mid-latitude northern Pacific basin they have termed the Pacific inter-Decadal Oscillation (PDO). The PDO signature appears to be an irregular, but robust pattern of climate variability that varies on interannual to interdecadal time scales that is clearly related to ENSO. However it appears to be a longer period oscillation that envelopes shorter-period ENSO events, such that the late 1970's shift to more frequent and intense ENSO's may be a manifestation of the PDO (Mantua, et al. 1998).

2.3. STORMS

In this section a distinction is made between tropical and extratropical cyclones. Although both systems are associated with regional-scale cyclonic surface wind circulations around a low pressure system, tropical cyclones are not associated with frontal systems whereas extratropical cyclones feed off such frontal boundaries. Tropical cyclones are very much dependent on evaporation and sensible heat fluxes from the oceans for their energy sources.

2.3.1. *Tropical cyclones*

Tropical cyclones are the costliest natural disasters around the world. Landsea et al. (1997) provide a comprehensive review of the Twentieth Century changes in tropical cyclone frequency and intensity around the world. IPCC (1996) and Landsea (1999) find that tropical cyclone frequency is not generally increasing or decreasing when considered across the globe, but there is significant decadal variability associated with both the number and intensity of tropical cyclones. On a regional basis there is evidence for a significant increase in tropical cyclone frequency in the Northwest Pacific since the 1970s, but records back to 1960 suggest that tropical cyclone frequency was also high during the 1960s (Landsea, 1999). On average, this basin is responsible for over 30% of the global tropical cyclones that form each year. In the Atlantic basin, which contributes to about 12% of the global total number of tropical cyclones each year, there has been a decrease in hurricane intensity since the mid-1940s (when reasonably reliable records begin), but since the turn of the Century there is no overall trend in hurricane frequency. Similarly, for land-falling hurricanes affecting the USA, there are large inter-decadal variations, but little evidence for any systematic trends. This decrease since the 1940s is consistent with the increase in ENSO activity that occurred in the late 1970s. Landsea and Gray (1992) point out that during an El Nino event tropical cyclone activity in the Atlantic tends to be suppressed due to increased shear in the lower and middle troposphere. In the North Indian Ocean, data suggest a significant downward trend in tropical cyclone frequency (Landsea, 1999). In the Australian region the data indicates little change in the number and frequency (Nicholls, 1999) of intense tropical storms since records began in the 1960s. However, less intense storms show a decline that is attributable, at least in part, to inhomogeneities in the record (Nicholls, 1999). Similarly, no change in tropical cyclone variations have been detected in the Southwest Indian Ocean or the Southwest Pacific since the 1960s.

2.3.2. Extratropical cyclones

Once again there are a dearth of analyses on the trends of intense extratropical cyclones, and some analyses are conflicting in terms of their result. Only recently, does there seem to be some consistency regarding an overall increase in the intensity of the strongest cyclones. For example, an increase in storm intensity during the late 1980s and first half of the 1990s has been found by a number of investigators for the North Atlantic Ocean, e.g., Stein and Hense(1994), Kushnir et al. (1997). There appears to be an abrupt shift toward more intense storms in the northern half of the basin (with the exception of the past two winters), but a decrease in intensity in the southern half over the past several decades. Lambert (1996) analyzed intense cyclones in the N. Pacific and Atlantic and found evidence to support a strong increase in intensity during the past several decades. Similarly, Bardin (1994) reported that the size and intensity of cyclones has increased since 1980. Davis and Dolan (1993) find an increase in the number of intense cyclones over eastern North America, but decadal variability in this region is great, making it difficult to separate out a statistically significant trend.

3. Loss Reduction Prospects

Reducing the impact of extreme weather and climate on human suffering and economic losses is a high priority for the International Decade of Natural Hazard Reduction. A critical step toward reaching this goal relates to better knowledge of what the future climate might bring in the way of these natural hazards. Improved projections and confidence in them are dependent on understanding the causes and variations within the modern instrumental record. This requires considerable attention be given to the linkages between observed changes and specific causes (the attribution issue). It is also important to ensure proper utilization of existing and new data and information about expected changes in climate extremes in long-range planning and infrastructure maintenance. Both of these activities are critically dependent on our ability to monitor changes in climate and weather extremes.

3.1. ATTRIBUTION OF TRENDS

Clearly many, although not all, of the changes in temperature extremes we have examined here are related to increases in the global mean temperature. Since the overall increase in global mean temperatures is likely to be at least partially a result of increases of greenhouse gases (IPCC, 1996) there is reason to believe that the changes in temperature extremes may be related to these increases as well. The anthropogenic greenhouse effect related to changes in temperature variability are less certain, but there is some suggestion from climate models with enhanced concentrations of atmospheric CO_2 that short-term temperature variability would decrease, and this has been detected in several regions (Karl and Knight, 1995). Nonetheless the models are not entirely consistent in this regard (IPCC, 1996).

Perhaps one of the most critical attribution issues for changes in climate and weather extremes relates to the hypothesis that the hydrologic cycle should intensify as global warming progresses. Trenberth and Shea (1996) provide a conceptual model for such a hypothesis (Fig. 1). There are some indications, although by no means is the argument unequivocal, that the hydrologic cycle is growing more intense. In many instances the data is based on just a few decades with incomplete global coverage, so it is difficult at this stage to be comprehensive. Nonetheless, it is important to consider the changes and variations that have been observed related to an intensification of the hydrologic cycle. These changes are summarized below:

An increase in evaporation from the tropics (IPCC, 1996)
- Increase in convective clouds and related cirrus (IPCC, 1996)
- Increased continental cloud cover contributing to reduced diurnal temperature range and reduced evaporation from water surfaces over land (IPCC, 1996; Dai et al., 1997)
- Increased precipitation in the mid-and high-latitude land areas contributing to enhanced evaporation and more runoff (Dai et al., 1997)
- Increased atmospheric water vapor over North America, China, and tropical regions (IPCC, 1996; Ross and Elliott, 1996; Zhai and Eskridge, 1997)
- Increased precipitation intensity in many portions of the Northern Hemisphere (Karl and Knight, 1997; Groisman et al., 1999; Karl et al., 1997)

Figure 1-Conceptual model of the effect of greenhouse gases and global warming on the hydrologic cycle and phenomena associated with many climate extremes (from Trenberth and Shea, 1996).

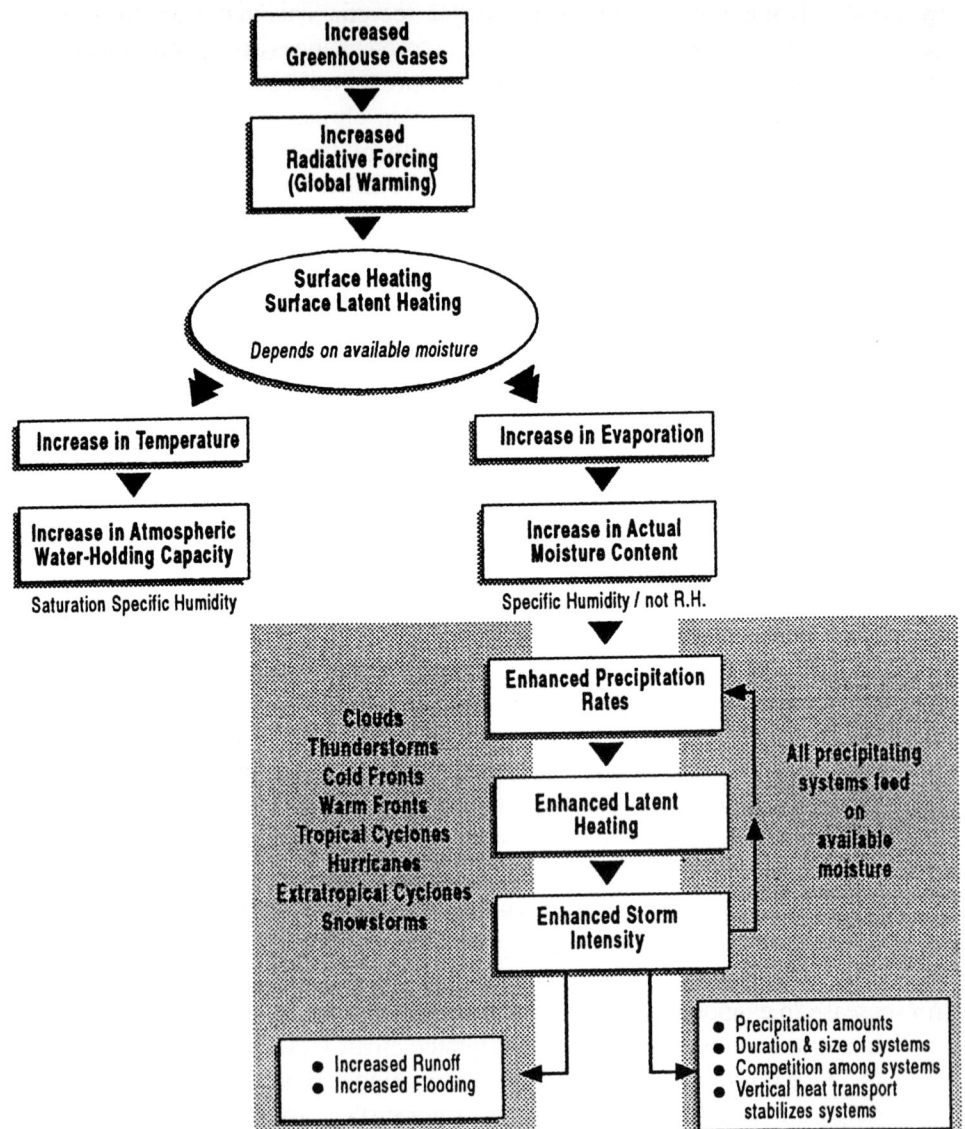

•   An increase in extratropical storm severity (Lambert, 1996)
Many of these changes are consistent with the conceptual model (Fig. 1) put forward
by Trenberth and Shea (1996), and many of the changes have been projected to
occur as global temperatures increase due to increases in atmospheric greenhouse
gases, e.g., increased precipitation intensity, more precipitation in the mid- and high-
latitudes, increased atmospheric water vapor etc.

Currently, there are a number of impediments preventing us from more
effectively understanding the linkages between changes in climate extremes and
natural hazards to anthropogenically-induced climate change. Certainly, model
deficiencies are high among the list, but just as important is our lack of long-term
reliable climate data. Time and time again, we find that our observing systems and
data sets often have large systematic biases of uncertain magnitude casting doubt on
our ability to detect multi-decadal changes. This is why efforts like the Global
Climate Observing System (GCOS) are so critical.

3.2. BETTER USE OF EXISTING DATA

It is becoming increasingly apparent that even in the absence of clear attribution of
the causes of the observed changes in the frequency and intensity of climate
extremes we do not know how to address the problem of designing infrastructure for
the next several decades in a climate that is clearly demonstrating that it is not
stationary, even on decadal time scales. For example, what guidance can the climate
community provide engineers who are designing for 100 and 200 year events?
Clearly, one responsibility of the climate community is to convey to users of climate
information that climate statistics, such as return period calculations, are based on
past climate, and in some instances on relatively short periods. Therefore, the
statistics do not contain any information on how these statistics may change in the
future. As already pointed out, the extremes are far more sensitive to changes than
changes in the mean. One can argue that it is now less desirable to design and plan
for climate extremes by assuming the Twentieth Century climate will be a useful
guide to the future, compared with projecting a different climate. Clearly, we have
not explored the implications of such scenarios in terms of cost-benefit ratios. When
does it pay to project modest changes versus strong changes or no change at all.
This is an area of research that has not received adequate attention.

3.3. IMPROVED MONITORING AND DETECTION OF CHANGES IN EXTREMES

Improved monitoring, data management, and data diagnostics, as discussed in both
IPCC (1990) and IPCC (1996) are critical to understand how the climate has changed
and is changing or varying. The problem is even more sensitive for changes in
climate extremes than changes in other climate statistics. There are a number of

chronic problems related to long-term climate monitoring (Karl, 1995) that are now becoming acute. These are described and summarized below:

- International data exchange is being hampered due to cost recovery policies for high resolution historical climate records needed to estimate global changes of climate extremes. A step toward resolving this issue was recently taken by the jointly sponsored GCOS/CLIVAR international workshop on 'Indicators and Indices for Changes in Climate Extremes'. There is now an incipient effort to build joint databases suitable for analysis of changes in climate extremes including natural disasters. An institutional framework to encourage this fledgling effort has been requested by the scientists involved. Another suggestion proposed that the list of GCOS Global Surface Stations be used as a basis to develop and update a set of indices. They would provide considerable information about changes in climate extremes. These stations would have to provide statistics on at least daily resolution, updated annually.

- Considerable data on a variety of short-term (less than one month) weather and climate events remains inaccessible due to an absence of electronic digital data. An increase in emphasis on rescuing past measurements, with appropriate resolution, from deteriorating manuscripts and other non-electronic media is required to adequately quantify past changes in climate extremes. There are enormous collections of high resolution data related to precipitation, temperature, freezes, sea level pressure, etc. in addition to the metadata required to interpret these data that still reside in inaccessible media. Closer scientific linkages with projects like WMO's Data Rescue Project are called for to improve our information about changes high resolution climate extremes.

- Whether tropical cyclones occurrences are changing or precipitation is becoming more intense, the major problem affecting virtually every analysis relates to undocumented or unknown effects of inhomogeneities in data sets. A greater emphasis on removing inhomogeneities in the instrumental record and ongoing weather monitoring programs (that provide much of our information about changes and variations of weather and climate extremes) should take a high priority. Until weather observing networks and data management systems adopt and adhere to a set of climate monitoring principles it is unlikely the situation will improve. Such a set was recently recommended at a GCOS in-situ/space-based calibration validation meeting (Sept. 1996). A set of climate monitoring principles might include the following characteristics:

(1) Prior to implementing any changes in existing observing or data processing and management systems, an assessment should be completed related to the impact on our ability to monitor environmental variations and changes.

(2) Overlapping measurements, both in time and space for old and new observing systems should be standard practice for critical environmental variables whenever implementing changes in order to develop appropriate transfer functions from one system's measurements to the other .

(3)Calibration, validation, processing algorithms, knowledge of instrument, station and/or platform history, and any other information relevant to interpreting what is being measured are essential for data interpretation and use. This information should be recorded as a mandatory part of the observing routine and be archived with the original data.

(4)Routine assessment of both random and systematic errors is necessary to adequately monitor environmental variations and change.

(5)Environmental assessments that require knowledge of environmental variations and change should be well integrated into strategies for development and maintenance of Global Observing Systems.

(6) Observations with a long uninterrupted record should be maintained, and every effort should be made to protect the data sets that document long-term homogeneous observations.

(7) The highest priority in the design and implementation of new environmental observing systems should be given to data poor regions, variables and regions sensitive to change, as well as key measurements with inadequate temporal resolution.

(8) Network designers, operators, and instrument engineers must be provided environmental monitoring requirements at the outset of network design. Instruments must have adequate accuracy with biases small enough to resolve environmental variations and changes of primary interest.

(9) Much of the development of new observation capabilities and much of the evidence supporting the value of these observations stem from research-oriented needs or programs. Stable, long-term commitments to these observations, and a clear transition plan from research to operations, are two requirements in the development of adequate environmental monitoring capabilities.

(10) Data management systems that facilitate access, use, and interpretation are essential. Freedom of access, low cost, mechanisms which facilitate use (directories, catalogs, browse capabilities, availability of metadata on station histories, algorithm accessibility and documentation, etc.) and quality control (both random errors and systematic biases) should guide data management. International cooperation in all these areas is critical.

Some of the most effective means to monitor extreme weather and climate events relate to more effective use of space-based measurements and reanalysis products derived from climate models. Re-analysis products may be quite effective in analyses of extreme extratropical cyclones. This is clearly critical in the area of monitoring tropical cyclones, but improved estimates of precipitation from satellite and radar coverage are likely with careful integration of in-situ measurements. If programs such as the Global Precipitation Project were encouraged to focus more on time series of high resolution precipitation events this would help in identifying changes in climate extremes.

At the present time there are very few analyses of local extreme weather events such as tornadoes, hail, lightning, and wind. This at least partially reflects the inattention that has been given to these phenomena as part of multi-decadal climate monitoring. Since these phenomena are of vital importance to society and ecosystems, they must receive greater climatological emphasis in routine weather monitoring. At the present time for example, it is impossible to ascertain whether there has been any change in tornado frequency because of the inhomogeneities in reporting tornadoes during the past several decades. New networks, such as lightning detection contain many time-related biases due to changing configurations.

At the present time there are few venues to integrate and communicate information about what we know and do not know about changes in climate extremes. A step forward has been taken in this area with the joint sponsorship (GCOS/CLIVAR) of the recent meeting on "Indicators and Indices for Changes in Climate Extremes" which brought together an international group of scientists, and representatives from industry, all focused on developing data and information to document changes in climate extremes. The continuation and development of this initial effort, and others like it, are critical for effective data and information exchange outside of specialty fields.

4. Conclusions

Several important questions have been posed regarding climate extremes and natural hazards, some of which we do not yet have satisfactory answers. Existing data indicates that the climate is becoming more extreme in some areas and for some variables, but at this time it is difficult to unambiguously link such changes to anthropogenic effects. Nonetheless, there is beginning to emerge some hints that the global hydrologic cycle may be intensifying in response to global temperature increases. Inadequate data access and poor climate monitoring practices are two primary issues that must be improved if we expect to make much progress in this area. Climate monitoring can no longer be relegated to weather operations, but the scientific basis, rationale, and oversight for long-term monitoring of climate and weather extremes must be given a high priority in future WCRP activities.

Acknowledgments

The comments of Dr. Roger Pielke, Jr. and an anonymous reviewer are gratefully acknowledged. Partial support for this work was provide by NOAA's Climate Change Data and Detection Program Element of the Office of Global Program's Climate and Global Change Program and by a U.S. Dept. of Energy Interagency Agreement.

References

Bardin, M. Yu., 1994: Parameters of cyclonicity at 500 mb in the Northern Hemisphere extratropics. In Proc. XVIII Climate Diagnostics Workshop, Boulder, CO., NTIS, US Dept. of Commerce, Sills, Building, 5285 Port Royal Road, Springfield, VA 22161, U.S.A., pp 397.

Changnon, S.A., Kunkel, K.E., and Reinke, B.C.: 1996, Impacts and responses to the 1995 heat wave: A call to action, *Bull. Amer. Meteor. Soc.*, 77, 1497-1506.

Cooter, E.J. and LeDuc, S.K.: 1995, 'Recent frost date trends in the northeastern United States', *Int. J. Climatology*, **15**, 65-75.

Dai, A., Trenberth,K.E., and Karl T.R.: 1997, 'Hydrological control of diurnal temperature range', *Geophys. Res. Letts.*, in press..

Dai, A., Trenberth, K.E., and Karl, T.R.: 1998, 'Global variations in droughts and wet spells', *Geophys. Res. Letts.*, **25,** 3367-3370.

Davis, R.E., and Dolan, R.: 1993, ' Nor'easters', *American Scientist,* **81**, 428-439.

Downton, M.W. and Miller, K.A.: 1993, 'The freeze risk to Florida citrus. Part II: Temperature variability and circulation patterns', *J. Climate,* **6**, 364-372.

Easterling, D.R., Horton, B., Jones, P.D., Peterson, T.C., Karl, T.R., Parker, D.E., Salinger, M.J., Razuvayev, V., Plummer, N, Jamason, P., and Folland, C.K.: 1997, 'Maximum and minimum temperature trends for the globe', *Science*, 277, 364-367.

Easterling, D.R.: 1999, 'Recent trends in temperature extremes and threshold exceedences in the United States', submitted to, *Bull. Amer. Meteor. Soc.*

Easterling, D. R., Diaz, H.F., Douglas, A.V., Hogg, W.D, Kunkel, K.E., Rogers, J.C., and Wilkinson, J.F.: 1999, 'Long-term observations for monitoring extremes in the Americas', Workshop Proceedings on Indices and Indicators for Climate Extremes, Asheville, NC. Sponsors, CLIVAR, GCOS and WMO. *Climatic Change,* this issue.

Groisman, P. Y., Karl, T.R., Easterling,D.R., Knight, R.W., Jamason, P.F., Hennessy, K.J., Suppiah, R., Page, C.M., Wibig, J., Fortuniak, K, Razuveav, V., Douglas, A, Forland, E, and Zhai, P-M.: 1999, 'Changes in the probability of heavy precipitation: Important indicators of climatic change', Workshop Proceedings on Indices and Indicators for Climate Extremes, Asheville, NC. Sponsors, CLIVAR, GCOS and WMO. *Climatic Change,* this issue.

Intergovernmental Panel on Climate Change (IPCC), 1990: *Climate Change, Working Group 1 to the Intergovernmental Panel on Climate Change Scientific Assessment.* Houghton, J.T., Jenkins G.J., and Ephraums, J.J., editors. Cambridge University Press.

Intergovernmental Panel on Climate Change (IPCC), 1996: *Climate Change, Working Group 1 to the Second Assessment Report of the Intergovernmental Panel on Climate Change.* Meira Filho, L. G., Callander, B. A., Harris, N., Kattenberg, A., and Maskell, K., editors. Cambridge University Press.

International Federation of Red Cross and Red Crescent Societies (IFRCRCS) , 1997: *World Disasters Report 1997*, Oxford University Press, p 116.

Iwashima, T. and Yamamoto, R.: 1993, 'A statistical analysis of the extreme events: Long-term trend of heavy daily precipitation'. *J. Met. Soc. Japan*, **71**, 637-640.

Karl, T. R., Kukla, G., Razuvaev, V.N., Changery, M.J., Quayle, R.G., Heim, R.R., Easterling,D.R., and Fu, C.B.: 1991, 'Global Warming: Evidence for asymmetric diurnal temperature change', *Geophys. Res. Lett.*, **18**, 2253-2258.

Karl, T.R., Knight, R.W., and Plummer, N.: 1995, 'Trends in high-frequency climate variability in the twentieth century', *Nature*, **377**, 217-220.

Karl, T. R., Nicholls, N., and Gregory, J.: 1997, 'The Coming Climate', *Scientific American*, **276**, 54-59.

Karl, T.R. and Knight, R.W.: 1997, 'The 1995 Chicago Heat Wave: How likely is a recurrence', *Bull. Amer. Meteor. Soc.*, **78**, 1107-1119.

Karl, T. R. and Knight, R.W.: 1998, 'Secular trends of precipitation amount, frequency, and intensity in the USA', *Bull. Amer. Meteor. Soc.*, 79, 231-242.

Katz, R.W. and Brown B.G.: 1992, 'Extreme events in a changing climate: Variability is more important than averages. *Climatic Change*, **21**, 289-302.

Knovitz, J.: 1990, Why cities don't die, *Amer. Hert. Inv. Tech.*, **5**, 59-63.

Kumar, K.K., Deshpande, N.R., and Kumar, K.R.: 1997, 'Long-term changes in the heavy rainfall events over India', Workshop Proceedings on Indices and Indicators for Climate Extremes, Asheville, NC. Sponsors, GCOS/CLIVAR and WMO.

Kushnir, Y., Cardone, V., Greenwood, J., and Cane, M.: 1997, ' On the recent increase in North Atlantic wave heights', *J. Climate* 2107-2113.

Lambert, S. J.: 1996, ' Intense extratropical Northern Hemisphere winter cyclone events: 1899-1991', *J. Geophys. Res.*, **101**, 21319-21325.

Landsea, C. W. and Gray, W.M.: 1992, Associations of Sahel monsoon rainfall and concurrent intense Atlantic hurricanes, *J. Climate*, **5**, 435-453.

Landsea, C.W.: 1999, 'Atlantic basin hurricanes: Indices of climatic changes', Workshop Proceedings on Indices and Indicators for Climate Extremes, Asheville, NC. Sponsors, CLIVAR, GCOS and WMO. *Climatic Change,* this issue.

Mantua, N.J., Hare, S.R., Zhang, Y., Wallace, J.M., and Francis, R.C. : 1998, 'A Pacific interdecadal climate oscillation with impacts on salmon production', *Bull. Amer. Meteor. Soc.*, 78, 1069-1079.

Mason, S.J., Waylen, P.R., Raiaratma, B., and Harrison, J. M.: 1998, 'Changes in extreme rainfall events in South Africa', *Climatic Change*. In press.

Mearns, L. O., Katz, R. W., and Schneider, S. H.: 1984, ' Extreme high-temperature events: Changes in their probabilities with changes in mean temperature', *J. Clim. Appl. Meteorol.*, **23**, 1601-1613.

Mitchell, J.K.: 1990, 'Natural hazards predictions and reponses in very large cities', in *Prediction and Perception of Natural Hazards*, Nemic, et al. (Eds.), Dordrecht, The Netherlands, pp. 29-37.

Munich Re, 1996: Annual review of natural catastrophes 1996. *Topics*, 1-16.

Nicholls, N.: 1997, 'CLIVAR and IPCC interests in extreme events', Workshop Proceedings on Indices and Indicators for Climate Extremes, Asheville, NC. Sponsors, CLIVAR, GCOS and WMO.

Parker, D.E., Legg, T.P., and Folland, C.K.: 1994, 'Interdecadal changes of surface temperatures since the late 19th century', *J. Geophys. Res.* 99, 14373-14399.

Pielke, R.A., Jr., and Landsea, C.W.: 1998, 'Normalized hurricane damages in the United States: 1925-1995', *Weather and Forecasting*, in press.

Plummer, N., Nicholls, N., Lavery, B.M., Leighton, R.M., and Trewin, B.C.: 1999, 'Twentieth century trends in Australian climate extremes indices', Workshop Proceedings on Indices and Indicators for Climate Extremes, Asheville, NC. Sponsors, GCOS/CLIVAR and WMO, *Climatic Change* this issue.

Rogers, J.C. and Rohli, R.V.: 1991, 'Atmospheric teleconnections and citrus freezes in the Great Plains', J. Climate, **4**, 1103-1113.

Rösner, S., Muller-Westermier, G., and Rosenhagen, G.: 1997, 'German data for indices/indicators for changes in extreme events', Workshop Proceedings on Indices and Indicators for Climate Extremes, Asheville, NC. Sponsors, GCOS/CLIVAR and WMO.

Ross, R.J. and Elliott, W.P.: 1996, 'Tropospheric water vapor trends over North America: 1973-93', Submitted *to J. Climate.*

Salinger, J.M.: 1997, 'Indices and indicators of changes in extreme events in the South Pacific preliminary results from New Zealand', Workshop Proceedings on Indices and Indicators for Climate Extremes, Asheville, NC. Sponsors, GCOS/CLIVAR and WMO.

Schneider, S.H.: 1998, 'Kyoto protocol: The unfinished agenda.' *Climatic Change*, **39,** 1-21.

Stein, O. and A. Hense: 1994, 'A reconstructed time series of the number of extreme low pressure events since 1880', *Meteorol Zeitschrift,* **N.F. 3**, 43-46.

Suppiah, R. and Hennessy, K. J.: 1996, 'Trends in intensity and frequency of heavy rainfall in tropical Australia and links with the Southern Oscillation', *Aust. Meteorol. Mag.*

Suppiah, R. and Hennessy, K. J.: 1997, 'Indices of changes in intense precipitation for Australia', Workshop Proceedings on Indices and Indicators for Climate Extremes, Asheville, NC. Sponsors, GCOS/CLIVAR and WMO.

Trenberth, K.E. and D.J. Shea: 1996, 'Atmospheric circulation changes and links to changes in rainfall and drought', 7th Conference on Climate Variations, American Meteorological Society, Long Beach, CA. J14-J19.

Zhai, P., S. Anjian, F. Ren, B. Gao, Q. Zhang, and X. Liu: 1999, 'Changes of climate extremes in China', Workshop Proceedings on Indices and Indicators for Climate Extremes, Asheville, NC. Sponsors, GCOS/CLIVAR. *Climatic Change,* this issue.

Zhai, P. and R. Eskridge: 1997, 'Atmospheric water vapor over China', *J. Climate*, 10, 2643-2652.

(Received 5 November 1997; in revised form 16 November 1998)

CONCEPTUAL FRAMEWORK FOR CHANGES OF EXTREMES OF THE HYDROLOGICAL CYCLE WITH CLIMATE CHANGE

KEVIN E. TRENBERTH

National Center for Atmospheric Research[1], *P. O. Box 3000, Boulder, CO 80307, U.S.A.*

Abstract. A physically based conceptual framework is put forward that explains why an increase in heavy precipitation events should be a primary manifestation of the climate change that accompanies increases in greenhouse gases in the atmosphere. Increased concentrations of greenhouse gases in the atmosphere increase downwelling infrared radiation, and this global heating at the surface not only acts to increase temperatures but also increases evaporation which enhances the atmospheric moisture content. Consequently all weather systems, ranging from individual clouds and thunderstorms to extratropical cyclones, which feed on the available moisture through storm-scale moisture convergence, are likely to produce correspondingly enhanced precipitation rates. Increases in heavy rainfall at the expense of more moderate rainfall are the consequence along with increased runoff and risk of flooding. However, because of constraints in the surface energy budget, there are also implications for the frequency and/or efficiency of precipitation. It follows that increased attention should be given to trends in atmospheric moisture content, and datasets on hourly precipitation rates and frequency need to be developed and analyzed as well as total accumulation.

1. Introduction

The character of precipitation, with highly variable rain rates and enormous spatial variability, makes simply determining mean precipitation difficult let alone how it will change as the climate changes. For instance, a detailed examination of spatial structure of daily precipitation amounts by Osborne and Hulme (1997) shows that in Europe the average separation distance between climate stations where the correlation falls to 0.5 is about 150 km in summer and 200 km in winter — the more convective nature of summer precipitation is responsible for the difference. In addition, this complexity also makes it difficult to model precipitation reliably, as many of the processes of importance can not be resolved by the model grid (typically 200 km) and so sub-grid-scale processes have to be parameterized. Yet there are some overall aspects of precipitation related to the hydrological

---

[1] The National Center for Atmospheric Research is sponsored by the National Science Foundation.

Climatic Change **42**: 327–339, 1999.
©1999 *Kluwer Academic Publishers. Printed in the Netherlands.*

cycle that can be clarified and for which expectations as to how they will change are physically based. Here the processes involved that influence precipitation and link it to evaporation and heating are outlined along with the importance of dealing not just with accumulated amounts, but also precipitation rates (or intensity) and precipitation frequency. The relative roles of moisture stored in the atmosphere, its advection, and resupply have been examined in detail in Trenberth (1998), and only a brief summary of those aspects are included here.

The term "global warming" is often taken to refer to global increases in temperature accompanying the increases in greenhouse gases in the atmosphere. In fact it should refer to the additional global heating (sometimes referred to as radiative forcing, e.g., by the IPCC (1996)) arising from the increased concentrations of greenhouse gases, such as carbon dioxide, in the atmosphere. Increases in greenhouse gases in the atmosphere produce global warming through an increase in downwelling infrared radiation, and thus not only increase surface temperatures but also enhance the hydrological cycle, as much of the heating at the surface goes into evaporating surface moisture. This occurs in all climate models regardless of feedbacks, although the magnitude varies substantially (see section 3).

Temperature increases signify that the water-holding capacity of the atmosphere increases and, together with enhanced evaporation, the actual atmospheric moisture should increase, as is observed to be happening in many places (Hense et al., 1988; Gaffen et al., 1991; Ross and Elliott, 1996; Zhai and Eskridge, 1997). Of course, enhanced evaporation depends upon the availability of sufficient surface moisture and over land, this depends on the existing climate. However, it follows that naturally-occurring droughts are likely to be exacerbated by enhanced potential evapotranspiration. Further, globally there must be an increase in precipitation to balance the enhanced evaporation but the processes by which precipitation is altered locally are not well understood.

It is shown that precipitating systems of all kinds feed mostly on the moisture already in the atmosphere at the time the system develops, and precipitation occurs through convergence of available moisture on the scale of the system. Hence, the atmospheric moisture content directly affects rainfall and snowfall rates, but not so clearly the precipitation frequency and thus total precipitation, at least locally. Thus, it is argued that global warming leads to increased moisture content of the atmosphere which in turn favors stronger rainfall events, as is observed to be happening in many parts of the world (Karl et al., 1995), thus increasing risk of flooding. It is further argued that one reason why increases in rainfall should be spotty is because of mismatches in the rates of rainfall versus evaporation. The arguments assembled here imply the need for new observations, datasets, and ways of analyzing both model and observed data. Trenberth (1998) discusses these aspects more fully.

2. Atmospheric Moisture Cycling

New estimates of the moistening of the atmosphere through evaporation at the surface and of the drying of the atmosphere through precipitation are given in Trenberth (1998). These are simple estimates based on the precipitable water and average local evaporation and precipitation rates, which ignore transport. Overall for the global annual mean, the e-folding residence time (the time for amounts to fall by a factor $e = 2.718$) for atmospheric moisture is just over 8 days. For precipitation, local values of e-folding residence time of the atmospheric depletion rate of moisture are less than a week in the tropical convergence zones but they exceed a month in the dry zones in the subtropics and desert areas. Time constants for depletion and restoration rates of atmospheric moisture are fairly similar overall, but this conclusion does not take account of the fact that rain falls only a small fraction of the time. In midlatitudes precipitation typically falls from zero to 30% of the time, and so rainfall rates, conditional on when rain is falling, are much larger than evaporation rates. The depletion rate time scale is about 4 hours in the tropics when rain is falling. In middle latitudes, typical unconditional rainfall rates are 3 mm/day, but with rain falling about 10% of the time and precipitable water amounts of 15 mm, the depletion rate time scale of 5 days drops conditionally on rain falling to about 12 hours (Trenberth, 1998). This inferred imbalance in the drying versus moistening of the atmosphere implies that most of the moderate and heavy rain that falls comes directly from the precipitable water already in the atmosphere at the time the storm responsible for the precipitation developed, not directly from evaporation, and so the lifetime of moisture in the atmosphere and its availability to rain systems is a limiting factor. However, atmospheric depletion of moisture by light rain could easily be restored by evaporation.

These above aspects do not take moisture transport into account. Therefore new estimates have also been made of how much precipitated moisture comes from evaporation from within versus transport from outside a domain, called recycling. Approximate values of recycling are computed following the approach of Brubaker et al. (1993), as detailed in Trenberth (1998). Equilibrium conditions are assumed, so that there are no changes in atmospheric moisture content but changes in moisture storage in the atmosphere do not impact the results for seasonal or longer averages. A domain of length L aligned along the trajectory of the air is considered. An important assumption is that the atmosphere is well mixed so that the ratio of precipitation that falls arising from advection versus local evaporation

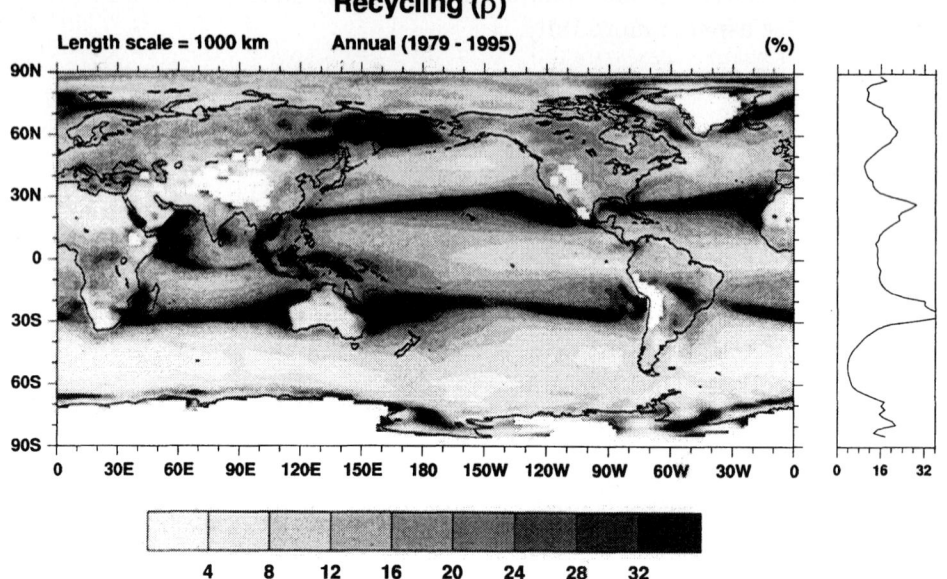

Figure 1. The recycling, for annual mean conditions, for length scales of 1,000 km, and using *E* and moisture flux from the NCEP reanalyses (Kalnay et al., 1996) and *P* from CMAP (Xie and Arkin, 1997). Values are set to missing (white) where the surface pressure is less than 800 mb.

is equal to the ratio of average advected to evaporated moisture in the air. While interest has often been on recycling estimates for large drainage basins, the heterogeneity of the land surface is such that the recycling clearly varies substantially over the basins. The regions of mountains (where surface pressures are less than 800 mb) are screened out from the calculation, as those are regions where the moisture flux is small and there are huge variations over short distances owing to orographic effects on rainfall.

In Trenberth (1998) recycling results for annual means are presented for $L = 500$ km. Here results presented for $L = 1000$ km (Fig. 1) reveal recycling percentages of about 8 to 20% over land typically. For 500 km scales the global mean is 9.6%, consisting of 8.9% over land and 9.9% over the oceans and for 1000 km scales the mean recycling is 16.8% globally, 15.4% over land and 17.3% over the oceans. Over the Amazon, the average is about 10% and over the Mississippi basin about 12%. These values prove to be compatible with most previous extimates (e.g., Brubaker et al., 1993) once the different scales of the basins are taken into account. It is worth pointing out that the larger values previously obtained for the Amazon versus the Mississippi are mostly a result of the scale of the domain.

The recycling fraction depends greatly on the magnitude of the total moisture flux. In the computations, this includes advection by the mean flow as well as the transient eddies. Relatively high values (>30%) of recycling occur either in the subtropical highs, where evaporation E is high and the advective moisture flux is small, or in convergence zones where, again, the advective moisture flux is small. Low values occur over the southern oceans, the North Pacific, and the eastern equatorial Pacific, where the moisture flux is at a maximum. All of these recycling values show that on average less than 20% of the precipitation that falls comes from evaporation within a distance of about 1000 km. Therefore the results reinforce the arguments given above concerning the importance of transport of moisture and local storage in feeding precipitating systems.

The dominant storm-scale process in both thunderstorms and extratropical storms is the convergence of moisture by the storm-scale circulation. The latter determines how much moisture is available to the system and can vary in size from a few tens of kilometers to over 2000 km spatial scales. The advected moisture may combine with the in situ moisture to feed the storm but it is not all available as the relative humidity can not be reduced to zero, except perhaps approximately in strong down drafts very locally. The efficiency of thunderstorms is observed to vary from about 20% to 50%. "Precipitation efficiency" is defined as the ratio of the water mass precipitated to the mass of water vapor entering the storm through its base (e.g., Fankhauser, 1988) or the ratio of total rainfall to total condensation in modeling studies (e.g., Ferrier et al., 1996).

In the United States, much of the moisture for precipitation, especially in the winter half year, comes from moisture transported out of the subtropics often in a southwesterly flow ahead of cold fronts. For storms east of the Rockies, moisture flows northwards from the Gulf of Mexico or subtropical Atlantic. At advection rates of 12 m s^{-1} (which is the standard deviation of the northward velocity component at 850 mb just north of the Gulf of Mexico in January), the moisture travels over 1,000 km in a day, so that moisture from the Gulf can be readily precipitated out over the Great Plains or Ohio Valley just a day or so later. In major storms, transient northward advection rates often exceed 20 m s^{-1} at 850 mb. In the western United States, the moisture comes from the subtropical Pacific. Therefore much of the extratropical precipitation originates from moisture advected from the Gulf of Mexico and subtropical Atlantic or Pacific a day or so earlier and it is estimated that about 70% to 75% of the moisture in an extratropical storm comes from moisture that was stored in the atmosphere at the beginning of the storm and brought into the region by the storm-scale circulation. For thunderstorms, whose life is a few hours, nearly all of the precipitated moisture comes from moisture that was already in the atmosphere at the time the storm began.

3. Relevance to Climate Change

The above discussion reveals the mismatch between precipitation rates and evaporation, so that moderate and heavy precipitation, which contributes most to the total accumulation, depends upon the moisture already in the atmosphere and the advection and resupply of moisture by the storm circulation. These points are pertinent to climate change experiments with global climate models. However, most climate model studies have not analyzed the results in a way that throws light on these aspects. The surface heat budget is especially relevant.

There are many feedback processes in nature that can either amplify or diminish the climate response to increases in greenhouse gases. The net radiative forcing or "warming" at the surface depends critically on these and the surface heat budget. In every case it seems that at the surface there is an increase in downwelling infrared radiation associated with both the greenhouse effect from carbon dioxide and other greenhouse gases, as well as changes in water vapor and clouds. In some models, changes in clouds produce an offset by reducing shortwave radiation, but the net energy available from radiation at the surface is increased in spite of the greater surface emissions associated with the higher temperatures. Moreover, changes in the sensible heat flux also act to warm the surface because of stabilization of the lower atmosphere (Boer, 1993; Roads et al., 1996).

This leaves only the latent heat flux through increased evaporation to compensate and balance the surface heat budget. The latent heat flux, which ranges from 3 to 10 W m^{-2} for CO_2 doubling for the four models considered by Boer (1993), determines the global enhancement of the hydrological cycle and average precipitation rate (of about 3 to 10%). However, the atmospheric moisture content increases by about 20% (Mitchell et al., 1987) or more (in the case of the CCM2, Roads et al., 1996) although with very little change in model relative humidity. With other things kept constant, moisture convergence would be enhanced by the same amount and should lead to similarly enhanced precipitation. But a 20% increase in precipitation cannot occur because of the limitations associated with the surface energy budget. Nevertheless such mechanisms should take place for individual storms, whether thunderstorms, or extratropical cyclones, leading to increased rainfall rates. If this is the case, however, there are implications for the frequency of storms or other factors that must come into play to restrict the total precipitation.

One factor clearly of importance is that the moisture increases are not uniform. Generally, evaporative cooling is more important in the tropics and subtropics. Bigger increases occur in lower latitudes because of the non-linear nature of the Clausius-Clapeyron equation in spite of larger increases in surface temperatures at high latitudes. Thus much of this moisture may not be within reach of many extratropical storms. Another factor is the precipitation efficiency, discussed above. How precipitation efficiency might change with climate change is not known and this is not a factor that can be dealt with by current climate models. Warmer conditions could imply that more moisture might remain in the atmosphere if this is determined by relative humidity, as is likely. Therefore the rainfall may not increase in direct proportion to the moisture convergence, because more moisture is left in the atmosphere.

In most models, surface temperature increases with increased greenhouse gases are greatest in the Arctic, in part because of ice-albedo feedback, so that the meridional surface temperature gradient and baroclinicity is reduced, although this may not be the case above the surface. Therefore another factor relates to extratropical storms and the overall baroclinicity, as argued by Held (1993). Held notes that one effect of increased moisture content in the atmosphere is to enhance the latent heating in such storms and thereby increase their intensity. But he also notes that more moist air would be transported polewards, reducing the required poleward energy transports normally accomplished by baroclinically unstable eddies and the associated poleward down-gradient heat transports. He therefore argues that this would contribute to "smaller eddies" and a decrease in eddy amplitudes. While recognizing that both effects are important, Held suspects that the latter is dominant. There are other possibilities not considered by Held. In particular, individual storms could be more intense from the latent heat enhancement, but fewer and farther between. Changes in the vertical temperature structure (the lapse rate) will also play a role in such storms.

Therefore the other major factor worth considering in more detail is the frequency of precipitation events. The above discussion suggests that for rain rates to increase faster than rain amounts, then the frequency of rain should decrease. However, this would only apply globally. A preliminary examination of trends in frequency of precipitation events for the United States computed over the period 1963 to 1994 in Trenberth (1998) shows that the most notable statistically significant trends are for increases in the southern United States in winter and decreases in the Pacific Northwest from November through January, which may be related to changes in atmospheric circulation and storm tracks associated with the trend toward more El Niño events (Trenberth and Hoar, 1996). For instance, an example has been the 1997-98 El Niño winter which featured heavy rains across the southern states from California to Florida, while somewhat drier conditions generally prevailed across the northern states.

These aspects have been explored only to a limited extent in climate models. None deal with true intensity of rainfall, which requires hourly (or higher resolution) data, as the analysis is of daily rainfall amounts. Cubasch et al. (1995) and Hennessy et al. (1997) have analyzed changes in intensity and frequency in coarse resolution models with increased CO_2. Cubasch et al. note that while precipitation change does not display a clear signal, increases in rain intensity and dry periods are simulated in the ECHAM3 model. The UKHI and CSIRO9 models (Hennessy et al., 1997) are consistent in showing heavier rainfall events with doubled CO_2, a general decrease in the probability of moderate precipitation, and an increase in no or light precipitation. Return periods for extreme events whose period is greater than one year decrease by factors of 2 to 5. Hennessy et al. further argue that the frequency of precipitation should be expected to decrease with increases in intensity, and find this to be true in the model simulations for the most part.

An analysis by Mearns et al. (1995) used a nested regional model with 60 km resolution for regions of the United States for control and doubled-carbon dioxide results. They explored the frequency and intensity of modeled precipitation but only for daily values, not the true precipitation rates. Results revealed increased daily rainfall variability under doubled CO_2. There are some areas where frequency of precipitation decreases but precipitation mean daily amounts increase. Overall, however, they find both increases and decreases of both precipitation frequency and intensity. Jones et al. (1997) produced results over Europe using a similar technique and a nested model with 50 km resolution. They find a substantial increase in precipitation intensity in extreme events, and were able to trace most of that increase simply to the increased atmospheric moisture concentrations in the models. While moderate precipitation decreased, the frequency of dry days also increased along with an increase in evaporation, and so these were all symptoms of an increased hydrological cycle.

4. Conclusions and Recommendations

The arguments on how climate change can influence moisture content of the atmosphere, and its sources and sinks are assembled in the schematic in Fig. 2. This provides the sequence described earlier. The sequence given is simplified by omitting some of the feedbacks that can interfere. For example, an increase in atmospheric moisture may lead to increased relative humidity and increased clouds, which could cut down on solar radiation (enhance shortwave cloud forcing) and reduce the energy available at the surface for evaporation. Those feedbacks are included in the climate models and alter the magnitude of the surface heat available for evaporation in different models but not its sign. Figure 2 provides the rationale for why rainfall rates and frequencies as well as accumulations are important in understanding

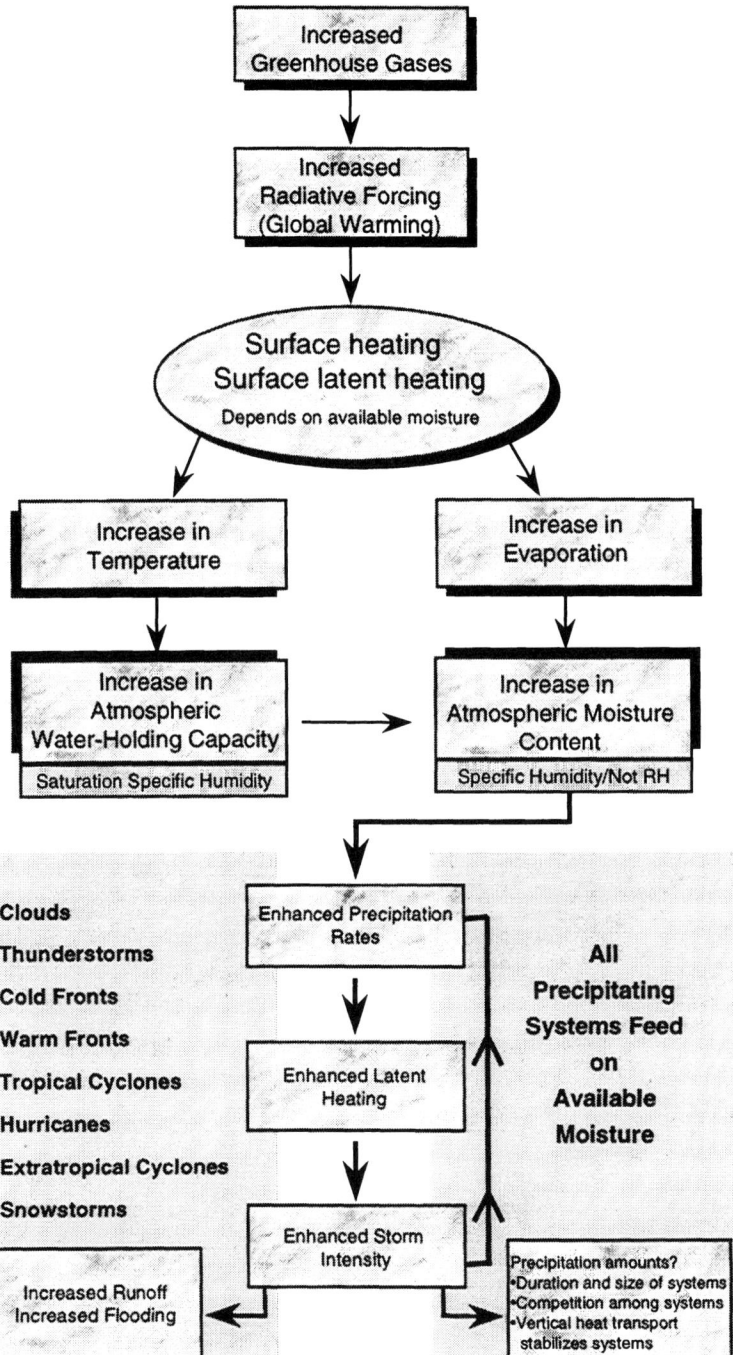

Figure 2. Schematic outline of the sequence of processes involved in climate change and how they alter moisture content of the atmosphere, evaporation, and precipitation rates. All precipitating systems feed on the available moisture leading to increases in precipitation rates and feedbacks.

what is going on with precipitation locally. The accumulations depend greatly on the frequency, size and duration of individual storms, as well as the rate (Byers, 1948) and these depend on static stability and other factors as well. In particular, the need to vertically transport heat absorbed at the surface is a factor in convection and baroclinic instability both of which act to stabilize the atmosphere. Increased greenhouse gases also stabilize the atmosphere. Those are additional considerations in interpreting model responses to increased greenhouse gas simulations.

Another clearly important factor in interpreting observed and modeled changes, not explored here, is the changes in atmospheric circulation which can alter the location and intensity of storm tracks and thereby lead to dipole structures in precipitation changes, with decreases in rainfall in some areas and increases in others. For example, Trenberth and Guillemot (1996) show how storm tracks changed across North America to help bring about the spring-summer 1988 drought and 1993 floods.

There is firm evidence that moisture in the atmosphere is increasing. In the Western Hemisphere north of the equator, annual mean precipitable water amounts below 500 mb are increasing over the United States, Caribbean and Hawaii by about 5% per decade as a statistically significant trend from 1973 to 1993 (Ross and Elliott, 1996), and these correspond to significant increases in relative humidities of 2 to 3% per decade over the Southeast, Caribbean and subtropical Pacific. Precipitable water and relative humidities are not increasing significantly over much of Canada, however, and are decreasing slightly in some areas. In China, recent analysis by Zhai and Eskridge (1997) also reveals upward trends in precipitable water in all seasons and for the annual mean from 1970 to 1990. Earlier, Hense et al. (1988) revealed increases in moisture in the western Pacific. A claim for recent drying in the tropics by Schroeder and McGuirk (1998) using TOVS data is questionable owing to the changes in instruments and satellites. Clearly, there is a need to obtain more reliable atmospheric moisture trends over the entire globe.

Moreover, there is clear evidence that rainfall rates have changed in the United States. The incidence of heavy rainfall events has steadily increased at the expense of moderate rainfall events throughout this century. This has been shown by an analysis of the percentage of the U.S. area with much above normal proportion of total annual precipitation from 1 day extreme events, where the latter are defined to be more than 2 inches (50.8 mm) amounts (Karl et al., 1996). The "much above normal proportion" is defined to be the upper 10%. This quantity can be reliably calculated from 1910, and the percentage has increased steadily from less than 9 to over 11%, a 20% change. Karl and Knight (1998) have provided further analysis of U.S. precipitation increases and show how it occurs mostly in the upper tenth percentile of the distribution and that the portion of total precipitation derived from extreme and heavy events is increasing at the expense of more moderate events. Other evidence for increasing precipitation rates occurs in Japan (Iwashima and Yamamoto, 1993) and Australia (Suppiah and Hennessy, 1996).

It has been argued that increased moisture content of the atmosphere favors stronger rainfall and snowfall events, thus increasing risk of flooding. As noted, there is a pattern of heavier rainfalls observed in many parts of the world where the analysis has been done. However, flooding records are confounded by changes in land use and increasing settlement of flood plains. Moreover, because there is a disparity between the rates of increase of atmospheric moisture and precipitation, there are implied changes in the frequency of precipitation and/or efficiency of precipitation (related to how much moisture is left behind in a storm).

These arguments may help to explain the exceptional rain and snow falls over the U.S. in the winter of 1996-97. These included heavy rains and flooding in the Pacific Northwest in early January, where observed increases in moisture content of the atmosphere at Hawaii and in the subtropical Pacific (Ross and Elliott, 1996) are especially pertinent. Also, heavy snowfalls in the Great Plains and Upper Mississippi Basin led to extensive flooding in the spring of 1997 as snows melted, and heavy rains in the Ohio River Valley which, along with snow melt, also produced extensive flooding. Note that the primary argument here is not that these flooding events would not have occurred but that they have probably been enhanced, perhaps by as much as 10%, because of the increased moisture in the atmosphere, over what would have occurred two decades ago.

The above arguments suggest that there is not such a clear expectation on how local total precipitation amounts should change, except as an overall global average. With higher average temperatures in winter expected, more precipitation is likely to fall in the form of rain rather than snow, which will increase both soil moisture and run off, as noted by the IPCC (1996) and found in many models. In addition, faster snow melt in spring is likely to aggravate springtime flooding. In other places, dipole-like structures of precipitation change should occur in places where storm tracks shift meridionally. Beyond this, it is suggested that examining moisture content, rainfall rates and frequency of precipitation and how they change with climate change may be more important and fruitful than just examining precipitation amounts in understanding what is happening in model projections. To be compatible with life times of significant rain events, yet still deal with whole storms rather than individual rain cells, hourly precipitation data are recommended. Such data are also retrievable from climate models.

Acknowledgments

This research is partly sponsored by NOAA under grant NA56GP0247 and by a joint NOAA/NASA grant NA56GP0576. I thank Dave Stepaniak for producing Fig. 1, and Liz Rothney for Fig. 2.

References

Boer, G. J.: 1993, 'Climate change and the regulation of the surface moisture and energy budgets.' *Clim. Dyn.* **8**, 225–239.

Brubaker, K. L., Entehabi, D., and Eagleson, P. S.: 1993, 'Estimation of continental precipitation recycling.' *J. Clim.* **6**, 1077–1089.

Byers, H. R.: 1948, 'The use of radar in determining the amount of rain over a small area.' *EOS Trans. AGU* **29**, 187–196.

Cubasch, U., Waszkewitz, J., Hegerl, G., and Perlwitz, J.: 1995, 'Regional climate changes as simulated in time-slice experiments.' *Clim. Change* **31**, 273–304.

Fankhauser, J. C.: 1988, 'Estimates of thunderstorm precipitation efficiency from field measurements in CCOPE.' *Mon. Wea. Rev.* **116**, 663–684.

Ferrier, B. S., Simpson J., and Tao, W-K.: 1996, 'Factors responsible for precipitation efficiencies in midlatitude and tropical squall simulations.' *Mon. Wea. Rev.* **124**, 2100–2125.

Gaffen, D. J., Barnett, T. P., and Elliott, W. P.: 1991, 'Space and time scales of global tropospheric moisture.' *J. Clim.* **4**, 989–1008.

Held, I. M.: 1993, 'Large-scale dynamics and global warming.' *Bull. Am. Meteorol. Soc.* **74**, 228–241.

Hennessy, K. J., Gregory, J. M., and Mitchell, J. F. B.: 1997, 'Changes in daily precipitation under enhanced greenhouse conditions.' *Clim. Dyn.* **13**, 667–680.

Hense, A., Krahe P., and Flohn, H.: 1988, 'Recent fluctuations of tropospheric temperature and water vapour content in the tropics.' *Meteorol. Atmos. Phys.*, **38**, 215-227

IPCC (Intergovernmental Panel of Climate Change): 1996, *Climate Change 1995: The Science of Climate Change.* Eds. J. T. Houghton, F. G. Meira Filho, B. A. Callander, N. Harris, A. Kattenberg, and K. Maskell, Cambridge Univ. Press, Cambridge, U.K., 572pp.

Iwashima, T., and Yamamoto, R.: 1993, 'A statistical analysis of the extreme events: Long-term trend of heavy daily precipitation.' *J. Met. Soc. Japan* **71**, 637–640.

Jones, R. G., Murphy, J. M., Noguer, M., and Keen, A. B.: 1997, 'Simulation of climate change over Europe using a nested regional-climate model II: Comparison of driving and regional model responses to a doubling of carbon dioxide.' *Quart. J. Roy. Met. Soc.* **123**, 265–292.

Kalnay E, Kanamitsu, M., Kistler, R., Collins, W., Deaven, D., Gandin, L., Iredell, M., Saha, S., White, G., Woollen, J., Zhu, Y., Chelliah, M., Ebisuzaki, W., Higgins, W., Janowiak, J., Mo, K-C., Ropelewski, C., Leetmaa, A., Reynolds, R., and Jenne, R. 1996, 'The NCEP/NCAR Reanalysis Project.' *Bull. Am. Meteorol. Soc.* **77**, 437–471

Karl, T. R., and Knight R. W.: 1998, 'Secular trends of precipitation amount, frequency and intensity in the USA.' *Bull. Am. Meteorol. Soc.* **79**, 231–242.

Karl, T. R., Knight, R. W., Easterling, D. R. and Quayle, R. G.: 1996; 'Indices of climate change for the United States. *Bull. Am. Meteorol. Soc.* **77**, 279–292.

Karl, T. R., Knight, R. W. and Plummer, N.: 1995, Trends in high frequency climate variability in the twentieth century.' *Nature* **377**, 217-220.

Mearns, L. O., Giorgi, F., McDaniel, L., and Shields, C.: 1995, 'Analysis of daily variability of precipitation in a nested regional climate model: comparison with observations and doubled CO_2 results.' *Global Planetary Change* **10**, 55–78.

Mitchell, J. F. B., Wilson C. A., and Cunnington, W. M.: 1987, 'On CO_2 climate sensitivity and model dependence of results.' *Quart. J. Roy. Met. Soc.* **113**, 293–322.

Osborn, T. J., and Hulme, M.: 1997, 'Development of a relationship between station and grid-box rainday frequencies for climate model evaluation.' *J. Clim.* **10**, 1885–1908.

Roads, J. O., Marshall, S., Oglesby R., and Chen, S-C.: 1996, 'Sensitivity of the CCM1 hydrological cycle to CO_2.' *J. Geophys. Res.* **101**, 7321–7339.

Ross, R. J., and Elliot, W. P. 1996, 'Tropospheric water vapor climatology and trends over North America: 1973-93.' *J. Clim.* **9**, 3561–3574.

Schroeder, S. R., and McGuirk, J. P.: 1998, 'Widespread tropical atmsopheric drying from 1979 to 1995.' *Geophys. Res. Lett.* **25**, 1301–1304.

Suppiah, R., and Hennessy, K. J.: 1998, 'Trends in the intensity and frequency of heavy rainfall in troipcal Australia and links with the Southern Oscillation.' *Aust. Meteorol. Mag.*, **45**, 1–17.

Trenberth, K. E.: 1998, 'Atmospheric moisture residence times and cycling: Implications for rainfall rates with climate change.' *Clim. Change*, **36**, (in press).

Trenberth, K. E., and Guillemot, C. J.: 1996, 'Physical processes involved in the 1988 drought and 1993 floods in North America.' *J. Clim.* **9**, 1288–1298.

Trenberth, K. E., and Hoar, T. J.: 1996, 'The 1990–1995 El Niño-Southern Oscillation event: Longest on record.' *Geophys. Res. Lett.* **23**, 57–60.

Xie, P, and Arkin, P. A.: 1997, 'Global precipitation: A 17-year monthly analysis based on gauge observations, satellite estimates and numerical model outputs.' *Bull. Am. Meteorol. Soc.* **78**, 2539–2558.

Zhai, P., and Eskridge, R. E.: 1997, 'Atmospheric water vapor over China.' *J. Clim.* **10**, 2643–2652.

(Received 5 November 1997; in revised form 11 August 1998)

ON TESTING FOR CHANGE IN EXTREME EVENTS

ANDREW R. SOLOW

Woods Hole Oceanographic Institution, Woods Hole, MA 02543, U.S.A.

Abstract. A common problem in climatology is detecting a change in the frequency or magnitude of extreme events in an historical time series. This paper compares the performance of two general approaches to this problem in a simple situation. The first approach is based on modelling the entire distribution, while the second approach focuses on the tail of the distribution. Although the second approach has clear advantages, the results of this paper suggest that it can also involve a substantial loss of information.

1. Introduction

As the papers in this collection attest, interest in climate extremes is growing. In many situations, particular interest centers on detecting a trend in the frequency or magnitude of extreme events in an historical time series (e.g., IPCC, 1995, Chapter 3). As illustrated below, there are two broad statistical approaches to this problem. Under the first approach, inference about extremes is based on modelling the entire time series. Under the second approach, inference is based on modelling the extreme events themselves. For economy, I will refer to these as the traditional and extreme approaches, respectively.

Methodological development under the extreme approach has been active. Recent reviews include Leadbetter, et al. (1983), Resnick (1987), and Davison and Smith (1990). As discussed in these reviews, the extreme approach is attractive for at least two reasons. First, a trend may be confined to one or both tails of a distribution, so that focusing the analysis in this region does not involve a substantial loss of information. Second, the approach enjoys certain robustness properties. Despite these undeniable attractions, in certain realistic situations, the use of this approach involves a loss of information. Although it may well be outweighed by increased robustness, it is still important to understand how large the information loss may be. It is natural to measure the loss of information by comparing the powers of the tests based on the two approaches under a specific model of the data-generating process. This is common practice in comparing alternative tests (e.g., Hodges and Lehmann, 1956).

The purpose of this paper is to present some results comparing the power of tests based on the two approaches in two different contexts: detecting a trend in the frequency of extreme events and detecting a trend in their magnitude. These results suggest that the loss of information from adopting the extreme approach can be substantial. However, it is important to emphasize that these results concern only the potential *costs* of adopting this approach and not the potential *benefits* that have been stressed in the references given above.

The paper is organized in the following way. Section 2 describes the basic

Climatic Change **42**: 341–349, 1999.
© 1999 *Kluwer Academic Publishers. Printed in the Netherlands.*

model under which the power of the tests are evaluated. Section 3 discusses the problem of testing for an increasing trend in the frequency of extreme events, while Section 4 discusses the problem of testing for an increasing trend in their magnitude. The results of a small simulation experiment are presented in Section 5. Section 6 contains some concluding remarks.

2. The Basic Model

Consider a continuous-valued, discrete time process Y_t, $t = 1, 2, ..., n$. The basic model considered in this paper is:

$$Y_t = \sigma_t \, \varepsilon_t \tag{1}$$

where $\varepsilon_1, \varepsilon_2, ..., \varepsilon_n$ is an independent sequence of normal random variables with mean 0 and variance 1. Suppose further that:

$$\sigma_t^2 = \exp(\sigma t) \tag{2}$$

so that the unknown growth rate of the variance of Y_t is σ. This form of trend in variability ensures that the variance of Y_t is positive. If $\sigma = 0$, then there is no change in the variability of Y_t. If σ is small, the trend in variance is approximately linear. This model might be appropriate for the logarithm of annual total precipitation after any estimated trend has been removed. Figure 1 shows a realization of this process with $n = 60$ and $\sigma = 0.0231$. For this value of σ, the variance of Y_t doubles in 30 years. It is important to note that, under this basic model, there is no change in the shape of the distribution of Y_t (although there is a change in the shape of the lognormal distribution of $\exp(Y_t)$).

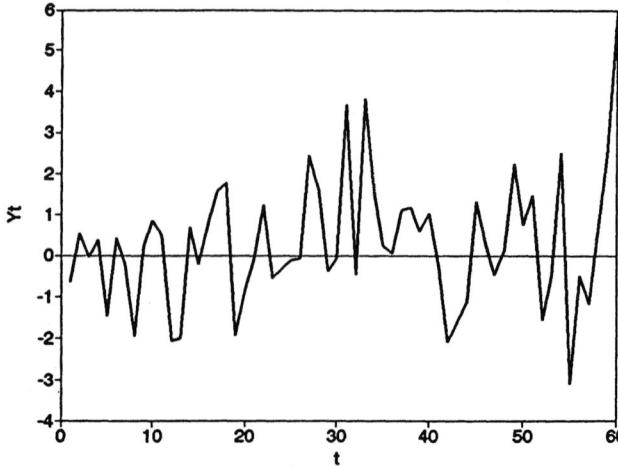

Figure 1. A simulated realization of the basic model with $n = 60$ and $\sigma = 0.0231$.

3. Testing for Increasing Exceedance Probability

Let $y > 0$ be a fixed level and suppose that interest centers on whether or not the exceedance probability:

$$p_t(y) = \text{prob}(Y_t > y) \tag{3}$$

increases with t. It is important to note that, while y may refer to a threshold of direct interest, this is not necessarily the case: y may be chosen to focus attention on the upper tail of the distribution of Y_t. For the model outlined in the previous section:

$$p_t(y) = 1 - \Phi(y \exp(-\sigma t/2)) \tag{4}$$

where Φ is the standard normal distribution function. The question of interest can be addressed formally by testing the null hypothesis H_o: $\sigma = 0$ against the one-sided alternative hypothesis H_1: $\sigma > 0$.

The test can be based on the likelihood ratio (LR) statistic:

$$\Lambda = -2 (\log L_o - \log L_1(\sigma^*)) \tag{5}$$

where L_o is the likelihood of the data under H_o and $L_1(\sigma)$ is the likelihood of the data evaluated at the maximum likelihood (ML) estimate σ^* of σ under H_1. In this simple case, the model is fully specified under H_o, while under H_1 there is a single parameter σ to estimate. The null hypothesis is rejected at approximate significance level α if $\lambda > \chi_1^2(2\alpha)$ where λ is the observed value of Λ and $\chi_1^2(2\alpha)$ is the upper 2α-quantile of the chi-squared distribution with 1 degree of freedom (Cox and Hinkley, 1974, p. 320).

The implementation of the LR test depends on the nature of the observations. Under the traditional approach, the process Y_t itself is observed. In this case:

$$\log L_o = -\Sigma \, y_t^2 / 2 \tag{6}$$

where here and below the summation runs from $t = 1$ to $t = n$ and y_t is the observed value of Y_t. Also:

$$\log L_1(\sigma) = -\Sigma \, (\sigma t + y_t^2 \exp(-\sigma t)) / 2 \tag{7}$$

The ML estimate of σ under H_1 can be found by maximizing (7) numerically subject to the constraint that the estimate is non-negative.

Under the simplest version of the extreme approach, the data consist of observations of whether or not an exceedance occurred in each period:

$$Z_t = 1 \text{ if } Y_t > y \tag{8}$$

0 otherwise

In this case:

$$\log L_o = \Sigma \ (z_t \log \ (1\text{-}\Phi(y)) + (1\text{-}z_t) \log \ \Phi(y)) \tag{9}$$

where z_t is the observed value of Z_t and:

$$\log L_1(\sigma) = \Sigma \ (z_t \log \ (1\text{-}\Phi(y \ \exp(\text{-}\sigma t/2)) +$$
$$(1\text{-}z_t) \log \ \Phi(y \ \exp(\text{-}\sigma t)/2)) \tag{10}$$

The ML estimate of σ under H_1 can again be found numerically by maximizing (10) subject to the non-negativity constraint.

In Section 5, some simulation results are presented comparing the power of these two tests.

4. Testing for Increasing Magnitude of Extremes

In some situations, interest centers on detecting a trend in the magnitude of extreme events. For example, Smith (1989) was interested in detecting a trend in the magnitude of extreme ground-level ozone events. Specifically, suppose that interest centers on the expected value of Y_t given that Y_t exceeds a threshold $y >$ 0. Under the basic model described in Section 2, this conditional expected value is:

$$E(Y_t \mid Y_t > y) =$$
$$\exp(\sigma t/2) \ \phi(y \ \exp(\text{-}\sigma t/2)) \ / \ (1\text{-}\Phi(y \ \exp(\text{-}\sigma t/2))) \tag{11}$$

where $\phi = \Phi'$ is the standard normal probability density function. The expression in (11) is an increasing function of σ. Thus, an increase in the magnitude of extreme events can again be detected by testing H_o: $\sigma = 0$ against H_1: $\sigma > 0$. Moreover, under the traditional approach, the LR test for an increase in the frequency of extreme events is also the LR test for an increase in their magnitude.

An extreme approach analysis could be based on the magnitudes of the observed exceedances of the threshold, discarding any observations of Y_t that do not exceed the threshold. This is closely related to the peaks-over-threshold approach that has long been used in hydrology. Suppose that exceedances occur at times $t(1)$, $t(2)$, ..., $t(M)$. The total number M of these exceedances is a random variable. In the sequel, I will condition on their observed number m. Let $Z_i = Y_{t(i)}$. The probability density function of Z_i is:

$$f_i(z) = \sigma_{t(i)}^{-1} \ \phi(z \ / \ \sigma_{t(i)}) \ / \ (1 - \Phi(y \ / \ \sigma_{t(i)})) \tag{12}$$

with $z > y$, where $\sigma_{t(i)}^2 = \exp(\sigma\, t(i))$. In this case, the log likelihood function:

$$\log L(\sigma) = \Sigma \, \log f_i(z_i) \tag{13}$$

(where i runs from 1 to m) does not simplify very much, particularly under H_1. However, it is straightforward to evaluate the LR statistic numerically and the LR test can be performed using the chi-squared approximation as in the previous section.

5. Some Results

This section presents the results of a simulation experiment aimed at comparing the power of the traditional and extreme approaches to detecting trends in the frequency and magnitude of extremes under the basic model described in Section 2. The experiment proceeded in the following way. For selected combinations of n, σ, and y, 1000 time series were simulated from the basic model. The traditional test (which serves as a test for trend in both frequency and magnitude) and the extreme approach tests were applied at the 0.05 significance level. The power of each test was estimated by the proportion of simulated time series for which H_o was rejected.

The values of n, σ, and y considered in this study were (30, 60, 90), (0.0077, 0.0116, 0.0231), and (0.5, 1.0, 1.5), respectively. These values of σ correspond to doubling times for variance of 90, 60, and 30 years, respectively. For illustration, the trends in exceedance probability (3) for these values of σ when y = 1.0 are shown in Figure 2. The trends have similar shape for the other values of y, although the exceedance probabilities are higher when y = 0.5 and lower

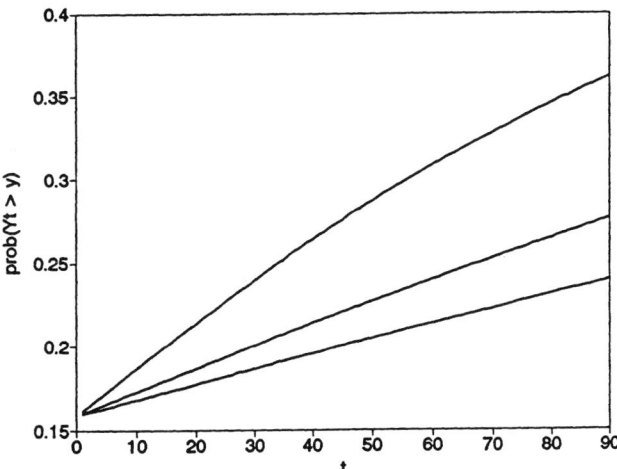

Figure 2. Exceedance probabilities with y = 1.0 and σ = 0.0077 (lower curve), σ = 0.0116 (middle curve), and σ = 0.0231 (upper curve).

when y = 1.5. The results of testing for a trend in the frequency of extreme events are given in Table 1. These results show that the power of the traditional test (which does not depend on the threshold y) is generally considerably higher than that of the test based on the extreme approach. The two exceptions are when the time series is short and the trend in frequency is weak, in which case neither test is powerful, and when the time series is long and the trend is strong, in which case both tests have high power.

Figure 3 shows the trends in mean exceedance (11) for the three values of σ when y = 1.0. Again, the trends for the other values of y have the same shape, but mean exceedance is lower when y = 0.5 and higher when y = 1.5. The extreme approach test for a trend in the magnitude of extreme events was also applied at the 0.05 significance level. As noted, the number M of exceedances is a random variable. When n is small and y is large, this number can be quite small. This calls into question the adequacy of the chi-squared approximation to the distribution of Λ under H_o. If this approximation is inadequate, significance can be assessed by a simple randomization procedure. In the experiment described here, only realizations with at least 5 exceedances were retained and the chi-squared approximation worked reasonably well, although it is somewhat conservative.

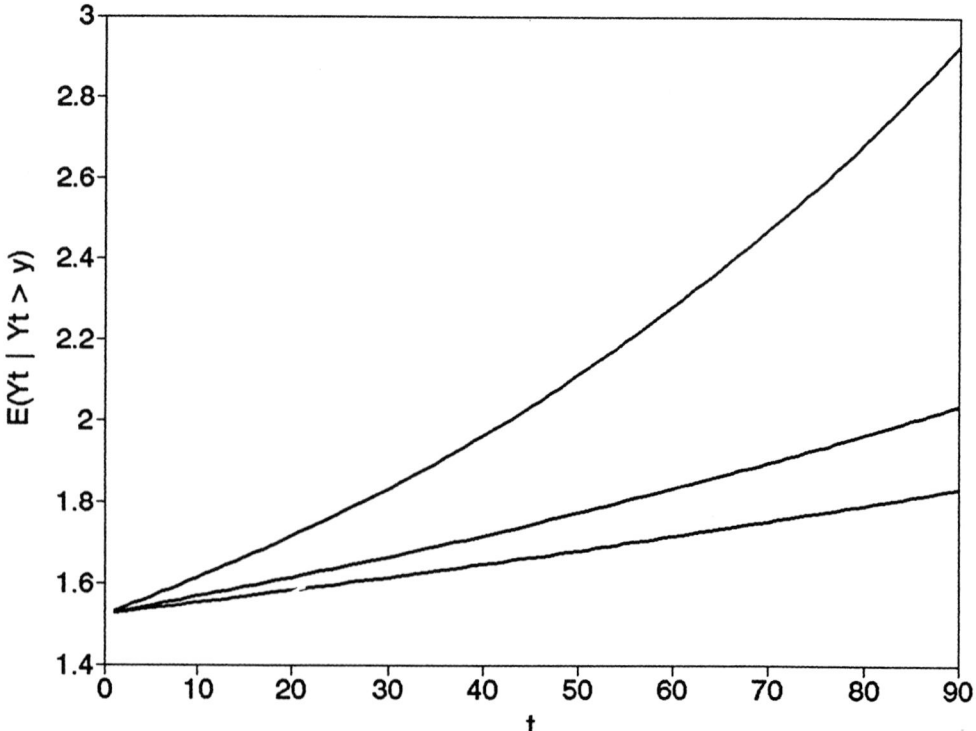

Figure 3. Mean exceedances with y = 1.0 and σ = 0.0077 (lower curve), σ = 0.0116 (middle curve), and σ = 0.0231 (upper curve).

Table I

Estimated power of the traditional test (Test 1), the extreme approach test for a trend in the frequency of extreme events (Test 2), and the extreme approach test for a trend in the magnitude of extreme events.

$\sigma = 0.0077$

n	y	Test 1	Test 2	Test 3
30	0.5	0.12	0.06	0.09
	1.0		0.08	0.06
	1.5		0.10	0.09
60	0.5	0.43	0.11	0.25
	1.0		0.17	0.17
	1.5		0.20	0.14
90	0.5	0.86	0.17	0.56
	1.0		0.33	0.45
	1.5		0.43	0.33

$\sigma = 0.0116$

n	y	Test 1	Test 2	Test 3
30	0.5	0.19	0.06	0.10
	1.0		0.09	0.09
	1.5		0.12	0.10
60	0.5	0.72	0.13	0.43
	1.0		0.26	0.32
	1.5		0.37	0.27
90	0.5	0.99	0.26	0.82
	1.0		0.52	0.72
	1.5		0.70	0.57

$\sigma = 0.0231$

n	y	Test 1	Test 2	Test 3
30	0.5	0.50	0.09	0.29
	1.0		0.18	0.22
	1.5		0.23	0.25
60	0.5	0.99	0.26	0.86
	1.0		0.51	0.81
	1.5		0.73	0.72
90	0.5	1.00	0.48	1.00
	1.0		0.88	1.00
	1.5		0.98	0.98

The results, which are also presented in Table 1 are mixed. When the number of exceedances is large, which occurs when n and σ are large and y is low, the loss of power is relatively modest. However, when the number of exceedances is small, the loss of power can be substantial. This underlines the importance of the selection of y. In particular, there is a trade-off between choosing y high enough to focus on tail behavior and choosing y low enough to ensure that a reasonable number of exceedances occur.

6. Discussion

The purpose of this paper has been to assess in a simple, but not unrealistic, situation the loss of power in applying the extreme approach to detect trends in the frequency and magnitude of extreme events. The limited results of the previous section indicate that this loss can be substantial. This suggests that a careful analysis along traditional lines may be preferable to an analysis based on tail behavior alone.

It is important to emphasize that the analysis presented here compared the *correct* traditional approach to the *correct* extreme approach in a situation in which the trend in extreme behavior arose from a change in scale. One of the clear benefits of the extreme approach is that it is robust against departures from the assumed model. For example, when y is high, the excess Y_t - y given that Y_t > y has an approximate generalized Pareto (GPD) distribution for a large family of parent distributions. Thus, while the traditional approach based on the wrong parent distribution may go wrong, the extreme approach based on the GPD distribution will often still be valid. There may also be situations in which trend is confined to the tails, so that the correct traditional approach is itself an extreme approach.

Acknowledgements

The helpful comments of two anonymous reviewers are acknowledged with gratitude.

References

Cox, D.R. and Hinkley, D.V.: 1974, *Theoretical Statistics*, Chapman and Hall, London, 511 p.

Davison, A.C. and Smith, R.L.: 1990, *'Models for exceedances over high thresholds (with discussion)'*, Journal of the Royal Statistical Society B52, 393-442.

Hodges, J.L. and Lehmann, E.L.: 1956, *'The efficiency of some nonparametric competitors of the t-test'*, Annals of Mathematical Statistics 27, 324-335.

IPCC: 1995, *Climate Change 1995: The Science of Climate Change,* Cambridge University Press, Cambridge, 572 p.

Leadbetter, M.R., Lindgren, G., and Rootzen, H.: 1983, *Extremes and Related Properties of Random Sequences and Processes,* Springer Verlag, Berlin, 488 p.

Resnick, S.I.: 1987, *Extreme Values, Regular Variation, and Point Processes,* Springer Verlag, New York, 462 p.

Smith, R.L.: 1989, *'Extreme value analysis of environmental time series: An application to trend detection in ground-level ozone'*, Statistical Science 4, 367-393.

(Received 5 November 1997; in revised form 10 August 1998)